Sensor Technology: Design and Analysis

Sensor Technology: Design and Analysis

Edited by **Marvin Heather**

CWILLFORD PRESS

New York

Published by Willford Press,
118-35 Queens Blvd., Suite 400,
Forest Hills, NY 11375, USA
www.willfordpress.com

Sensor Technology: Design and Analysis
Edited by Marvin Heather

International Standard Book Number: 978-1-68285-032-9 (Hardback)

The publisher's policy is to use permanent paper from mills that operate a sustainable forestry policy. Furthermore, the publisher ensures that the text paper and cover boards used have met acceptable environmental accreditation standards.

Printed in the United States of America.

Contents

Preface

Sensor technologies are ubiquitous in the modern era. They are used extensively in process control, robotics, measurement technology, etc. This book traces the progress of sensor technology and highlights some of its key concepts and applications. The topics covered in this profound book deal with the core subjects of sensor technology like sensor principles, sensor materials, measurement systems, modeling and simulation, etc. This text is appropriate for students seeking detailed information in this area as well as for experts.

All of the data presented henceforth, was collaborated in the wake of recent advancements in the field. The aim of this book is to present the diversified developments from across the globe in a comprehensible manner. The opinions expressed in each chapter belong solely to the contributing authors. Their interpretations of the topics are the integral part of this book, which I have carefully compiled for a better understanding of the readers.

At the end, I would like to thank all those who dedicated their time and efforts for the successful completion of this book. I also wish to convey my gratitude towards my friends and family who supported me at every step.

Editor

Improvement of the sensitivity of a conductometric soot sensor by adding a conductive cover layer

P. Bartscherer[1] and R. Moos[2]

[1]Robert Bosch GmbH, Robert-Bosch-Platz 1, 70839 Gerlingen-Schillerhöhe, Germany
[2]Bayreuth Engine Research Center (BERC), University of Bayreuth, 95447 Bayreuth, Germany

Correspondence to: R. Moos (functional.materials@uni-bayreuth.de)

Abstract. Diesel particulate filters are emission-relevant devices of the exhaust gas aftertreatment system. They need to be monitored as a requirement of the on-board diagnosis. In order to detect a malfunction, planar sensors with interdigital electrodes on an insulating substrate can be installed downstream of the filter. During the loading phase, soot deposits onto the electrodes, but the sensor remains blind until the percolation threshold has been reached (initiation time) and the sensor current starts to flow. In order to detect small soot concentrations downstream of the filter from small defects, this initiation time needs to be as low as possible. One may reduce the initiation time by covering the interdigital electrodes with an electrically conductive layer. Using finite element method (FEM) simulations, the influence of conductivity and thickness of such a coating on the initiation time are determined. It is found that a thin, screen printable coating with a thickness of $20\,\mu m$ and a conductivity in the range of 10^{-3} to $10^{-1}\,S\,m^{-1}$ may reduce the initiation time by about $40\,\%$. The FEM results were verified by a commercially available thick film resistor paste with a conductivity of $0.45\,mS\,m^{-1}$, showing an improvement of about $40\,\%$ compared to an uncoated sensor.

1 Introduction

Together with the constantly tightening limits of harmful exhaust gas emissions of internal combustion engines, a permanent monitoring of all emission-relevant devices is required (Europäisches Parlament, 2007; European Union, 2008). For Diesel engines, NO_x abatement and particulate matter (PM) reduction are of special importance (Johnson, 2007; Twigg, 2007). PM consists of aggregated carbon soot particles, covered with organics like (polycyclic aromatic) hydrocarbons, and inorganic oxides (e.g., ashes originating from wear of the engine or engine oil additives) (Spears, 2008; U.S. Environmental Protection Agency, 2002). Not only due to the small size of these particles, ranging between only a few tens and some hundreds of nanometers (Harris and Maricq, 2001), but also because of the soluble organic fraction (SOF) covering the surface of these small particles (Otto et al., 1980; U.S. Environmental Protection Agency, 2002), serious concerns exist about the effect of PM to human health (Geiser, 2005; Adar et al., 2010; Grahame and Schlesinger, 2010; U.S. Environmental Protection Agency, 2002).

In order to reduce the mass and the number of PM in the exhaust, diesel particulate filters (DPF) – typically porous ceramic wall-flow filters (Fino, 2007; Twigg and Phillips, 2009) – are installed in the exhaust line. When soot is deposited in the filter, the pores get clogged and the exhaust backpressure increases with increasing soot load (Alkemade and Schumann, 2006; Duvinage et al., 2001). Therefore, the DPF must be regenerated from time to time, typically by heating the filter to several hundred °C and oxidizing the soot. Currently the approach is to use the pressure difference up- and downstream of the DPF together with an engine map-oriented soot load model to estimate the soot load and the point in time when the regeneration needs to be initiated (Alkemade and Schumann, 2006; Rose and Boger, 2009). Recent research ideas try to determine the soot content by measuring the electrical impedance of the DPF itself during operation (Feulner et al., 2013) or by evaluating the soot load-dependent perturbation of the electrical resonance behavior in the GHz range (Moos et al., 2013; Fischerauer et al., 2010).

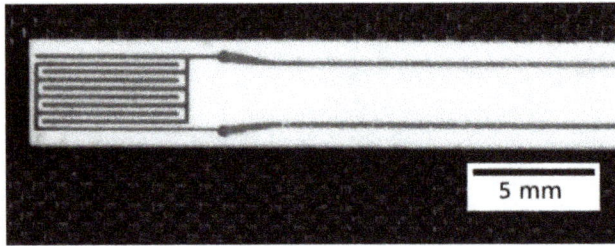

Figure 1. Image of an uncoated blank sensor with interdigital electrodes (IDEs) and feed lines.

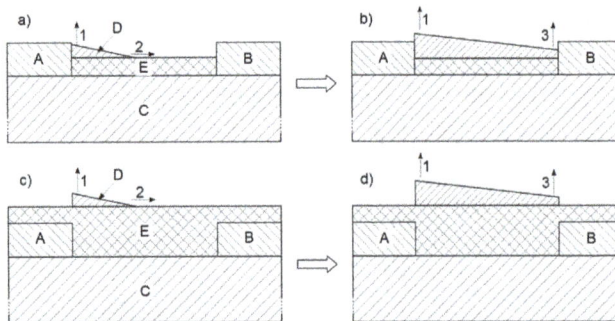

Figure 2. Geometry used for the simulations. The upper row (**a** and **b**) depicts the setup for a coating (E) being thinner than the electrodes (A) and (B). The lower row (**c** and **d**) depicts the case of a coating (E) being thicker than the electrodes (coating covers the electrodes). 1, 2 and 3 indicate the growth direction of the soot path (D). C indicates the electrically insulating substrate.

Table 1. Electrical conductivities of the conductive layer in mS m^{-1} used for different layer thicknesses and base currents/resistances.

Wanted sensor properties		Conductivity for different thickness b of the conductive layer			
$R_{no\,soot}$	$I_{no\,soot}$	1 μm	11 μm	20 μm	40 μm
300 kΩ	0.1 mA	9.01	0.82	0.46	0.25
30 kΩ	1 mA	90.09	8.19	4.61	2.49
3 kΩ	10 mA	900.90	81.93	46.08	24.96

initiation time increases with decreasing PM concentration. Therefore, it will take some time to detect a DPF defect, especially for very small defects with very low soot concentration downstream of the DPF. To overcome this and to reduce the blind time, we suggest applying an electrically conductive layer that connects both electrodes and covers all the space in between them. By adding such a layer, a soot-dependent current can be measured even if no soot percolation paths have reached the counter electrode.

It is the object of this study to demonstrate the feasibility of this idea. The first part of this article describes an FEM study to estimate suitable parameters with respect to conductivity and thickness of the conductive film. Later on, sensors coated with a commercial ruthenium oxide (RuO$_2$)-based thick-film paste were used to verify the results of the simulations. RuO$_2$ was only selected to prove the concept. For a reliable application in the exhaust, material with better long-term stability have to be applied.

2 FEM Modeling

2.1 Setup and modeling

One part of an IDE (as depicted in Fig. 2) was FEM modeled (FEM software Comsol Multiphysics). To comply with Fig. 1, the electrode width (w) and spaces (s) between the electrodes were set to 150 μm each and the electrode thickness to $h = 10$ μm. The substrate thickness was set to 500 μm. Further elements of the sensor, like the embedded heater (Ochs et al., 2012), have not been considered in the model. All free edges of the sensors were set to an electric insulation (boundary condition). For the substrate conductivity, 10^{-12} S m^{-1} was assumed, which is a typical value for technical alumina at 200–400 °C (Evans, 1995). Four different thicknesses (b) of the conductive layer were modeled, each one with three different electrical conductivities (for values of the conductivities see Table 1). The conductivities were adjusted in a way that the basic current of the sensor without soot load, $I_{no\,soot}$, was 0.1 mA, 1 mA, or 10 mA at an applied voltage of $U = 30$ V. Therefore, the conductivity for a thin layer had to be higher; the conductivity for a thicker layer had to be lower in order for the four different layers to exhibit the same base currents for the three different resistance levels.

The first method is serial standard, but it has not been clarified yet whether this method is precise enough to detect reliably small malfunctions of DPFs, like small cracks or holes that lead to soot slip. Therefore, sensors measuring the amount of soot in the exhaust downstream of a DPF are in discussion. Many principles for such sensors have been investigated, but the conductometric approach is considered the best choice for PM detection, especially with respect to the low-cost of a sensor (Riegel and Klett, 2008). A typical conductometric PM sensor is shown in Fig. 1. It consists of two electrodes that are typically applied as interdigital electrodes (IDEs) on an electrically insulating substrate. Recent approaches are given by Ochs et al. (2010), Lloyd Spetz et al. (2012), Groß et al. (2012), Hagen et al. (2010), Bartscherer et al. (2007), Bartscherer and Schmidt (2008), or Bartscherer et al. (2008).

Soot is deposited by thermophoresis on the IDE structure. If a voltage is applied between both electrodes, electrophoresis also occurs. It even prevails over the thermophoretic effect.

If one neglects the substrate conductivity, the current does not flow before the first percolation path forms between the electrodes. In other words, the sensor remains "blind" at the beginning of a loading cycle. This blind time as well as the

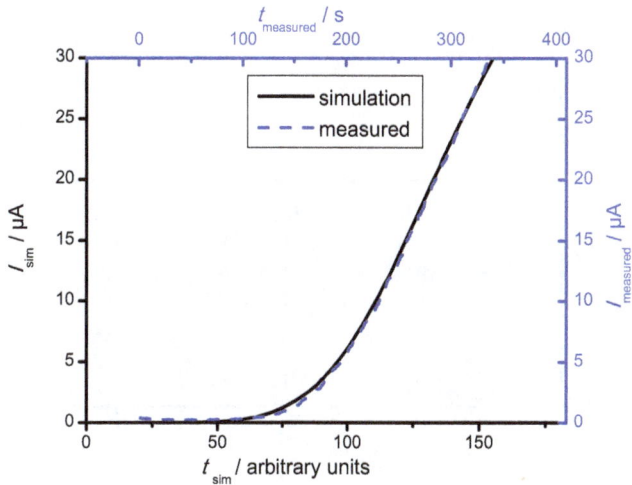

Figure 3. Simulation and real measurement of soot signal for a blank sensor.

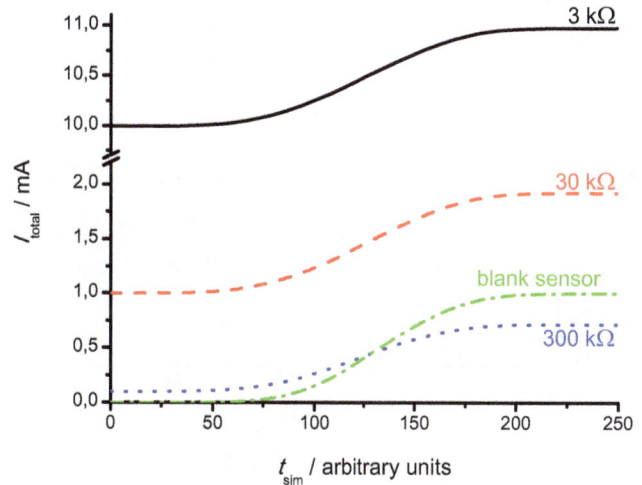

Figure 4. Simulation of the total current, I_{total}, for sensors with different conductive coatings leading to base resistances (without soot) of $3\,k\Omega$, $30\,k\Omega$, and $300\,k\Omega$ (see Table 1) compared to a blank sensor. Coating thickness $b = 11\,\mu m$.

Two geometries of the conductive layer have to be considered. In one case, the layer is thinner than the electrodes (i.e. $b < h = 10\,\mu m$) and the electrodes are not covered by the applied conductive layer (Fig. 2a and b; top row). In this case, the soot can contact the electrodes directly. In the other case, the conductive layer is thicker than the electrodes ($b > h$) and covers them (Fig. 2c and d; bottom row). In this case, soot has no direct contact to the electrodes.

Since there is not much known about the growth of soot paths in an electric field, some assumptions had to be made for the simulation. Light optical microscopy (Fig. 8) revealed that the soot paths are denser at one electrode – presumably at the electrode at which the growth is initiated – than on the other one. Therefore, the growing soot path was assumed to be wedge-shaped (D in Fig. 2), initially growing vertically at position 1 and horizontally at position 2 at the same time. When it reaches the counter electrode (position 3), it stops growing horizontally and the growth is vertical also at this position. From this point on, the the soot layer would contribute also to the current of a sensor without a conductive layer. At the end of the growth, the height of the wedge is $5\,\mu m$ at position 1 and $2.5\,\mu m$ at position 3. Since the electrical conductivity of soot may vary with the type of soot, a macroscopic approach was conducted. A blank sensor without conductive layer but fully covered with soot shows a current of approximately $1\,mA$, if a dc voltage of $30\,V$ is applied to the electrode (at a sensor temperature of about $50\,°C$). This corresponds to a resistance of $30\,k\Omega$. The electrical conductivity of the soot in the simulations was chosen in a way that the resistance of the soot wedge amounted to this value at the end of the growth. Hence, the electrical conductivity of the soot should be about $25\,mS\,m^{-1}$.

In reality, there is not only one soot path growing from one electrode to the other but instead many soot paths start to grow at different times. All paths are parallel connected and contribute to the total current according to their state of growth. To represent that, a Gaussian distribution (Eq. 1) of the number of paths, n_{paths}, starting to grow at a certain time, t_s, in the simulation was assumed.

$$n_{paths}(t_s) = \frac{1}{25 \cdot \sqrt{2\pi}} \cdot e^{-\frac{1}{2}\left(\frac{t_s - 75}{25}\right)^2} \cdot 15\,000 \qquad (1)$$

Since the simulation should reveal only relative results for the different varied parameters, and due to the many uncertainties of the soot growth and its nature, the simulations have been done using arbitrary units for the time, t_s. Different growth rates of the soot paths were not considered. Using this approach, only one path had to be calculated, but the output current of the whole sensor was obtained by superposing the currents of each single path according to the Gaussian distribution. Calculated and measured current progressions for a blank sensor without conductive layer agree quite well (Fig. 3) if one assumes appropriate scaling factors for current and time.

It should be annotated here that this model for the soot path growth does not consider the physical deposition mechanism of the soot. However, it will be shown that it is a very useful approximation that describes the current response quite well.

The measurement procedures for the experiments leading to Fig. 3 are described in the experimental Sect. 3.

2.2 Variations of the layer parameters

In the following, I_{total} refers to the total current of a sensor, including the current through the conductive cover layer (if present) and the collected soot (the current through the substrate can be neglected because the substrate is insulating), and $I_{no\,soot}$ representing I_{total} without collected soot.

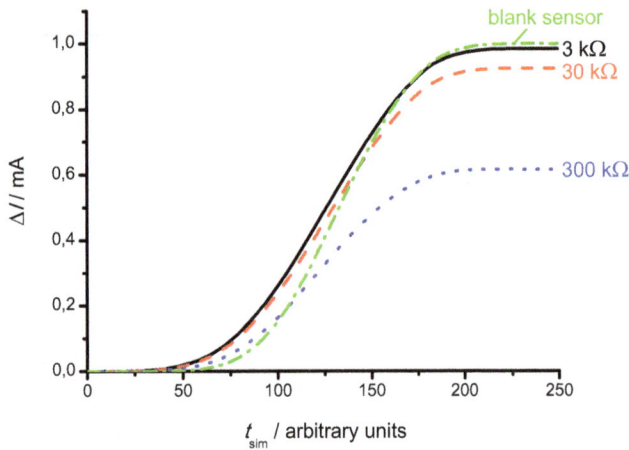

Figure 5. Current deviation $\Delta I = I_{total} - I_{no\,soot}$, for coatings with base resistances (without soot) of $3\,k\Omega$, $30\,k\Omega$ and $300\,k\Omega$, compared to a blank sensor. For raw data see Fig. 4.

Table 2. Total current and current deviation ΔI of a simulated conductive layer ($b = 11\,\mu m$ thickness) with different electrical conductivities.

$R_{no\,soot}$	$I_{no\,soot}$	I_{total}	ΔI
$300\,k\Omega$	$0.1\,mA$	$0.716\,mA$	$616\,\mu A$
$30\,k\Omega$	$1\,mA$	$1.924\,mA$	$924\,\mu A$
$3\,k\Omega$	$10\,mA$	$10.977\,mA$	$977\,\mu A$

Examples for the simulated total current, I_{total}, and the soot-related current deviation $\Delta I = I_{total} - I_{no\,soot}$ for a blank sensor and sensors with a conductive layer of $b = 11\,\mu m$ (see Fig. 2c and d) are shown in Figs. 4 and 5 for three different electrical conductivities in a way that $R_{no\,soot}$ corresponds to $3\,k\Omega$, $30\,k\Omega$, or $300\,k\Omega$. Table 2 lists the numerical values. Figure 6 shows the magnification of Fig. 5 to extract the initiation time t_i for these four sensors at a trigger limit of $\Delta I_{trigger} = 3\,\mu A$.

The differences in the total current arise from the different resistances of the conductive layers (as they depend on the cover layer materials conductivity and on the thickness of the cover layer), which add up to the resistance of the soot path. A low conductivity of the conductive layer yields high serial resistances, and therefore limits the current. It appears implausible at first glance that the current of the fully soot loaded but uncovered sensor (Fig. 4) should be higher than the $300\,k\Omega$-covered one. However, one has to consider that the conductive layer (which in case of a $300\,k\Omega$ layer exhibits poor conductivity) also covers the electrodes, prevents direct contact between soot and electrodes, and therefore limits the current.

The improvements that can be obtained with the conductive layers are summarized in Fig. 7. Here, $(1 - t_i/t_{ref})/\%$ indicates the percentaged reduction of the initiation time, t_i, compared to the initiation time, t_{ref}, of the uncoated blank sensor.

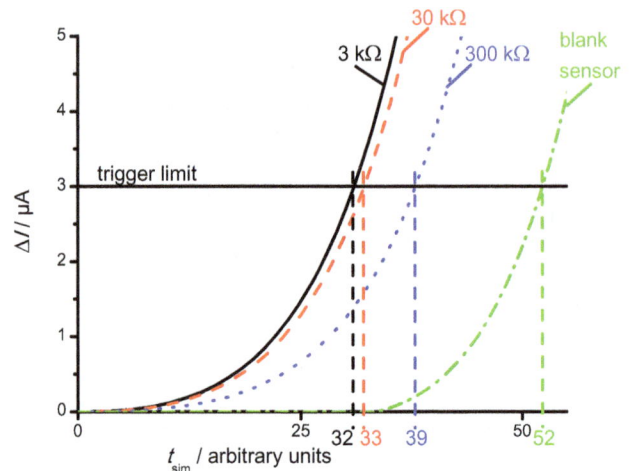

Figure 6. Magnification of Fig. 5 to determine the initiation time, t_i, for coatings with different base resistances (without soot).

As shown by the simulations, an improvement of about 40 % should be possible.

Thin layers with a high conductivity show the best performance, especially for the $1\,\mu m$ layer. Since such thin layers with $b < h$ (Fig. 2a and b) do not cover the electrodes, soot can contact the electrodes directly without any limiting serial resistances between electrode and soot. However, the calculated improvements in reduction of the blind time between coatings of $1\,\mu m$ and $20\,\mu m$ thickness as well as with coatings between a resistance of $3\,k\Omega$ and $30\,k\Omega$ (for the data of the sensor with respect to conductivity and film thickness, see Table 1) can be considered insignificant, so that the already mentioned disadvantages of thin layers or low resistance without soot outweigh the slight advantages regarding the current signal. Considering an economical screen printing process, typical layer thicknesses of $5\,\mu m$ or more can be achieved. Hence, a functional layer with an electrical resistance of $30\,k\Omega$ and a thickness of 5 to $20\,\mu m$ seems to be a good tradeoff between improved measurement duration, manufacturability and usability of inexpensive measurement equipment.

3   Experimental verification

In Sect. 2, the advantages of a conductive layer were worked out using simulation. In this section, we will describe experimental results and compare them with our simulations.

Experiments

For this study, the platinum IDEs were screen-printed on alumina film-covered yttria stabilized zirconia (YSZ) substrates. Due to the setup in planar tape technology, a heater could be integrated into the YSZ monolithic substrate. Further details can be obtained from Ochs et al. (2012).

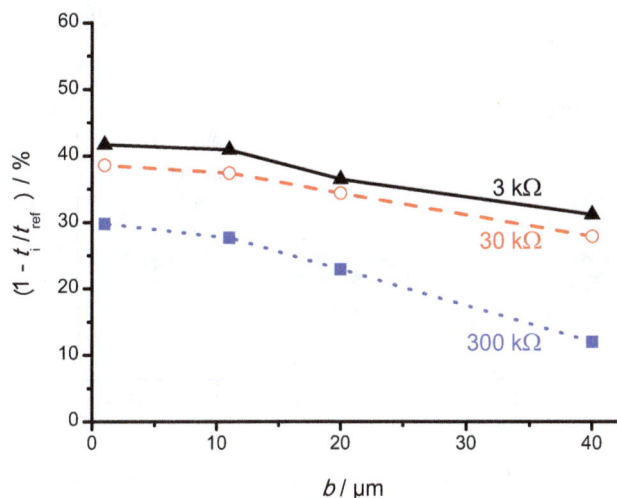

Figure 7. Simulated percentaged reduction of the initiation time t_i for different layer thicknesses and conductivities, see Table 1. t_{ref} is the initiation time of the uncoated sensor.

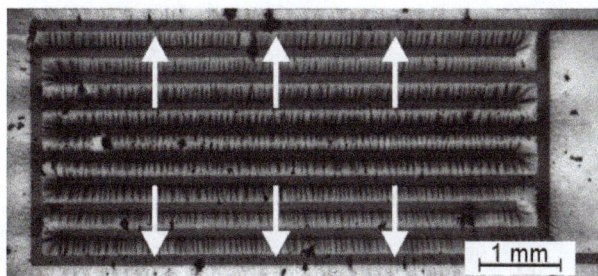

Figure 8. Light optical microscope image of a soot loaded interdigital electrode structure. The soot accumulates in form of percolation paths, growing from one electrode to the other. Exhaust gas hits the sensor perpendicular to the surface and spreads as indicated by the arrows.

Figure 8 shows a typical IDE (without additional coating) during electrophoretic soot deposition. Electrically conductive soot paths grow from one electrode to the other, resulting in an ohmic current (Ochs et al., 2012; Hagen et al., 2010). The initiation time, t_i, i.e., the time the current needs to reach a trigger limit (for instance a current of 3 μA) can be used as a measure for the amount of PM in the exhaust gas. After a deposition cycle, the sensor is heated to above 600 °C to burn the soot on the IDE. After cooling, the sensor is regenerated and the subsequent loading cycle can start.

To verify the results from the simulation and to proof the concept of improving the particle sensor by applying a conductive layer, a commercial resistor paste which is typically used in LTCC and thick-film technology, was applied (Heraeus, type R8281). It consists of RuO_2 embedded in a glass matrix (for literature see, e.g., Pike and Seager, 1977, or Nicoloso et al., 1995). The nominal square resistance according to the data sheet was 100 MΩ for a fired film thickness of 22 μm. This leads to a conductivity of 0.45 mS m^{-1}, which is close to the optimum determined in the simulation. For the test sensor, this small deviation is not important, because even for a slightly deviating electrical conductivity, the reduction of the initiation time should be significant. It should be clarified that RuO_2 embedded is a glass matrix is a good fabrication approach for thick-film resistors in electronics but it is not suitable to be applied in automotive exhausts at elevated temperatures. It is known that RuO_2-based materials show reversible as well as irreversible changes in resistance, depending on temperature, mechanical stress and electrical field (Pike and Seager, 1977). Since the temperature during regeneration can reach or even exceed the firing temperature of the RuO_2/glass layer, the glass matrix may soften each time the soot is burned off at regeneration temperatures. This may result in plastic deformation of the conductive layer that

changes the electrical characteristic. Furthermore, high temperatures may affect the conductivity of RuO_2 due to oxidation and loss of volatile reaction products (Colomer and Jurado, 1997). All these factors affect the conductivity of RuO_2-based resistor pastes and may lead to a drift of the sensor current that is not tolerable for a sensor that should operate for many years in the exhaust gas. However, in order to verify the concept, these shortcomings are not relevant and the already mentioned paste was selected despite all its disadvantages.

For the test sample, blank sensors were manufactured as described briefly above and in detail in Ochs et al. (2012). Some of the sensors were coated with the screen-printed paste, dried for 10 min at 80 °C and fired at 850 °C according to the manufacturer's specifications. A fired film thickness of about 35 μm was obtained. The sensors were built into the exhaust pipe of an artificial soot source (CAST2, Matter Engineering) together with an uncoated blank sensor serving as a reference. The gas stream was 5 L min^{-1} with an average particle size of 85 nm. During soot deposition, the sensors were not heated, i.e. soot collection occurred at about 35 °C.

Figure 9 shows a typical measuring cycle. At t_1, the IDE voltage (30 V dc) was applied to determine a possible temperature-based drift of the sensor base signal. At t_2, about 2 min later, the artificial soot source was switched on. After the current of both sensors (coated sensor and blank sensor) exceeded the triggering limit (3 μA) significantly, the artificial soot source was switched off at t_3. At t_4, the IDE voltage was also switched off and shortly after that, the regeneration was initiated (at t_5). Using the applied heater, a temperature above 750 °C was achieved and the collected soot was oxidized. After turning off the sensor heating at t_6, the sensor was allowed to cool down to a stable temperature before the subsequent cycle started.

The raw signal of a coated and an uncoated, blank sensor is shown in Fig. 10. Due to the reasons mentioned above, the coated sensors showed a baseline drift even if no soot was present. A linear extrapolation of the drift (line A in Fig. 10)

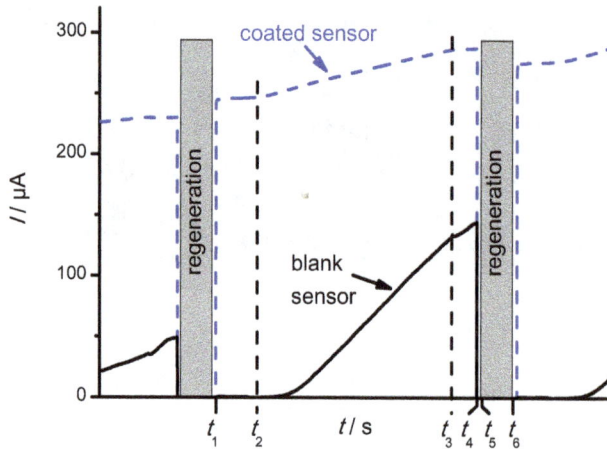

Figure 9. Typical measurement and regeneration cycle. t_1: sensor voltage over the IDE is applied, t_2: artificial soot source is switched on, t_3: artificial soot source is switched off, t_4: sensor voltage is switched off, t_5 to t_6: heater current for regeneration is applied.

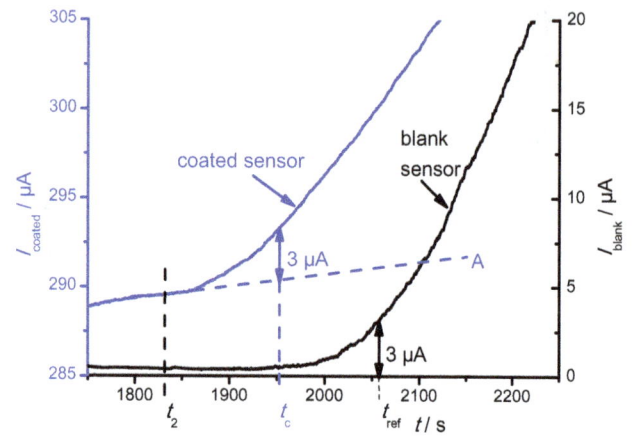

Figure 10. Raw signal for a coated (left axis) and an uncoated (right axis) sensor. The current drift for the coated sensor was compensated using linear extrapolation. The artificial soot source is switched on at t_2. The coated sensor reaches the trigger limit of $3\,\mu\mathrm{A}$ at t_c, the uncoated sensor reaches the trigger limit at t_{ref}.

yielded an increasing baseline. It was used to correct the initiation time t_c as depicted in Fig. 10. The coated sensor showed a baseline of almost $300\,\mu\mathrm{A}$, which is equivalent to a resistance of about $100\,\mathrm{k\Omega}$ in the soot-free state. Furthermore, the coated sensor reached the trigger limit of $\Delta I \approx 3\,\mu\mathrm{A}$ about $100\,\mathrm{s}$ earlier than the uncoated sensor, and, as predicted from the simulations, the coated sensor showed a smaller slope $\mathrm{d}I/\mathrm{d}t$. This behavior can be explained by the serial resistance of the conductive coating that reduces the contribution of the soot to the total current.

The very low (but not zero) current of the reference sensor before applying the soot source may originate from some soot that had been deposited between the feed lines during antecedent cycles (see Fig. 1). Since the sensor feeds are not heated, this part was not regenerated. However, this leakage current can be neglected.

The tests were repeated with five different sensors with 1 to 5 cycles for each sensor. The results for each sensor with the standard deviation of the percentaged reduction of the initiation time are shown in Fig. 11. Despite not being optimized, the initiation time with a conductive layer can be reduced by between 25 and 50 % compared to the uncoated sensor. The results are even slightly better than predicted by the simulations (25 to 40 %).

4   Concluding remarks

The simulations showed that the conductivity of a conductive coating of the electrode area of a PM sensor greatly influences the sensor performance. Conductivity and layer thickness can be further optimized. On the one hand, higher conductivities are preferred because they reduce the current through the serially connected soot only a little. However, the benefit of a high conductivity gets smaller for higher conduc-

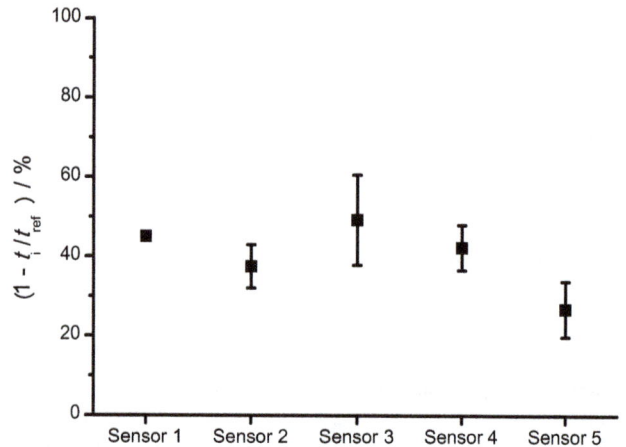

Figure 11. Reduction of the initiation time due to coating, proven for five different sensors.

tivities. Comparing the simulated low and medium conducting layers, an improvement of the initiation time of about 12 % can be obtained with a conductivity change of a factor of 10. Increasing further the conductivity of the layer by a factor of 10 yields an additional reduction of only about 2 %.

Besides the sensor performance with respect to a shorter initiation time, system aspects as well as manufacturing process issues deserve consideration. How can such a conductive coating be applied, what electronic measurement equipment is needed? Both aspects may limit the application.

The conductive layers are preferably applied in thick-film technology. This technique is process compatible, inexpensive and well-known for ceramic exhaust gas sensors (Riegel et al., 2002). Layer thicknesses between about 5 to $50\,\mu\mathrm{m}$ can be manufactured reliably. Thin film techniques like CVD or PVD are more expensive and are not preferred.

Cost issues of the measurement circuit play a crucial role. The higher the required resolution, the more expensive it will be[1]. If the conductivity of the layer is too high, a 12 bit A/D converter may not be sufficient due to the high base current without soot. In case of the highly conducting coating, a resolution of 15 bit would be needed. For the medium conductivity, 12 bit would be sufficient and for the lowest conductivity 9 bit would suffice. With respect to the required resolution, a low conductivity would be preferable.

As a tradeoff, medium conductive layers are considered best. The demands on the A/D converters are moderate, and the initiation time is short. A higher conductivity does not show a huge advantage regarding the initiation time, but will result in much higher cost of the measuring equipment. A material with an electrical conductivity of 10^{-3} to 10^{-1} S m^{-1} would be suitable for a functional layer of $20\,\mu m$ thickness and 12 bit resolution.

The functional demonstration with a RuO$_2$-based coating showed even more potential to reduce the initiation time than expected from the simulation. But it also showed the current limiting effect of a serial connection between soot and functional coating. With a reduction of the initiation time of about 45 % with a non-optimized system, an even higher effect can be expected by using an optimized material and coating, therefore a continuation of these investigations seems promising.

A reduction in the initiation time enhances the accuracy of the sensor and therefore allows a much better prediction of the particle filter condition, meaning that smaller defects may be more reliably detected. However, it is challenging to find a material with a suitable electrical conductivity and the robustness to withstand the harsh conditions in automotive exhausts.

Recently, dosimeter-type resistive gas sensors are suggested for gas concentration measurements (Geupel et al., 2010; Groß et al., 2012). Both soot sensor and dosimeter-type resistive gas sensor rely on the same principle: with sorption of gas (or soot) the resistance changes and as soon as saturation effects occur, sensor regeneration is required, typically induced by heating the sensitive film far above working temperature. Recently, Hennemann et al. (2012), even suggested a dosimeter-type gas sensor based on percolation effects. H$_2$S percolation paths are formed on nanofibers leading to a steep resistance decrease when a defined dose is applied. This behavior is very similar to the conductometric soot sensor where a defined amount of soot yields a similar behavior. It is suggested to transfer the idea of a conductive layer also to such percolation-type conductometric dosimeter gas sensors.

Acknowledgements. The authors thank Ralf Schmidt, Frank Rettig and Gerd Teike for discussions and their help on the simulation and model building, as well as Ulrich Hasenkox and Helmut Marx for the support during the measurements. The authors also thank Robert Bosch GmbH, where most of the work have been done, for supporting this work and for the permission to publish the results.

Edited by: A. Lloyd Spetz
Reviewed by: two anonymous referees

References

Adar, S. D., Klein, R., Klein, B. E. K., Szpiro, A. A., Cotch, M. F., Wong, T. Y., O'Neill, M. S., Shrager, S., Barr, R. G., Siscovick, D. S., Daviglus, M. L., Sampson, P. D., and Kaufman, J. D.: Air Pollution and the Microvasculature: A Cross-Sectional Assessment of In Vivo Retinal Images in the Population-Based Multi-Ethnic Study of Atherosclerosis (MESA), PLOS Med., 7, 1–11, 2010.

Alkemade, U. G. and Schumann, B.: Engines and exhaust after treatment systems for future automotive applications, Solid State Ionics, 177, 2291–2296, 2006.

Bartscherer, P. and Schmidt, R.: Sensor and method for detecting particles in a gas flow, Patent Application WO002008138659A1, 2008.

Bartscherer, P., Grabis, J., and Schmidt, R.: Verfahren zum Betrieb eines Partikelsensors, German Patent Application DE102007060939A1, 2007.

Bartscherer, P., Hasenkox, U., and Roesch, S.: Sensor for resistively determining concentrations of conductive particles in gas mixtures, Patent Application WO002008025602A1, 2008.

Colomer, M. T. and Jurado, J. R.: Preparation and characterization of gels of the ZrO$_2$-Y$_2$O$_3$-RuO$_2$ system, J. Non-Cryst. Solids, 217, 48–54, 1997.

Duvinage, F., Nolte, A., Paule, M., Schommers, J., and Brueggemann, H.: Dieselpartikelfilter für PKW – gestern, heute und morgen, 10. Aachener Kolloquium Fahrzeug- und Motorentechnik, 2001.

Europäisches Parlament: Verordnung (EG) Nr. 715/2007 über die Typgenehmigung von Kraftfahrzeugen hinsichtlich der Emissionen von leichten Personenkraftwagen und Nutzfahrzeugen (Euro 5 und Euro 6) und über den Zugang zu Reparatur- und Wartungsinformationen für Fahrzeuge, Amtsblatt der Europäischen Union, L171, 1–16, 2007.

European Union: Communication on the application and future development of Community legislation concerning vehicle emissions from light-duty vehicles and access to repair and maintenance information (Euro 5 and 6), Official Journal of the European Union, 2008/C 182/08, 2008.

Evans, B. D.: A review of the optical properties of anion lattice vacancies, and electrical conduction in Al$_2$O$_3$: their relation to radiation-induced electrical degradation, J. Nucl. Mater., 219, 202–223, 1995.

---

[1] If one accepts a resolution for the A/D converter of 12 bit, than the measuring range can be resolved in $2^{12} = 4096$ equidistant steps. The steps should be about 10 times smaller than the value to be measured. Hence, a trigger limit of 3 A requires steps of at least $0.3\,\mu A$. If the layer conductivity is too high, a 12 bit A/D converter may not be sufficient due to the high base current without soot.

Feulner, M., Hagen, G., Piontkowski, A., Müller, A., Fischerauer, G., Brüggemann, D., and Moos, R.: In-Operation Monitoring of the Soot Load of Diesel Particulate Filters – Initial Tests, Top. Catal., 56, 483–488, 2013.

Fischerauer, G., Förster, M., and Moos, R.: Sensing the Soot Load in Automotive Diesel Particulate Filters by Microwave Methods, Meas. Sci. Technol., 21, 035108, doi:10.1088/0957-0233/21/3/035108, 2010.

Fino, D.: Diesel emission control: Catalytic filters for particulate removal, Sci. Technol. Adv. Mat., 8, 93–100, 2007.

Geiser, M.: Distribution and Clearance of Inhaled Ultrafine TiO_2 Particles in Rat Lungs, 9th ETH Conference on CGP, 2005.

Geupel, A., Schönauer, D., Röder-Roith, U., Kubinski, D. J., Mulla, S., Ballinger, T. H., Chen, H.-Y., Visser, J. H., and Moos, R.: Integrating nitrogen oxide sensor: a novel concept for measuring low concentrations in the exhaust gas, Sensor. Actuat. B-Chem., 145, 756–761, 2010.

Grahame, T. J. and Schlesinger, R. B.: Cardiovascular Health and Particulate Vehicular Emissions: a Critical evaluation of the Evidence, Air Quality, Atmosphere and Health, 3, 3–27, 2010.

Grob, B., Schmid, J., Ivleva, N., and Niessner, R.: Conductivity for Soot Sensing: Possibilities and Limitations, Anal. Chem., 84, 3586–3592, 2012.

Groß, A., Beulertz, G., Marr, I., Kubinski, D. J., Visser, J. H., and Moos, R.: Dual Mode NO_x Sensor: Measuring Both the Accumulated Amount and Instantaneous Level at Low Concentrations, Sensors, 12, 2831–2850, 2012.

Hagen, G., Feistkorn, C., Wiegärtner, S., Heinrich, A., Brüggemann, D., and Moos, R.: Conductometric Soot Sensor for Automotive Exhausts: Initial Studies, Sensors, 10, 1589–1598, 2010.

Harris, S. J. and Maricq, M. M.: Signature size distributions for diesel and gasoline engine exhaust particulate matter, Aerosol Science, 32, 749–764, 2001.

Hennemann, J., Sauerwald, T., Kohl, C.-D., Wagner, T., Bognitzki, M., and Greiner, A.: Electrospun copper oxide nanofibers for H_2S dosimetry, Phys. Status Solidi A, 209, 911–916, 2012

Johnson, T. V.: Diesel emission control in review, SAE Paper 2007-01-0233, doi:10.4271/2007-01-0233, 2007.

Lloyd Spetz, A., Huotari, J., Bur, C., Bjorklund, R., Lappalainen, J., Jantunen, H., Schütze, A., and Andersson, M.: Chemical sensor systems for emission control from combustions, Sensor. Actuat. B-Chem., doi:10.1016/j.snb.2012.10.078, in press, 2012.

Moos, R., Beulertz, G., Reiß, S., Hagen, G., Fischerauer, G., Votsmeier, M., and Gieshoff, J.: Overview: Status of the microwave-based automotive catalyst state diagnosis, Top. Catal., 56, 358–364, 2013.

Nicoloso, N., LeCorre-Frisch, A., Maier J., and Brook, R. J.: Conduction mechanisms in RuO_2-glass composites, Solid State Ionics, 75, 211–216, 1995.

Ochs, T., Schittenhelm, A., Genssle, A., and Kamp, B.: Particulate Matter Sensor for On Board Diagnostics (OBD) of Diesel Particulate Filters (DPF), SAE paper 2010-01-0307, doi:10.4271/2010-01-0307, 2010.

Otto, K., Sieg, M. H., and Zinbo, M.: The Oxidation of Soot Deposits from Diesel Engines, SAE paper 800336, doi:10.4271/800336, 1980.

Pike, G. E. and Seager, C. H.: Electrical properties and conduction mechanisms of Ru-based thick-film (cermet) resistors, J. Appl. Phys., 48, 5152–5169, 1977.

Riegel, J. and Klett, S.: Sensors for modern exhaust gas after-treatment systems. In: Proceedings of the 5th International Exhaust Gas and Particulate Emissions Forum, Ludwigsburg, Germany, February 2008, 84–97, 2008.

Riegel, J., Neumann, H., and Wiedenmann, H.-M.: Exhaust gas sensors for automotive emission control, Solid State Ionics, 152–153, 783–800, 2002.

Rose, D. and Boger, T.: Different Approaches to Soot Estimation as Key Requirement for DPF Applications, SAE paper 2009-01-1262, doi:10.4271/2009-01-1262, 2009.

Spears, M.: Particulate Matter and Adsorption: The First 100 Seconds, 5. Internationales Forum Abgas- und Partikelmesstechnik, Ludwigsburg 2008.

Twigg, M. V.: Progress and future challenges in controlling automotive exhaust gas emissions, Appl. Catal. B-Environ., 70, 2–15, 2007.

Twigg, M. V. and Phillips, P. R. Cleaning the air we breathe – Controlling diesel particulate emissions from passenger cars, Platinum Metals Review, 53, 27–34, 2009.

U.S. Environmental Protection Agency (EPA): Health assessment document for diesel engine exhaust, EPA/600/8-90/057F, 2002.

On-board hybrid magnetometer of NASA CHARM-II rocket: principle, design and performances

C. Coillot[1], J. Moutoussamy[1], G. Chanteur[1], P. Robert[1], and F. Alves[2]

[1]LPP/CNRS/UPMC/Ecole Polytechnique, Route de Saclay, 91128 Palaiseau, France
[2]LGEP/CNRS/Paris XI, 11 rue Joliot Curie, 91192 GIF sur Yvette, France

Correspondence to: C. Coillot (christophe.coillot@lpp.polytechnique.fr)

Abstract. We present a hybrid tri-axes magnetometer designed to measure weak magnetic fields in space from DC (direct current) up to a few kHz with a better sensitivity than fluxgate magnetometers at frequencies above a few Hz. This magnetometer combines a wire-wound ferromagnetic ribbon and a classical induction sensor. The nature of the wire-wound ferromagnetic ribbon sensor, giant magneto-impedance or magneto-inductance, is discussed. New configurations of wire-wound ferromagnetic ribbon sensors based on closed magnetic circuits are suggested and the hybrid sensor is described. The electronic conditioning of the wire-wound ribbon makes use of an alternating bias field to cancel the offset and linearize the output. Finally we summarize the main performances of the hybrid magnetometer and we discuss its advantages and drawbacks. A prototype has been built and was part of the scientific payload of the NASA rocket experiment CHARM-II (Correlation of High Frequency and Auroral Roar Measurements) launched in the auroral ionosphere. Unfortunately the launch campaign ended without any noticeable magnetic event and the rocket was eventually launched on 16 February 2010, through a very quiescent arc in the magnetic cusp and no wave activity was detected at frequencies observable by the hybrid magnetometer.

1 Introduction

The hybrid magnetometer aims to cover a frequency range of magnetic field measurement from DC up to a few kHz for plasma waves study in space and especially the investigation of the Langmuir and upper hybrid waves in the auroral ionosphere (Samara et al., 2004) in the context of the NASA rocket CHARM-II (Correlation of High Frequency and Auroral Roar Measurements). This magnetometer extends the magnetic field measurement of the induction sensor to the low frequencies (< 1 Hz) in order to ensure a redundancy with fluxgate magnetometers which are onboard rockets and spacecraft for plasma wave studies. The ultimate objective is to reach a noise equivalent magnetic induction comparable to the one of the fluxgate used in space experiments (about 5 pT Hz/ $\sqrt{\text{Hz}}$ at 1 Hz for Oersted magnetometer from Hika et al., 1996). This novel hybrid sensor combines induction sensor and a magneto-impedance sensor and covers magnetic field measurement from DC up to 20 kHz. In this paper we will focus on the DC part of the hybrid sensor which is ob-

tained using a kind of giant magneto-impedance (GMI). The GMI effect is the large variation of the AC impedance of a magnetic conductor with an applied DC or ultra-low frequency magnetic field. It has been evidenced in ferromagnetic wires by Harrison et al. (1936), in microwires by Panina and Mohri (1994) and Panina et al. (1994), in ribbons and sandwiches by Hika et al. (1996), Panina et al. (1994) and Morikawa et al. (1997), and more recently in transverse or longitudinal wire-wound ferromagnetic ribbons by Moutoussamy et al. (2009). Sensors making use of the GMI effect are characterized by their robustness, small mass and high sensitivity to weak magnetic fields which make them suitable for many applications. The ambient magnetic field to be measured modifies the dynamic magnetization of the magnetic material through its magnetic susceptibility tensor. At high frequency the modification of susceptibility due to the magnetic field is accompanied by a skin depth change resulting in a strong impedance variation.

2 Principle of the wire-wound ferromagnetic ribbon magnetic sensor

2.1 Discussion about the nature of the wire-wound ferromagnetic ribbon sensor: magneto-inductance or giant magneto-impedance?

The skin effect is usually invoked to explain the GMI effect but, in fact, the skin effect in sandwich GMI is more complicated than usually thought due to the lateral skin effect which occurs first in the copper conductor as described by Belevitch (1971). Simulation results in the middle section of a ferro/copper/ferro sandwich, presented in Fig. 1, illustrates the lateral skin effect at 1 MHz on the cross section of an infinite copper ribbon (width = 200 μm, thickness = 20 μm); the current density is expelled from the centre of the conductor to the lateral edges of the ribbon. However the skin effect in a ferromagnetic ribbon occurs as expected (the current density is confined into the skin thickness) when the relative permeability is sufficiently large, due to the edge effect discussed by Garcia-Arribas et al. (2008). Nevertheless the real GMI sandwich exhibits a poor current conduction between the copper ribbon and the magnetic material ribbons. The current flowing through the ferromagnetic ribbon is negligible, this experimental fact has led Moutoussamy et al. (2009) to implement an excitation by coils. However it should be noticed that the magnetic field radiated by the current flowing through the copper ribbon generates eddy currents inside the ferromagnetic ribbon, these are neglected in this article.

Let us now examine the effect of an external magnetic field on the magnetization of the ferromagnetic ribbon. When a magnetic field H is applied to a ferromagnetic ribbon, as shown in Fig. 2, the magnetization M, which makes an angle θ with the x axis, is modified in two ways, by its rotation and through the displacement of the domain walls. When considering only the rotation of the magnetization the susceptibility tensor is obtained by solving the Landau–Lifschitz–Gilbert (LLG) equation governing the evolution of the magnetization M:

$$\frac{d\boldsymbol{M}}{dt} = -\gamma\mu_0\boldsymbol{M}\times\boldsymbol{H}_{\text{eff}} + \alpha\frac{\boldsymbol{M}}{M_s}\times\frac{d\boldsymbol{M}}{dt}, \qquad (1)$$

where γ corresponds to the gyromagnetic ratio, α measures the damping, and M_S is the saturated magnetization. The use of the LLG equation is based on the hypothesis of a single magnetic domain structure which is a rough assumption as the magnetic structure of the ribbon is obviously multidomain.

The effective magnetic field $\boldsymbol{H}_{\text{eff}}$ is derived, accordingly to Eq. (2), by minimizing the free energy, defined by Eq. (3), which is the sum of the magneto-crystalline anisotropic energy, the Zeeman energy, and the demagnetizing energy.

$$\boldsymbol{H}_{\text{eff}} = -\frac{1}{\gamma\mu_0 M_S}\frac{\partial W}{\partial\theta} \qquad (2)$$

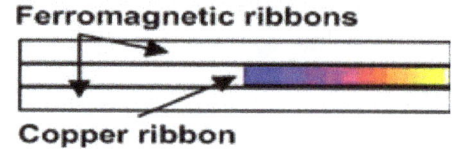

Figure 1. Current density distribution inside the middle half cross section of a ferro/copper/ferro sandwich at 1 MHz (blue to yellow corresponds to a ratio on the current density magnitude from 1 to 4).

$$W = K\sin^2\theta - \mu_0\boldsymbol{H}\cdot\boldsymbol{M} - \frac{1}{2}\mu_0\boldsymbol{H}_{\text{d}}\cdot\boldsymbol{M} \qquad (3)$$

K is the anisotropy constant, $\boldsymbol{H}_{\text{d}}$ the demagnetizing field, and μ_0 the vacuum permeability. The resolution of these equations is beyond the scope of this article but is detailed for ferromagnetic ribbons in Moutoussamy (2009); it allows one to compute the susceptibility tensor χ linking the fluctuation m of the magnetization to the driving magnetic field $\boldsymbol{h}_{\text{exc}}$ at pulsation ω_{exc}.

$$m = \chi\boldsymbol{h}_{\text{exc}} \qquad (4)$$

In the case of a ferromagnetic ribbon, the susceptibility tensor will have the following components in the $x - y$ plane defined in Fig. 2:

$$\chi = \begin{vmatrix} \chi_{xx}(H,\omega_{\text{exc}}) & \chi_{xy}(H,\omega_{\text{exc}}) \\ \chi_{yx}(H,\omega_{\text{exc}}) & \chi_{yy}(H,\omega_{\text{exc}}) \end{vmatrix}. \qquad (5)$$

The dependency of each component of the susceptibility tensor upon the ambient magnetic field H gives rise to a possible measurement of the magnetic field, as demonstrated by Moutoussamy et al. (2009), even if there is no skin effect when the frequency of the excitation current is as low as a few kHz. The current excitation at high frequency used in classical GMI discussed by Panina and Mohri (1994) and Hika et al. (1996) allows the onset of the skin effect which enhances the impedance variation caused by the susceptibility variation. The ferromagnetic ribbon of the hybrid sensor is made of Ultraperm manufactured by VAC (2002), a high permeability soft ferromagnetic; it has a thickness of 40 μm, and is supplied at a frequency higher than 10 kHz. At this frequency the skin depth of the ribbon is much smaller than its middle thickness and we can consider that eddy currents inside the ferromagnetic ribbon will be confined into the skin depth that will be modified by the magnetic field.

Moreover, by considering the frequency regime distinction between magneto-inductance and GMI in ferromagnetic wire proposed by Panina et al. (1994), the wire-wound ferromagnetic ribbon could be categorized as a GMI sensor. Nevertheless the use of a winding appears as an alternative way to enhance the susceptibility variation caused by the magnetic field. The diagonal components of the susceptibility tensor

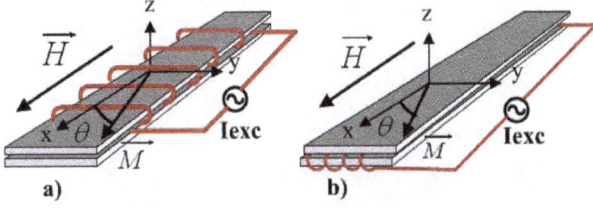

Figure 2. Transverse **(a)** and longitudinal **(b)** wire-wound ferromagnetic ribbon constituting the magnetic sensor under the ambient magnetic field (**H**).

are involved in the transverse and longitudinal wire-wound ferromagnetic ribbons constituting the magnetic sensor (cf. Fig. 2a, b) meanwhile its off-diagonal components are involved in GMIs using a pickup coil and named off-diagonal in Malatek and Kraus (2010). It should be noted that one or two ribbons can be used in both cases, transverse or longitudinal. The main advantage of using two ferromagnetic ribbons is to close the driving magnetic field lines thus allowing us to reduce the driving current, but in each case a magneto-impedance effect will occur. Let us notice the availability of miniaturized commercial magneto-inductive sensors apparently similar to the wire-wound transverse GMI with a single ribbon, although Leuzinger and Taylor (2010) gave no detail on their design.

2.2 Impedance behaviour of the transversal wire-wound GMI

The coil of the wire-wound GMI sandwich is driven by an excitation current I_{exc} at frequency f_{exc}. When a magnetic field **H** is applied to a transverse wire-wound GMI, the modulus of the impedance $|Z(H, f_{exc})|$ starts to increase smoothly before reaching a maximum and then starts to decrease asymptotically to zero as shown in Fig. 3a. The intrinsic sensitivity of the magnetic sensor defined by

$$S_I = \frac{\partial |Z(H, f_{exc})|}{\partial H} \qquad (6)$$

is plotted in Fig. 3b. The value of the applied magnetic field corresponding to the largest negative slope of $|Z(H, f_{exc})|$, i.e. to the maximal intrinsic sensitivity, is close to the anisotropy magnetic field H_{ani} resulting from a combination between the magneto-crystalline and shape anisotropies (Moutoussamy, 2009) of the ferromagnetic ribbon, where the shape anisotropy dominates because of the advantageous shape ratio of the ribbon.

2.3 Elements of impedance modelling

Let us consider a single ferromagnetic ribbon wound by a N turns coil. When the coil is driven by a current I_{exc} at pulsation ω_{exc}, the impedance modulus can be roughly expressed

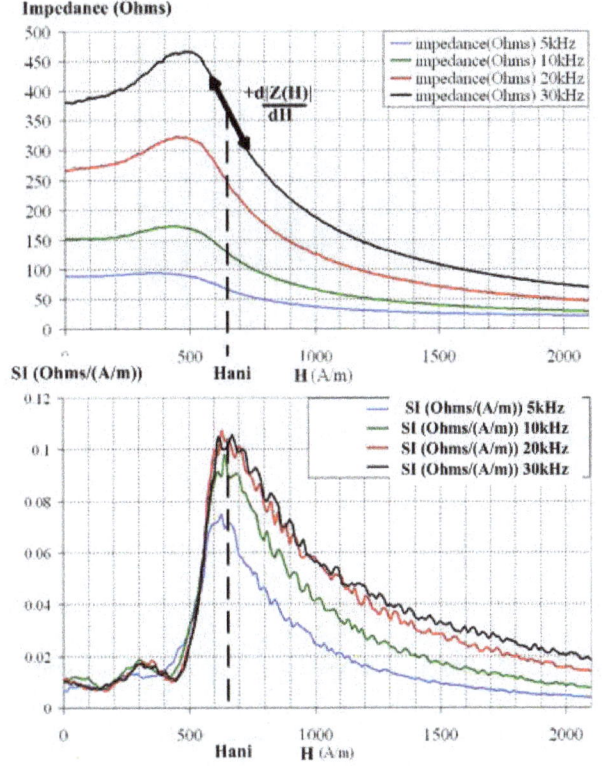

Figure 3. Variations with the magnetic field H of the modulus of the impedance (upper frame) and intrinsic sensitivity (lower frame) of a longitudinal wire-wound magnetic ribbon.

by the following equation if the resistance of the coil can be neglected:

$$|Z(H, \omega_{exc})| = \omega_{exc} \frac{N^2 S_{ribbon} \mu_{app}(H, \omega_{exc})}{l_{ribbon}}, \qquad (7)$$

where S_{ribbon} is the section of the ferromagnetic ribbon, l_{ribbon} its length, and $\mu_{app}(H, \omega_{exc})$ is the apparent permeability of the ribbon along the x direction which can be derived from component $\chi_{xx}(H, \omega_{exc})$ of the susceptibility tensor (Eq. 5) as the external magnetic field to be measured and the excitation magnetic field are both aligned with x. The susceptibility χ_{xx} is assumed to be complex in order to take into account losses in the ferromagnetic ribbon.

$$\mu_{app}(H, \omega_{exc}) = \frac{\chi_{xx}(H, \omega_{exc}) - 1}{1 + N_x \chi_{xx}(H, \omega_{exc})} \qquad (8)$$

The apparent permeability μ_{app} depends upon the susceptibility χ_{xx} but also on the shape of the ferromagnetic ribbon through its demagnetizing factor N_x in the sensing direction. Exact calculations of magnetometric demagnetizing factors of ribbons are given by Aharoni (1998).

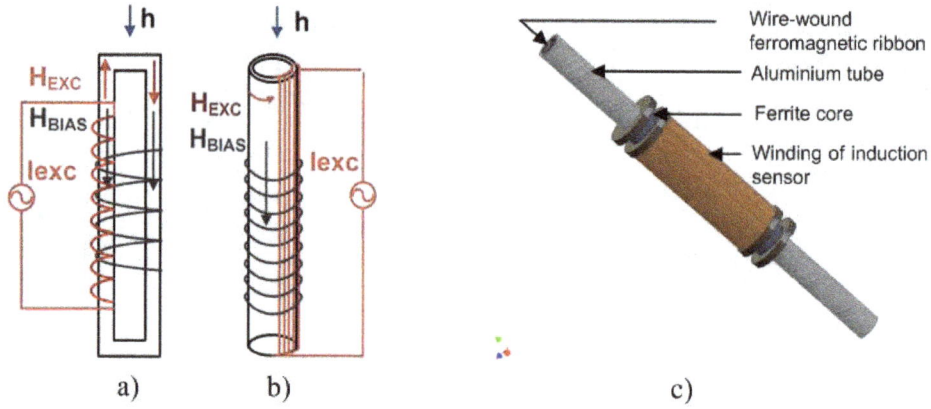

Figure 4. Racetrack (**a**) and cylindrical (**b**) wire-wound ferromagnetic core. (**c**) Hybrid sensor: induction sensor and wire-wound GMI inside an aluminium tube.

2.4 Intrinsic sensitivity

The full expression of the impedance requires a cumbersome analysis to design the sensor, nevertheless Moutoussamy et al. (2007) have made a linear analysis by considering the magnetic field to be measured (h) as a small fluctuation around a magnetic field bias $H = H_{bias}$. The optimal value of intrinsic sensitivity is obtained when $H_{bias} = H_{ani}$. Thus the impedance modulus, which is a function of both the bias magnetic field and the magnetic field to be measured (h), can be expressed by the equation below.

$$|Z((H+h), f_{exc})| = |Z(H_{bias}, f_{exc})| + \frac{\partial |Z(H, f_{exc})|}{\partial H}\Big|_{H=H_{bias}}$$

$$\times h\sin(\omega t) \qquad (9)$$

The optimal intrinsic sensitivity, (at $H = H_{bias}$), defined by

$$S_I = \frac{\partial |Z(H, f_{exc})|}{\partial H}\Big|_{H=H_{bias}} \qquad (10)$$

has been used by Moutoussamy et al. (2007) and Dumay et al. (2012) to characterize the ability of the GMI sensor to measure a weak magnetic field, it can be expressed as follows by combining Eqs. (7) and (10):

$$S_I = \omega_{exc}\frac{N^2 S_{ribbon}}{l_{ribbon}}\frac{\partial \mu_{app}(H, f_{exc})}{\partial H}\Big|_{H=H_{bias}}. \qquad (11)$$

It results from the partial derivative of the apparent permeability with respect to H that the lower the demagnetizing factor is, the higher the sensitivity that will be achieved. This basic fact explains why most of the efficient GMI sensors reported in the literature exhibit advantageous demagnetizing factors due to a large length/diameter ratio. The design of a GMI sensor, looking for the best material and shape of the ferromagnetic ribbon, and the optimal parameters (N, I_{exc}, ω_{exc}, H_{bias}) to get the maximal intrinsic sensitivity, is an experimental task as emphasized by Moutoussamy et al. (2009).

3 Wire-wound GMI ribbon for hybrid sensor design

New configurations of wire-wound GMI sensors based on closed magnetic circuits, presented in Fig. 4a and b, reduce the leakage of the magnetic flux created by the excitation current. This aspect is of great importance in order to combine the GMI sensor with a search coil. The GMI in Fig. 4a reminds the classical racetrack fluxgate reminded by Ripka (2003), while the GMI in Fig. 4b, which consists of a wire-wound sheet of ferromagnetic ribbon rolled around to make a cylindrical muff, reminds the orthogonal fluxgate studied by Paperno et al. (2008) even if the wire-wound GMI does not use a pickup coil since the measurement is done directly on the driving coil. Moreover the ferromagnetic material is not driven at saturation. Even if these "closed magnetic path configurations" exhibit interesting properties, the transverse configuration of GMI illustrated in Fig. 2a has been preferred for the rocket instrument due to its simplicity of manufacturing. The transverse wire-wound GMI sensor used for the experiment had 100 turns both for excitation and bias coils, the ribbon made of Fe-Ni alloy (Ultraperm manufactured by VAC, 2002) had a length of 120 mm, a width equal to 4 mm, and a thickness of 40 µm. The excitation coil was wound around one of the two ribbons, while the bias coil surrounded the two ribbons.

In the transverse wire-wound configuration with a closed magnetic path, the excitation and bias driving current will create an AC magnetic field. This magnetic field will have leakages flux which are reduced by the addition of an aluminium tube of 1mm thickness since the eddy current generated by the AC magnetic field expels the magnetic field outside the tube (this effect is stronger at frequencies for which the skin depth is comparable to the thickness of the tube). Finally, the GMI sensor surrounded by an aluminium tube is inserted inside a hollow ferrite magnetic core on which the induction sensor (also known as search coil) is built as shown in Fig. 4c. Thus, the AC magnetic field expelled from the aluminium tube (120 mm length, 6 mm internal diameter and 7.95 mm external diameter) is caught by the surrounding ferrite core of the induction sensor (Tumanski, 2007 and Coillot and Leroy, 2012). The ferromagnetic core of the induction sensor is a hollow tube made of a high permeability Mn-Zn ferrite ($\mu_r > 2500$) having a length of 50 mm, an internal diameter of 8 mm, and an external diameter of 10 mm. The winding consists of 9600 turns of a copper wire having a diameter equal to 70 µm in order to fulfill the detection requirement of in situ electromagnetic waves specified by a noise equivalent magnetic induction (NEMI) lower than 100 fT/$\sqrt{\text{Hz}}$ at few kHz. The induction sensor was combined with a low-noise feedback flux preamplifier (voltage and current input noise were respectively 6 nV/$\sqrt{\text{Hz}}$ and 100 fA/$\sqrt{\text{Hz}}$) to achieve the required NEMI. The low-noise amplifier implemented for the experiment was similar to the one described by Seran and Fergeau (2005), making use of a dual JFET transistor type LS-U404 from Linear Systems. A usual flux feedback, see for example Seran and Fergeau (2005) or Coillot and Leroy (2012), uses the output of the low-noise amplifier to generate a flux opposite to the measured one in order to flatten the transfer function. It has been checked experimentally that the internal aluminium tube does not increase the NEMI of the induction sensor above the specified level of 100 fT/$\sqrt{\text{Hz}}$.

4   Signal conditioning principle

Let us consider a bias field H_{bias} applied to the ferromagnetic ribbon. Accordingly to Eq. (9) the output voltage of the wire-wound transverse GMI supplied with a current I_{exc} is written

$$V_{\text{GMI}} = -[|Z(H_{\text{bias}})| + S_I(H_{\text{bias}})\, h \sin(\omega t)] I_{\text{exc}} \sin(\omega_{\text{exc}} t) \quad (12)$$

The implemented signal conditioning takes advantage of the symmetry of the modulus of the impedance, $|Z(-H)| = |Z(H)|$. For two opposite bias fields $\pm H_{\text{bias}}$ the offsets $Z(\pm H_{\text{bias}}) I_{\text{exc}} \cos(\omega_{\text{exc}} t)$ are equal meanwhile the slopes $S_I(\pm H_{\text{bias}})$ are opposite. If the bias field is modulated at a bias frequency f_{bias} then the magnetic field $h \sin(\omega t)$ to be measured is modulated twice: at the excitation frequency and at the bias frequency (plus bias harmonics) while the impedance will not be modulated by the bias field. We chose $f_{\text{exc}} = 100\,\text{kHz}$ and $f_{\text{bias}} = 100\,\text{Hz}$. If we assume a sinusoidal waveform of the excitation I_{exc} at f_{exc} and a square waveform of the bias field, the voltage of the GMI can be written

$$V_{\text{GMI}} = (I_{\text{exc}} \cos(\omega_{\text{exc}} t))$$

$$\times \left((Z(H_{\text{bias}}) + \sum_{k}^{\infty} S_I(H) \times h \sin(\omega t) \frac{4}{k\pi} \cos(k\omega_{\text{bias}} t) \right). \quad (13)$$

Consequently, the magnetic field signal (h) can be retrieved by using a double demodulation at both f_{exc} and f_{bias}

while the offset value (corresponding at $Z(H)$) will be easily suppressed. It should be noticed that the phase of demodulation at f_{exc} should be adjusted experimentally to get the maximum sensitivity. For wire-wound GMI supplied at low frequency, through the voltage source in series with the resistance Rs (cf. Fig. 5), the demodulation in phase with the inductive part is optimum. A schematic explanation of the electronic principle that is used to cancel the offset is presented in Fig. 5.

The demodulation is realized using multipliers based on a switch matrix (cf. Fig. 5). A switch matrix, based on DG419 switches, is used to multiply the signal by ± 1 with low added noise thanks to their extremely low R_{DSON}. This way of demodulating is more efficient in terms of power consumption than demodulation using analog multipliers (for low-noise multipliers, like AD835, current consumption is about 16 mA). One can notice that demodulation using peak detector, used by Dumay et al. (2012), offers also an efficient way to demodulate the GMI signal even if it is not applicable to the GMI with alternating bias. We will now expose the principle of this electronic conditioning. For this purpose, we will consider only the fundamental of the bias magnetic field. If we examine the GMI voltage after the primary amplification ($A1$) in the frequency domain (Fig. 6), it appears that the signal (h) is twice modulated around $f_{\text{exc}} \pm f_{\text{bias}}$ while the offset remains at f_{exc}.

Then, after a first demodulation at f_{exc} (with an optimal phase difference with excitation current determined experimentally) the offset signal is moved at null frequency while the signal (h) is moved around f_{bias} (cf. Fig. 7).

Next the offset is removed using simple high-pass filtering (Fig. 8) and finally the signal is moved in its original band (Fig. 9) thanks to the demodulation at f_{bias}.

Surprisingly, this offset cancellation technique seems very similar to the one used by Sasada (2002) for fluxgate in orthogonal mode.

5   Hybrid magnetometer performances

The characterization in terms of intrinsic sensitivity has been achieved for the GMI part of each hybrid sensor (Fig. 10a) in order to find the best compromise between a high sensitivity S_I, a high dynamic range and a low current consumption. In order to determine the minimum dynamic range in the Earth's magnetic field, characterizations of S_I have been done in 3 directions (Fig. 10a): parallel to Earth's magnetic field, anti-parallel and orthogonal. This explains the shift of the curves around the centered S_I curve which corresponds to orthogonal position. The electronic conditioning, which includes an offset cancellation technique offers an intrinsic linearization, demonstrated in Fig. 10b. The linearity range, approximatively $\pm 25\,\mu\text{T}$ would be significantly improved by using a flux feedback. The transfer function (Fig. 10c) obtained for the GMI part of the hybrid sensor (picture label in

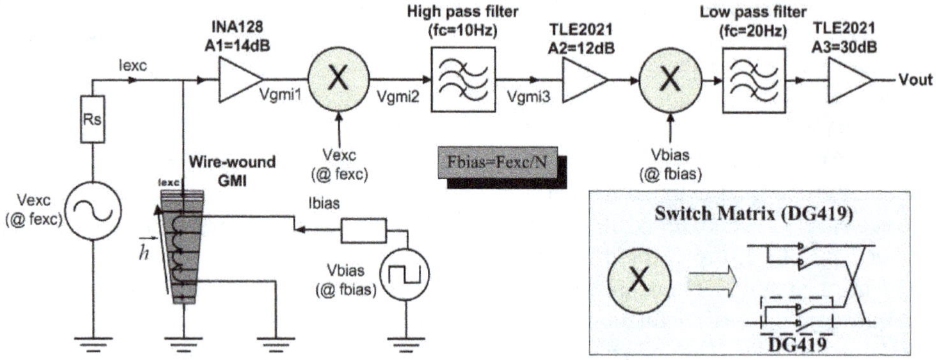

Figure 5. Signal conditioning principle.

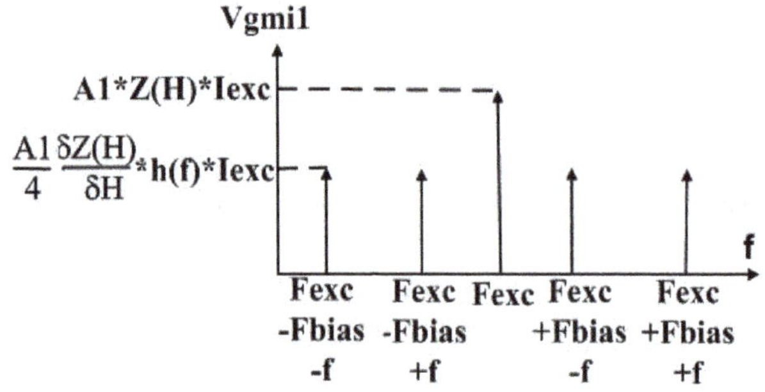

Figure 6. Frequency domain of GMI signal (Vgmi1) after amplification A1.

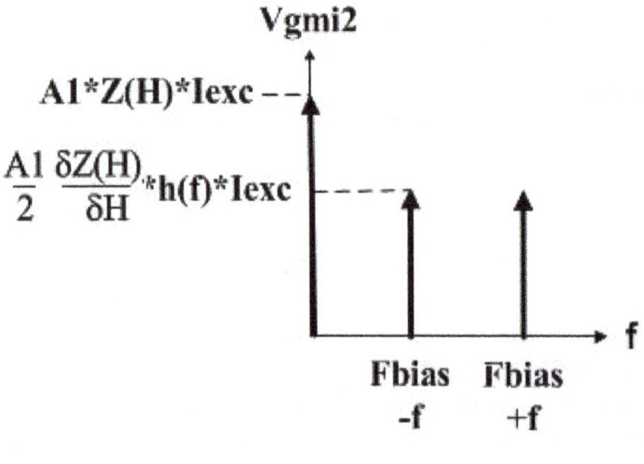

Figure 7. Frequency domain of GMI signal (Vgmi2) after multiplication by clock at f_{exc}.

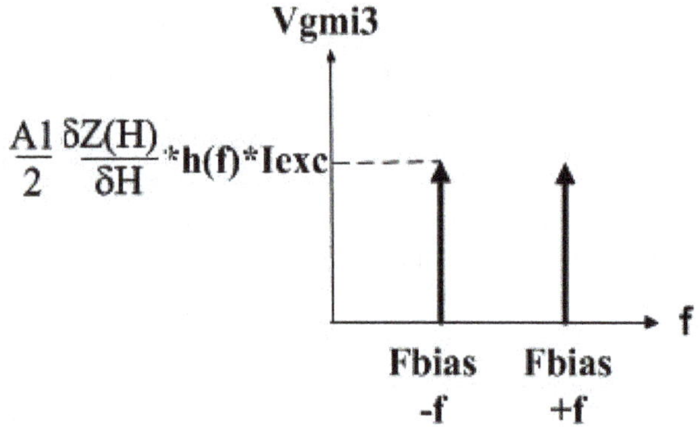

Figure 8. Frequency domain of GMI signal (Vgmi3) after high-pass filter.

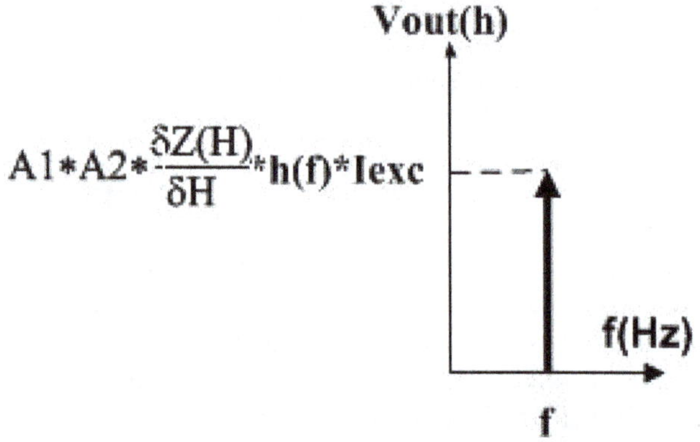

Figure 9. Frequency domain of GMI signal (Vgmi4) after multiplication by clock at f_{bias}, amplification ($A2$) and low-pass filtering.

Fig. 10), using the electronic principle described previously, exhibits a constant gain from DC up to 20 Hz. The NEMI of the GMI part (green plot in Fig. 10d) is about 600 pT/$\sqrt{Hz}$ at 1 Hz, far from the best fluxgate one, mentioned by Ripka (2003), while the NEMI for the inductive part of the hybrid sensor when the GMI is switched OFF (blue plot in Fig. 10d) reaches 100 fT/$\sqrt{Hz}$ at 4 kHz. Concerning the GMI part, many possibilities of improvement have to be investigated, some of them could be inspired from Dumay et al. (2012); namely, turn number, increase of excitation frequency and

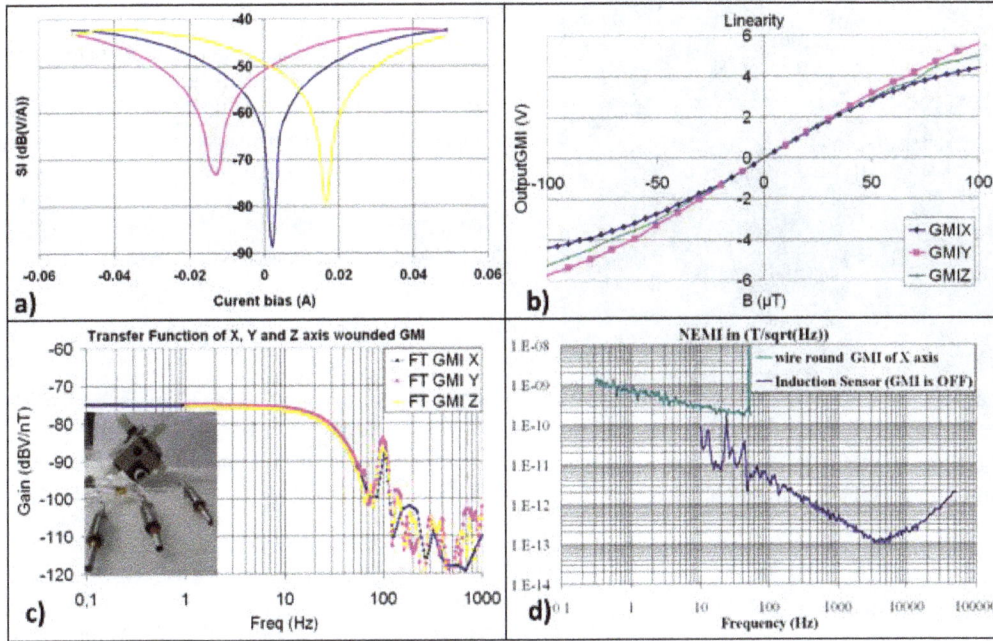

Figure 10. (a) Performances of the tri-axes hybrid magnetometer. The intrinsic sensitivity S_I has been measured when the sensor axis is perpendicular to an external DC magnetic field (black curve) and when it is either parallel or anti-parallel (yellow and magenta curves) , more precisely $20\log(S_I)$ is plotted in frame **(a)**, frame **(b)** demonstrates the linearity of the sensor with respect to the external DC magnetic field, frame **(c)** shows the transfer functions $20\log(\text{Gain})$, Gain being measured in V/nT, frame **(d)** is a plot of the noise equivalent magnetic induction in $T/\sqrt{Hz}$ between 0.3 and 50 Hz (in green) for the GMI part (the peak at 50 Hz being due to the network power supply) and between 10 Hz and 50 kHz (in blue) for the induction sensor part.

input voltage noise reduction. The induction sensor part of the hybrid sensor becomes more sensitive than the GMI sensor from a few Hz. One should notice an interesting property of the electronic principle which is the intrinsic linearization of the output voltage despite a strong variation of the intrinsic sensitivity around a bias field value. Thanks to the symmetry of S_I curves (Fig. 10a) and the alternating biasing, the variation of S_I due to the shift of the positive bias field when H field is measured is compensated by an opposite variation of S_I for the negative bias field. Concerning the induction sensor it should be noticed that the strong signature of the alternating bias of the wire-wound GMI sensor remains at the fundamental of bias magnetic field ($f_{\text{bias}} = 100\,\text{Hz}$) but it does not affect too much the induction sensor transfer function above a few 100 Hz since the reduction of the AC disturbance from the wire-wound GMI is more efficient when frequency increases thanks to the aluminium tube which acts as a high-pass filter magnetic shielding. However the induction sensor NEMI was worst in the presence of the GMI because of residual disturbing fields (mainly lines at the bias field harmonics frequencies). The way to reduce it would be to increase bias frequency out of the induction sensor frequency range. The performance of the hybrid magnetometer is summarized in Table 1.

Table 1. Performance of the tri-axes hybrid magnetometer combining a search coil and a wire-wound GMI.

Frequency range	DC–20 kHz
Sensitivity	200 kV T^{-1}
Linearity range	±25 µT
NEMI (1 Hz)	600 pT/ $\sqrt{Hz}$
NEMI (4 kHz)	100 fT/ $\sqrt{Hz}$
Power consumption	500 mW
Mass (preamplifier box)	550 gr
Mass (sensors)	500 gr
Size (tri-axis sensors)	120 mm × 120 mm × 120 mm

6 Instrument and measurement discussion

The NASA sounding rocket CHARM-II was launched successfully from Poker Flat, Alaska, on 16 February 2010. Figure 11 shows the launch of the rocket, where the green light in the lower right is a wonderful aurora borealis. The prototype embedded in the scientific payload has operated nominally. The measured DC magnetic field from the wire-wound GMI has been used to reconstruct the magnetic field line along the rocket trajectory. The measured magnetic field line fits well with the computed Earth magnetic field (using the

Figure 11. The launch of the CHARM-II rocket from Poker Flat (Alaska) during an aurora borealis (Copyright Micah P. Dombrowski).

International Geomagnetic Reference Field model) except at the time of the boom deployment where a jump of about 20 % of the Earth's magnetic field magnitude is obtained. The origin of this jump has not been elucidated and for this reason the measurement of the Earth's magnetic field is not reported in the paper.

The combination of the inductive sensor with a wire-wound racetrack GMI sensor allows one to cover a frequency range measurement from quasi-DC up to a few kHZ. The major drawback of this solution of hybrid sensor is the residual flux leakage of both excitation and bias fields which disturbs the induction sensor part by adding lines at harmonics of the bias frequency. That problem could be overcome by investigating an increase of the bias frequency. Moreover, the flux of the magnetic field to be measured is divided into two parts: one part is diverted into the ferromagnetic ribbon of the GMI while the other part is diverted into the ferromagnetic core of the search coil. That reduces the efficiency if we compare to the GMI or induction sensor working separately. For these reasons, possibly combined to a magnetically quiet launch period, the instrument did not measure the magnetic component of electromagnetic waves during the flight. Finally, the hybridization of the induction sensor with a wire-wound GMI sensor is not well suited for the induction sensor and the NEMI at 1 Hz remains much higher than the one of a fluxgate. Nevertheless the principle of the wire-wound GMI sensor and its electronics, presented in this paper, are easy to implement and allow to build, in a pedagogic way, an efficient magnetometer.

Acknowledgements. This research work has been funded by the Research and Development Programme of CNES, the Space French Agency. The authors are grateful to J. Labelle for their participation in the NASA CHARM-II rocket experiment. The authors would also like to thank Rabah Ikhlef and Zaki Grig Ahcene, master students who have participated with great enthusiasm on the design, tests and manufacturing of the magnetic sensors.

Edited by: A. Schütze
Reviewed by: two anonymous referees

References

Aharoni, A.: Demagnetizing factors for rectangular ferromagnetic prisms, J. Appl. Phy., 83, 3432–3434, 1998.

Belevitch, V.: The lateral skin effect in a flat conductor, Philips tech Rev., 32, 221–231, 1971.

Coillot, C. and Leroy, P.: Induction Magnetometers: Principle, Modeling and Ways of Improvement, Magnetic Sensors – Principles and Applications, edited by: Kuang, K., ISBN: 978-953-51-0232-8, InTech, 2012.

Dufay, B., Saez, S., Dolabdjian, C., Yelon, A., and Ménard, D.: Characterizaion of an optimized off-diagonal GMI based magnetometer, IEEE Sensors, 99, 379–388, 2012.

Garcia-Arribas, A., Barandiaran, J. M., and de Cos, D.: Finite element method calculations of GMI in thin films and sandwiched structures: size and edge effects, Journal of Magnetism and Magnetic Materials, 320, e4–e7, 2008.

Harrison, H. P., Turney, G. P., Rowe, H., and Gollop, H.: The electrical properties of High permeability Wires Carrying Alternating Current, Proc. Royal. Soc., 157, 451–479, 1936.

Hika K. et al., Magneto-Impedance in Sandwich Film for Magnetic Sensor Heads, IEEE Trans. Magn, Vol. 32, pp. 4594–4596, 1996.

Leuzinger, A. and Taylor, A.: Magneto-Inductive Technology Overview, PNI white paper, February 2010.

Malatek, M. and Kraus, L.: Off-diagonal GMI sensor with stress-annealed amorphous ribbon, Sensors and Actuators, 64, 41–45, 2010.

Moutoussamy, J.: Nouvelles solutions de capteurs effet de magneto-impdance gé ante: Principe, Modélisation et Performances, PhD dissertation, Ecole Normale Supérieure de Cachan (France). http://tel.archivesouvertes.fr/docs/00/50/57/44/PDF/Moutoussamy2009.pdf, 2009.

Moutoussamy, J., Coillot, C., Alvès, F., and Chanteur, G.: Feasibility of a giant magneto-impedance sandwich magnetometer for space applications, IEEE Sensors Conference (Atlanta-USA), 1013–1016, October 2007.

Moutoussamy, J., Coillot, C., Alvès, F., and Chanteur, G.: Longitudinal and transverse coiled giant magnetoimpedance transducers: principle, modelling and performances, Transducer 09 (Colorado-USA), June 2009.

Morikawa, T., Nishibe, Y., and Yamadera, H.: Giant Magneto-Impedance Effect in Layered Thin Films, IEEE Transactions on Magnetics, 33, 4367–4372, 1997.

Panina, L. V. and Mohri, K.: Magneto-impedance effect in amorphous wires, Appl. Phys. Lett., 65 pp., 1189–1191, 1994a.

Panina, L. V., Mohri, K., Bushida, K., and Noda, M.: Giant magneto-impedance and magneto- inductive effects in amorphous alloys, J. Appl. Phys., 76, 6198–6203, 1994b.

Paperno, R., Weiss, E., and Plotkin, A.: A Tube-Core Orthogonal Fluxgate Operated in Fundamental Mode, IEEE Transactions on Magnetics, 44, 4018–4021, 2008.

Ripka, P.: Advances in Fluxgate sensors, Sensors and Actuators A, 106, 8–14, 2003.

Sasada, I.: Symmetric response obtained with an orthogonal fluxgate operated in fundamental mode, IEEE Trans. Magn., 38, 3377–3379, 2002.

Samara, M., LaBelle, J., Kletzing, C. A., and Bounds, S. R.: Electrostatic upper hybrid waves where the upper hybrid frequency matches the electron cyclotron harmonic in the auroral ionosphere, Geophys. Res. Lett., 31, L22804, doi:10.1029/2004GL021043, 2004.

Seran, H. C. and Fergeau, P.: An optimized low frequency three axis search coil for space research, Rev. Sci. Instrum., 76, pp. 044502-1 044502-11, 2005.

Tumanski, S.: Induction coil sensors – A review, Meas Sci. Technol., 18, R31–R46, 2007.

VAC: Vacuumschmelze GMBH and Co Kg, Soft Magnetic Materials and Semi-Finished Products, Edition 2002.

3

Characterization and simulation of peristaltic micropumps

M. Busek, M. Nötzel, C. Polk, and F. Sonntag

Fraunhofer IWS Dresden, Dresden, Germany

Correspondence to: M. Busek (mathias.busek@iws.fraunhofer.de)

Abstract. The aim of the work is to find an analytical model of a pneumatic micropump which was integrated into a cell-culture system. This allows the estimation of peak velocities and wall-shear stress influencing the cultured cells in our multi-organ-chip (MOC) with respect to the applied pressure and the geometric properties of the pump. By adjusting those parameters, one can imitate physiological or pathological heart activity. The calculated flow within the MOC was compared to experimental results obtained via the non-invasive micro-PIV method.

1 Introduction

Within the last few years, a flexible lab-on-a-chip system with an integrated peristaltic micropump has been developed (Marx et al., 2012; Sonntag et al., 2010; Wagner et al., 2013; Winkelmann et al., 2013). The pump is pneumatically actuated and establishes a pulsatile fluid flow within a closed fluidic circuit of several cell-culture segments. Each segment corresponds to a certain organ (liver, skin, kidney, etc.), forming a human-on-a-chip. This so-called multi-organ-chip (MOC) can be used by the cosmetic and pharmaceutical industries and may replace animal testing within the next few years, as predicted by Baker (2011). Each organ has a specific media uptake and should be perfused with a well-defined flow rate. Therefore a mathematical model will be used to calculate the channel distribution. Furthermore, the wall-shear stress to the adherent cells should be known. In a previous work (Schimek and Busek, 2013) the chip was cultured with endothelial cells to determine the influence of different shear stress values on the cell formation. The flow velocity was measured using micro-PIV. Those experimental results enabled us to create a mathematical model of the pump specified in this work. The model is useful especially when micro-PIV measurement cannot be applied, e.g. when the chip is filled with blood or the walls of channels are fully covered with a cell layer.

2 Material and methods

2.1 Device design and fabrication

We designed and fabricated a microfluidic multi-organ-chip (MOC) device accommodating two separate microvascular circuits, each operated by a separate peristaltic on-chip micropump. Figure 1 illustrates the system at a glance. The cover plate accommodates six air pressure fittings and four inserts forming 300 µL compartments, each for media exchange and later integration of organ equivalents. The MOC holder supports controlled constant tempering of the MOC at 37 °C (Fig. 1a). Peristaltic on-chip micropumps (Fig. 1b) were modified from Wu et al. (2008). Micropump software control facilitates both clockwise and anticlockwise fluid flow. The flow rate (Q) can be varied by the adjustment of the pumping frequency. Each circuit (Fig. 1c) comprises a channel volume of 10 µL.

Standard soft lithography and replica moulding of PDMS (Sylgard 184, Dow Corning, Midland, MI, USA) have been applied for MOC fabrication. To fabricate the microsystem, the cover plate (CP) was treated with a silicon rubber additive (WACKER® PRIMER G 790; Wacker Chemie, Munich, Germany) at 80 °C for 20 min. The prepared cover plate was plugged into the master mould (channel height 100 µm, width 500 µm) and PDMS (10 : 1 v/v ratio of PDMS to curing agent) was injected into this casting station. The setup was incubated at 80 °C for at least 60 min. Teflon screws were

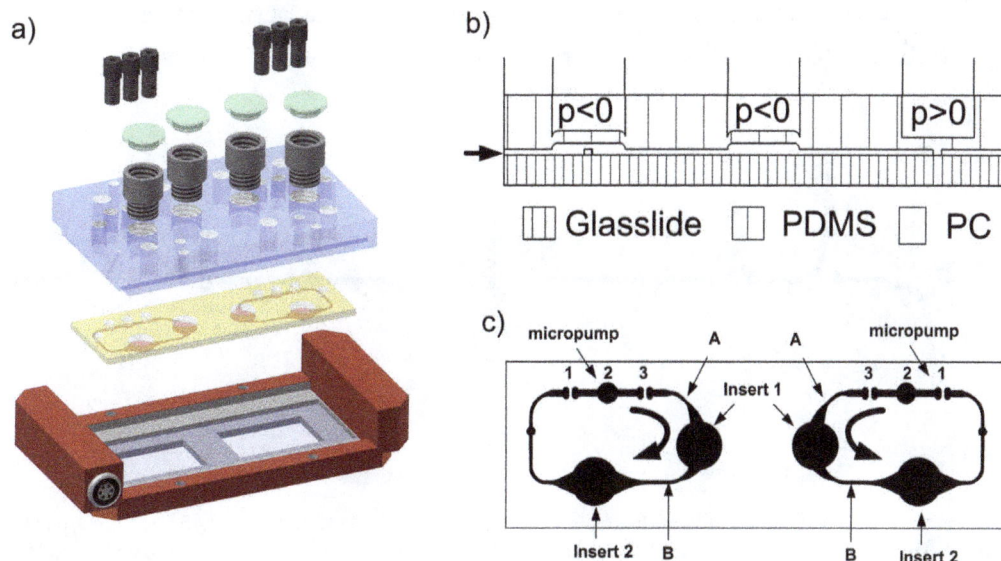

Figure 1. The microfluidic MOC device at a glance. **(a)** Exploded view of the device. **(b)** Cross-section of a peristaltic on-chip micropump; the arrow indicates the direction of flow. **(c)** Top view of the MOC layout illustrating the two separate microfluidic circuits. Spots A and B of each circuit designate the position of non-invasive fluid flow measurement.

used to generate the four PDMS-free culture compartments and the six 500 µm-thick PDMS membranes constituting the two on-chip micropumps (three membranes per micropump). The cast PDMS slice bonds fluid-tight to the CP. Thereafter, the PDMS slice attached to the CP was irreversibly bonded by low-pressure plasma oxidation treatment (Femto; Diener, Ebhausen, Germany) to a microscope slide. A sterile medium was injected immediately into the two microvascular circuits to avoid surface neutralization.

2.2 Characterization of fluid dynamics with µPIV

We applied non-invasive micro-particle image velocimetry (µPIV), reviewed in detail by Lindken (2009), to characterize the fluid flow in spots A and B (cf. Fig. 1c) of the microfluidic circuit. In brief, a Zeiss Primovert inverting microscope (Zeiss, Jena, Germany) with a standard halogen lamp as a continuous light source, coupled to a CMOS camera (Baumer Optronic HXC40, resolution: 2048×2048 pixels, interface: CameraLink; Baumer Optronic, Radeberg, Germany) was used to track the movement of 16 µm polystyrene beads (4×10^4 g mL^{-1}; Life Technologies, Darmstadt, Germany) at an exposure time of 4 µs per single image. A low magnification (4×) was chosen to constrain the shift between two frames to approximately 50 pixels (1 pixel = 3.2 µm). The z focus was set to the centre of the fluidic channel in the respective spot (50 µm above the glass slide) to detect the peak velocity. An interrogation window at the centre of the fluidic channel (1024×100 pixels, 3.28×0.32 mm^2) was observed, achieving frame rates up to 3200 fps. Finally, the correlation was carried out with a software program (Fraunhofer

IWS, Dresden, Germany) which analyzes an image stack of 15 000 frames, calculating the correlation maximum for the x component of the displacement in a specified area. The calculated values of five successive frames are averaged to minimize artefacts. The following pump configuration was used for all experiments: pressure: 500 mbar; vacuum: -520 mbar; and air flow: 1.5 L min^{-1} at 350 mbar. Figure 1c shows the different spots A and B where the velocity was measured.

3 Results

3.1 Time-resolved micro-PIV

We successfully applied µPIV at the two different spots of the MOC circuits to examine the transient behaviour of the pump (Fig. 2). Figure 2 shows a full pumping cycle consisting of four steps at a frequency of 0.48 Hz. Huang et al. (2006, 2008) also used µPIV to characterize the peristaltic pump at different pumping frequencies. Their results show a good correlation with the previously obtained results (Schimek and Busek, 2013).

3.2 Mathematical model of the pump

The mathematical model of the pump, schematically shown in Fig. 3, is an analytical approach consisting of the following two parts:

1. Airflow from the outputs of the control unit to the deflection of the membrane

2. Liquid transport from the pumping chamber to a reservoir.

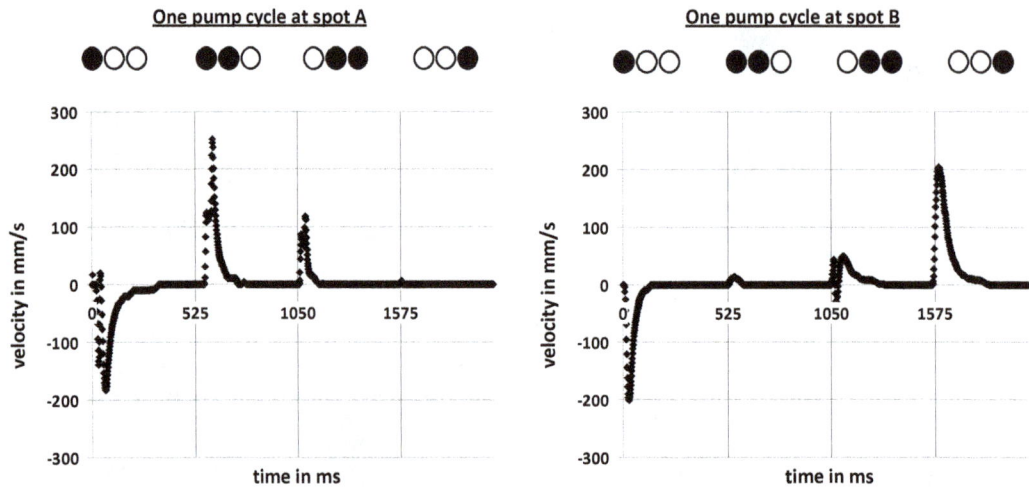

Figure 2. Exemplary velocity profiles throughout the four stages of a full pumping cycle (frequency: 0.48 Hz) measured at the two discrete fluid flow analysis spots. The state of the three pump valves is represented by the circles (black = valve closed, white = valve open).

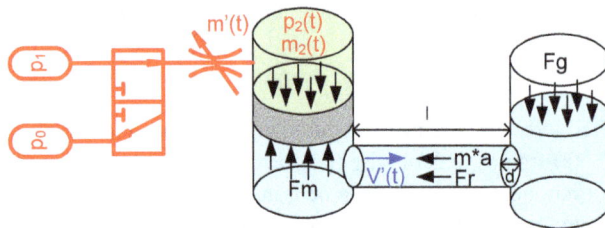

Figure 3. Schematic of the pump model with pneumatic–fluidic coupling.

By combining both parts to one differential equation, the velocity–time dependency for the liquid flowing in the channel when pressure is applied to the membrane can be calculated.

On the left side of the picture one can see the pneumatic switching circuit and the throttle which regulates the mass air flow $m'(t)$ to the pneumatic chamber with the dead volume V_0 at the membrane. The volume V_0 is constant. Described is the moment when the valve switches from p_0 to p_1. At this moment the air inside the pneumatic chamber is compressed, which leads to a pressure and density increase. The universal gas law describes this behaviour:

$$p_2(t) = p_0 + R_s \cdot T \cdot (\rho(t) - \rho_0) = p_0 + \frac{R_s \cdot T}{V_0} \cdot \int_0^t m'(t)\,\mathrm{d}t \quad (1)$$

The mass air flow through the throttle is represented by the Hagen–Poiseuille law:

$$m'(t) = \frac{\rho_1 \cdot \pi \cdot a^4 \cdot \Delta p}{8 \cdot \eta_L \cdot b} = \rho_1 \cdot k \cdot (p_1 - p_2(t)) \quad (2)$$

The geometrical properties of the throttle (a...radius, b...length) as well as the dynamic viscosity of air η_L are

summarized as the coefficient k. The universal gas constant R_s, the temperature T and the density ρ_1 are equal to p_1. The air flow through the throttle is therefore directly proportional to the pressure loss between both sides. The following linear differential equation describes the time-dependent pressure behaviour:

$$p_2'(t) = k \cdot \frac{p_1}{V_0} \cdot (p_1 - p_2(t)) \quad (3)$$

The following solution can be assumed:

$$p_2(t) = e^{-\frac{k \cdot p_1 \cdot t}{V_0}} \left[p_0 + p_1 \left(e^{\frac{k \cdot p_1 \cdot t}{V_0}} - 1 \right) \right] \quad (4)$$

Under constant pressure the membrane is deflected. The following formula describes the deflection of a round membrane with clamped edges (Timoshenko and Woinowsky-Krieger, 1959):

$$z(r) = z_{max} \left(1 - \frac{r^2}{r_0^2} \right)^2 \quad (5)$$

where z_{max} describes the maximum of the elongation at the centre, r is the position at the membrane and r_0 represents the membrane radius. The flow within the microchannel is equal to the displaced volume per time which can be determined by integrating this equation over the two cylindrical-space dimensions r and ϕ:

$$\begin{aligned} V(t) &= \iint z(r,t)\,\mathrm{d}\phi\,\mathrm{d}r = 2\pi \cdot z_{max}(t) \int_0^{r_0} r \cdot \left(1 - \frac{r^2}{r_0^2}\right)^2 \mathrm{d}r \\ &= \frac{\pi}{3} \cdot r_0^2 \cdot z_{max}(t) \end{aligned} \quad (6)$$

$$Q(t) = V'(t) = \frac{\pi}{3} \cdot r_0^2 \cdot z_{max}'(t) = \frac{\pi}{3} \cdot r_0^2 \cdot c \cdot p_2'(t) \quad (7)$$

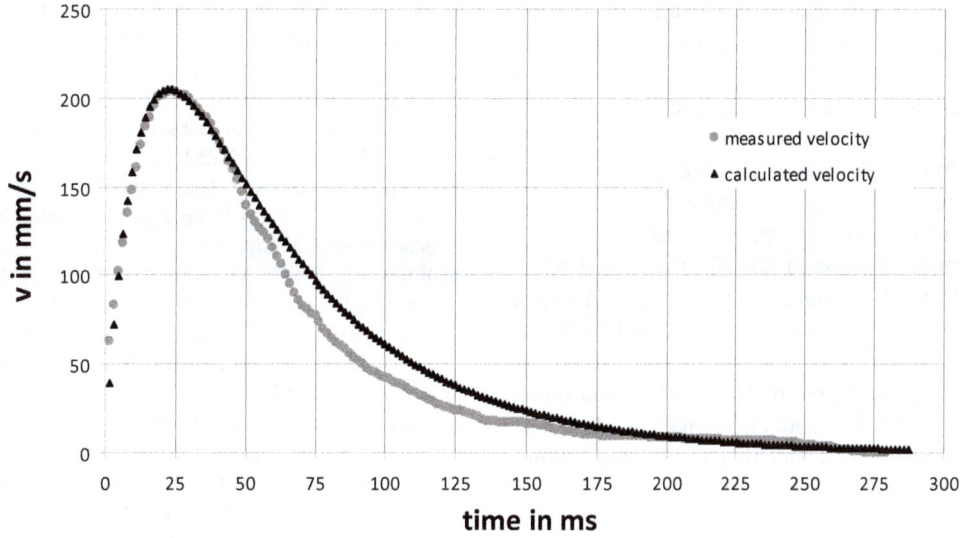

Figure 4. Calculated vs. measured velocity for one pump pulse.

As shown in Fig. 3, there are several counterforces working against the applied pressure:

1. The stiffness of the membrane, represented by the factor c (Thangawng, 2007).

2. The friction force F_r and the mass inertia $m \cdot a$ caused by the fluid flowing from the pump chamber to the reservoir through the tube-like channel.

3. The weight force F_g counteracting by the fluid level of the reservoir.

The effective driving pressure p_{2e} is represented by:

$$p_{2,e}(t) = p_2(t) - \frac{1}{r_0^2 \cdot \pi}\left(F_r + m \cdot a + F_g\right) \tag{8}$$

with the transported fluid mass m and the fluid acceleration a. If we assume that the displaced volume is very low, then the weight force F_g of the fluid could be neglected.

For a tube-like channel the friction force could be calculated by:

$$F_r = 8 \cdot \pi \cdot \eta \cdot l \cdot \bar{v} \tag{9}$$

where the dynamic viscosity of the media is represented by η, the channel length is represented by l and the mean velocity of the fluid in the tube is expressed by v.

The mass inertia can now be calculated with the fluid density ρ_w and the channel radius R:

$$m \cdot a = \rho_w \cdot \pi \cdot R^2 \cdot l \cdot v'(t) \tag{10}$$

If we differentiate Eq. (8) with respect to t and insert the solution in formula (7), the volumetric flow rate can be calculated. By dividing by the channel area ($\pi \cdot R^2$), one can formulate the following differential equation for the flow velocity

inside the channel:

$$v(t) = \frac{c}{3} \cdot \left[\frac{r_0^2}{R^2} \cdot \frac{k \cdot p_1}{V_0} \cdot \left\{ p_1 - e^{-\frac{k \cdot p_1 \cdot t}{V_0}} \left\langle p_1 \cdot \left(e^{\frac{k \cdot p_1 \cdot t}{V_0}} - 1\right) + p_0\right\rangle\right\} \right. $$
$$\left. - \frac{8}{r_0^2} \cdot \eta \cdot l \cdot v'(t) - \rho_w \cdot l \cdot v''(t)\right] \tag{11}$$

If we assume the following start and stop condition: $v(t = 0) = 0$; $v(t \to \infty) = 0$, the general solution of this differential equation is:

$$v(t) = v_{max} \cdot \left[2 \cdot e^{(-C_1 \cdot t)} - e^{(-C_2 \cdot t)} - e^{(-C_3 \cdot t)}\right] \tag{12}$$

To calculate the coefficients C_1, C_2 and C_3, a substitution and a numerical fitting algorithm were used. The exponents C_2 and C_3 could therefore be expressed with respect to C_1 with the three sampling points (t_0, v_{max}), $(t_1 = 2t_0, v_1)$ and $(t_2 = 3t_0, v_2)$. The velocity–time dependency shown in Fig. 2 at spot B is used to fit the parameters of the model to the experimental data. Figure 4 shows the measured and calculated velocity for one pump pulse:

The fitted parameters for this pressure pulse are $C_1 = 19$, $C_2 = 74.5$, $C_3 = 97$ for $t_0 = 0.025$ s, $v_{max} = 204$ mm s^{-1} and $v_1 = 121$ mm s^{-1}. This leads to the following approximation formula to calculate the pressure pulse for one pump cycle:

$$v(t) = 204 \frac{mm}{s} \cdot \left[2 \cdot e^{(-19 \cdot t)} - e^{(-74.5 \cdot t)} - e^{(-97 \cdot t)}\right] \tag{13}$$

4   Conclusions

Within this work the developed peristaltic micropump was characterized with the help of non-invasive μPIV measurements and modelled using an analytical approach. In contrast to the model described by Nedelcu et al. (2007), the mass inertia and friction force of the liquid are also to be considered. Furthermore, the calculated results for a pump pulse of the periodic pulsatile fluid flow fit the experimental data very well. Previous findings reported by Schimek and Busek (2013) show that the described pump is suitable for the cultivation of human endothelial cells; the produced wall-shear stress is within the physiological range. The developed model can be used to calculate the peak velocity with respect to the pumping parameters (pressure, diameter, etc.). In further works the influence of the elastic channel walls should be considered. It is planned to develop a network model of the MOC with a peristaltic pump based on simulation platform SimulationX, similar to the electrical equivalent network for a peristaltic pump described by Bourouina and Grandchamp (1996). This tool allows the calculation of the flow in a branched fluidic device with several cell-culture systems connected in parallel.

Acknowledgements. The authors thank the Federal Ministry of Economics and Technology for funding this work within the ZIM SimFluNet project.

Edited by: N.-T. Nguyen
Reviewed by: two anonymous referees

References

Baker, M.: A living system on a chip, Nature, 471, 661–665, 2011.

Bourouina, T. and Grandchamp, J.-P.: Modeling micropumps with electrical equivalent networks, J. Micromech. Microeng., 6, 398–404, 1996.

Huang, C.-W., Huang, S.-W., and Lee, G.-B.: Pneumatic micropumps with serially connected actuation chambers, J. Micromech. Microeng., 16, 2265–2272, 2006.

Huang, S.-B., Wu, M.-H., Cui, Z., Cui, Z., and Lee, G.-B.: A membrane-based serpentine-shape pneumatic micropump with pumping performance modulated by fluidic resistance, J. Micromech. Microeng., 18, 045008, doi:10.1088/0960-1317/18/4/045008, 2008.

Lindken, R.: Micro-particle image velocimetry (μPIV): Recent developments, applications, and guidelines, Lab Chip, 9, 2551–2567, 2009.

Marx, U., Walles, H., Hoffmann, S., Lindner, G., Horland, R., Sonntag, F., Klotzbach, U., Sakharov, D., Tonevitsky, A., and Lauster, R.: "Human-on-a-chip" Developments: A Translational Cutting-edge Alternative to Systemic Safety Assessment and Efficiency Evaluation of Substances in Laboratory Animals and Man?, ATLA, 40, 235–257, 2012.

Nedelcu, O. T., Morelle, J.-L., Tibeica, C., Voccia, S., Codreanu, I., and Dahms, S.: Modelling and simulation of a pneumatically actuated micropump, Semiconductor Conference, 2007.

Schimek, K., Busek, M., Brincker, S., Groth, B., Hoffmann, S., Lauster, R., Lindner, G., Lorenz, A., Menzel, U., Sonntag, F., Walles, H., Marx, U., and Horland, R.: Integrating biological vasculature into a multi-organ-chip microsystem, Lab Chip, 13, 3588–3598, 2013.

Sonntag, F., Schilling, N., Mader, K., Gruchow, M., Klotzbach, U., Lindner, G., Horland, R., Wagner, I., Lauster, R., Howitz, S., Hoffmann, S., and Marx, U.: Design and prototyping of a chip-based multi-micro-organoid culture system for substance testing, predictive to human (substance) exposure, J. Biotechnol., 148, 70–75, 2010.

Thangawng, A. L., Ruoff, R. S., Swartz, M. A., and Glucksberg, M. R.: An ultra-thin PDMS membrane as a bio/micro–nano interface: fabrication and characterization, Bomed. M., 9, 587–595, 2007.

Timoshenko, S. and Woinowsky-Krieger, S.: Theory of plates and shells, 2nd Edn., McGraw-Hill Book Company, 1959.

Wagner, I., Materne, E.-M., Brincker, S., Süßbier, U., Frädrich, C., Busek, M., Sonntag, F., Sakharov, D., Trushkin, E., Tonevitsky, A., Lauster, R., and Marx, U.: A dynamic multi-organ-chip for long-term cultivation and substance testing proven by 3D human liver and skin tissue co-culture, Lab Chip, 13, 3538–3547, 2013.

Winkelmann, C., Luo, Y., Lode, A., Gelinsky, M., Marx, U., Busek, M., Schmieder, F., and Sonntag, F.: Charakterisierung von in Lab-on-a-Chip Systemen eingebetteten Hohlfasern, Tech. Mess., 80, 147–154, 2013.

Wu, M.-H., Huang, S.-B., Cui, Z., Cui, Z., and Lee, G.-B.: A high throughput perfusion-based microbioreactor platform integrated with pneumatic micropumps for three-dimensional cell culture, Biomed. M., 10, 309–319, 2008.

4

Polymer composite based microbolometers

A. Nocke

Solid-State Electronics Laboratory, Technische Universität Dresden, Dresden, Germany

now at: Institute of Textile Machinery and High Performance Material Technology,
Technische Universität Dresden, Dresden, Germany

Correspondence to: A. Nocke (andreas.nocke@tu-dresden.de)

Abstract. This work focuses on the basic suitability assessment of polymeric materials and the corresponding technological methods for the production of infrared (micro-) bolometer arrays. The sensitive layer of the microbolometer arrays in question is composed of an electrically conductive polymer composite. Semiconducting tellurium and vanadium dioxide, as well as metallic silver, are evaluated concerning their suitability as conductive filling agents. The composites with the semi-conducting filling agents display the higher temperature dependence of electrical resistance, while the silver composites exhibit better noise performance. The particle alignment – homogeneous and chain-shaped alike – within the polymer matrix is characterized regarding the composites' electrical properties. For the production of microbolometer arrays, a technology chain is introduced based on established coat-forming and structuring standard technologies from the field of polymer processing, which are suitable for the manufacture of a number of parallel structures. To realize the necessary thermal isolation of the sensitive area, all pixels are realized as self-supporting structures by means of the sacrificial layer method. Exemplarily, 2×2 arrays with the three filling agents were manufactured. The resulting sensor responsivities lie in the range of conventional microbolometers. Currently, the comparatively poor thermal isolation of the pixels and the high noise levels are limiting sensor quality. For the microbolometers produced, the thermal resolution limit referring to the temperature of the object to be detected (NETD) has been measured at 6.7 K in the superior sensitive composite layer filled with silver particles.

1 Introduction

Microbolometers belong to the group of thermal infrared detectors and are used as sensor arrays primarily in thermal imaging devices. In accordance with Planck's law, the measuring of infrared radiation permits passive target analysis as well as non-contact temperature measurements of solid bodies, thus opening a large number of possible applications. In microbolometer detectors, the absorbed infrared radiation causes a change in temperature, triggering a local alteration in resistance within the sensitive area (thermoresistive effect). Conventionally, sensitive resistor elements include vanadium oxide (VO_x), amorphous silicon (a:Si) and ceramic semiconductors (YBCO), which are not usually used in semiconductor production and are difficult to deposit (Ambrosio et al., 2010). Additionally, such microbolometer arrays are manufactured in an elaborate technological process chain, as individual pixels have to be realized as self-supporting structures. This is necessary to ensure a high thermal insulation of the sensitive area and thus great thermal and electrical responsivity. The high cost of microbolometer production is still limiting their widespread use, particularly civilian use, which is highly price-dependent. This motivates the basic research of alternative, economically viable materials and the corresponding technical methods for the production of microbolometer arrays.

The focus of this contribution is on polymer-based materials, as they possess advantages over conventional inorganic materials with regards to both the wider and more cost-efficient material range, and the large number of simple and parallel processing possibilities. A number of publications suggest this approach. Kaufmann et al. describe a possible method for the manufacture of electrically conductive sensitive bolometer layers consisting of ion-implanted polymer

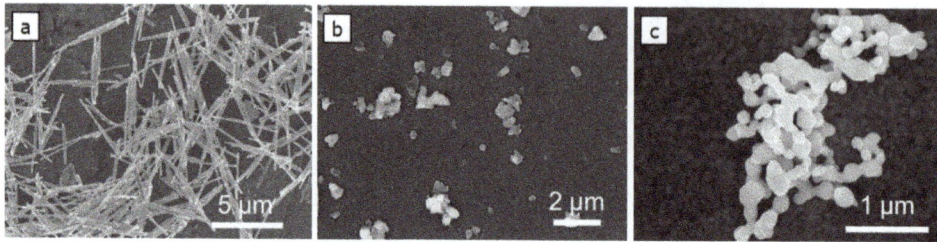

Figure 1. Scanning electron microscope (SEM) views of **(a)** synthesized tellurium needles, **(b)** vanadium dioxide particles, and **(c)** silver particles.

layers (Kaufmann et al., 1996). The electrical properties of the modified polymer layer depend on the ion dose, the ion energy and the ion current density. Liger et al. describe an approach in which the sensitive layer is formed by pyrolyzing the parylene C polymer (Liger, 2006). The pyrolysis of the pre-deposited parylene layer is performed in two stages at temperatures ranging from 660 to 800 °C and causes a share of the benzene rings contained in the parylene C (Liger, 2006) to transform into graphite-like areas. The resistance of the pixel is determined by the rate of graphitization, and thus by the temperature during pyrolysis. Liger (2006) gives a calculated noise equivalent temperature difference (NETD) value for his microbolometer pixels of 31–109 mK. In these works, polymer-based technologies for the manufacture of self-supporting bolometer-pixels are presented. The considerable energy inputs required for the manufacturing methods described limits the application spectrum, particularly with regards to flexible polymeric substrates. Additional works, addressing in particular the production and characterization of polymer-based sensitive bolometer layers, examine the suitability of the intrinsically conductive polymer Poly(3,4-ethylenedioxythiophene)/Poly(styrenesulfonate) (Son et al., 2009) and of carbon nanotubes with (Aliev, 2008) and without (Zeng et al., 2012) surrounding matrix polymer. Best sensor performance for these sensitive materials is given by Zeng et al. (2012) with a calculated detectivity D^* of 1.09×10^7 cm Hz$^{1/2}$ W^{-1}.

In this work, sensitive layers consisting of electrically conductive polymer composites composed of an insulating matrix polymer and a conductive filling material are used. When using such polymer composites, the mechanical, chemical and electrical properties of an individual layer can be adjusted and optimized separately. Especially chemical and thus technological properties are determined by the polymer matrix, as long as the proportion of the filling agent is small enough. The electrical properties are given by the type, structure and distribution of the conductive filling material.

2 Experimental

2.1 Materials

When selecting the materials to be used, the essential technological and electrical requirements of the polymer-based microbolometer arrays have to be taken into account. For the sensitive polymer composites, individual conditions apply for both the polymer matrix and the filling agents. Furthermore, both components have to be chemically compatible in order to form a stable suspension at least for the duration of processing.

2.1.1 Filling materials

Crucial criteria for material choice are the electrical parameters resistivity and temperature coefficient α_R of resistance. Semiconducting materials with a comparatively high α_R of $-(2\text{–}5)$ % K^{-1} as well as metallic materials with exceptionally good noise performance show great potential as filling materials. Another requirement results from the individual pixel element's geometry given by the sensitive layer's maximum layer thickness, which should be as small as possible (< 2 μm) to achieve the required low heat capacity of the microbolometer pixel. Therefore, only sufficiently small filling particles, which are also synthesizable in the desired geometric form while still meeting high quality standards, can be used. Considering these criteria, particles composed of tellurium (Te), vanadium dioxide (VO$_2$) and silver (Ag) have been used within the framework of this research (Fig. 1).

Monocrystalline tellurium is a semiconductor with an anisotropic trigonal crystal structure, giving it a predominant growth direction along the main axis, with the tellurium particles growing as needles. Another effect of its anisotropic structure is an anisotropic behavior of electrical conductivity, which is $\sigma_c = 2$ S cm^{-1} along the main axis at room temperature, while being lower by magnitudes along the other axes (Nussbaum, 1954). Te-needles were synthesized using chemical reduction of telluric acid (H$_6$TeO$_6$) with hydrazine (N$_2$H$_4$) as presented by Mayers and Xia (2002). The chosen synthesis procedure leads to a clean surface of the particles, which is desired for good electrical contacts. The resulting tellurium needles display a homogeneous size distribution at

Table 1. Electrical properties of the examined materials at room temperature (300 K): band gap E_g; temperature coefficient α_R of resistance and resistivity ρ.

Material	E_g [eV]	α_R [% K^{-1}]	ρ [Ωcm]	Reference
Te (c axis)	0.33	−2.45	2.0	Loferski (1954)
VO$_2$	0.65	−4.19	≈ 100	Berglund and Guggenheim (1969)
Ag	–	0.41	1.5×10^{-6}	Ashcroft and Mermin (2001)

diameters of 200–250 nm and length of 5–6 µm. The particles' aspect ratio, therefore, is ca. 25.

Polycrystalline vanadium dioxide particles with a size distribution from a few hundred nanometers to ca. 10 µm are commercially acquired with a purity of 99.9 % based on trace metals analysis, according to manufacturer information (manufacturer: Aldrich). By means of a sedimentation process, particles are separated depending on their size. The resulting particles have a maximum size of 2 µm, as can be seen in Fig. 1b), and thus meet the above-mentioned geometrical requirements of filling material.

The silver particles used (manufacturer: Aldrich) were also acquired commercially and have a silver content of at least 99 % and a diameter of ca. 150 nm, according to manufacturer information. The particles have a high defect structure and internal energy, causing a metastable, energetic, activated powder, which may form agglomerates of a size of approximately 1–2 µm.

The electrical properties of these materials are summarized in Table 1.

2.1.2 Polymers

The aspired requirement of realizing the individual sensor structures with just a few simple process steps is most easily attained by using filled photoresists. The structure transfer is performed by UV exposure through a photomask. The respective process step is reproducible and applicable in large scale. Furthermore, many photoresists can be cross-linked, making them chemically stable against solvents and giving them higher mechanical solidity. The cross-linking reaction takes place either directly under UV exposure or by subsequent heat input.

One photoresist with the above-named properties is the AZ 1514 positive photoresist by Clariant, with the polymeric main component being Novolak. This photoresist is used as the polymer matrix of the microbolometer pixel's sensitive layer. The special suitability of the AZ 1514 photoresist stems from its capability to be thermally cross-linked into a phenolic resin in the temperature range 120–160 °C after its lithographic structuring (Roy et al., 2003), which gives it extremely stable mechanical and chemical properties.

Another structuring polymer used is the Pyralin 2722 photoresist by HD MicroSystems. It is a negative photoresist with polyimide (PI) for its polymeric main component. Poly-

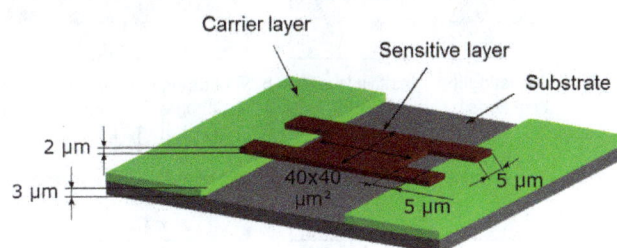

Figure 2. Geometric target parameters for the microbolometer pixel to be realized.

imides are very stable against most solvents and high temperature strains (> 300 °C) (Fukukawa and Ueda, 2008) and therefore suitable for use as permanent, structuring carrier layers, which are created in the first process step of the microbolometer array manufacture.

Other polymers used in this work are the heat-cured two-component Sylgard 184 (manufacturer: Dow Corning), with the silicone polydimethylsiloxane (PDMS) being the main component, and a paraffin wax purchased from Aldrich, with a melting range of 70–80 °C.

2.2 Technology

A technology chain was developed for the production of microbolometer arrays whose sensitive layers are made from electrically conductive polymer composites. As the thermal insulation of the sensitive area is key, any pixels have to be realized as self-supporting structures. The construction scheme of a self-supporting pixel is portrayed in Fig. 2.

For the cost-effective production of microbolometer arrays, established coating and structuring standard technologies for polymer processing were used, all of which are suitable for the simultaneous creation of several structures. In particular, the following aspects had to be considered for these polymer technologies:

- It is necessary that the individual polymer layers can be coated homogeneously in the desired thickness range, and easily structured laterally.

- The solvents and developers to be used must not dissolve or macerate the previously produced layers.

- The thermal stability of existent polymer layers has to be ensured at any exposure to heat, which often occurs

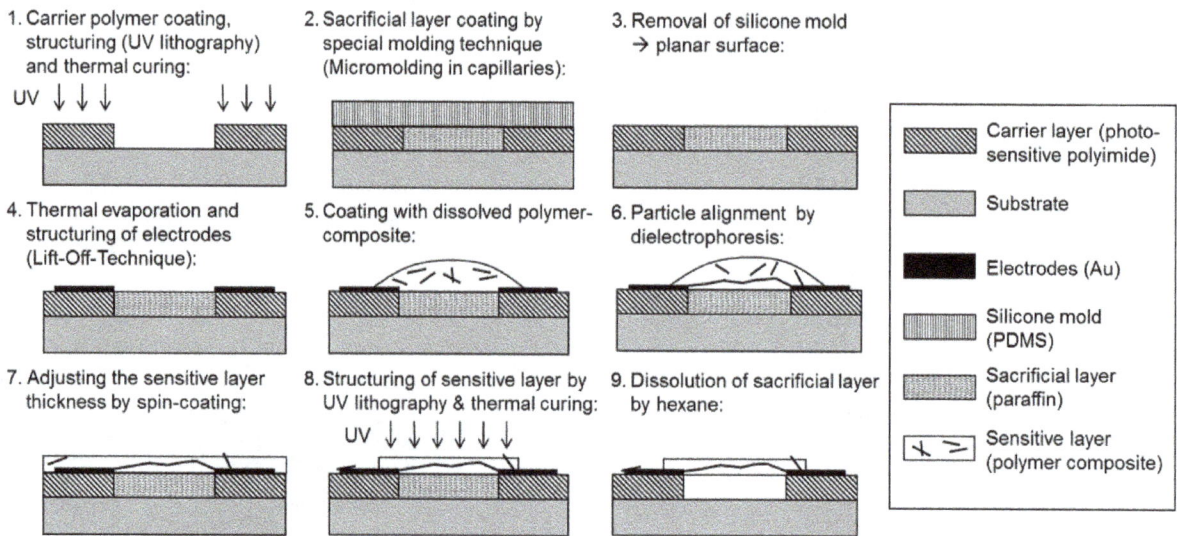

Figure 3. Schematic portrayal of the individual process steps for the manufacture of microbolometer arrays under principal use of polymeric materials and technologies; UV – ultraviolet, PDMS – polydimethylsiloxane.

in the coating of inorganic materials (e.g., the deposition of electrodes).

The derived technology chain consists of seven elementary process steps (Fig. 3). Additional information can be found in Nocke (2011). The technological process is based primarily on the sacrificial layer technology, in which the sacrificial layer is removed during the final process step, creating the self-supporting structure. The micromolding in capillaries (MIMIC) stamping technology was chosen for layer formation and simultaneous structuring of the sacrificial layer (Moonen et al., 2012). The sacrificial layer of heated, liquid paraffin wax fills a pre-structured channel network, owing to capillary action. Due to its low surface tension and viscosity, molten paraffin wax is particularly suitable for this structuring process. Furthermore, it displays favorable solubility properties, as it is only soluble in alkanes and few other solvents. Therefore, it barely limits the choice of possible organic materials in the next process steps.

The electrically conductive networks in the sensitive polymer composites are realized by means of a dielectrophoretic alignment of the filling particles. This effect is based on the particles' polarization in electrical fields and the resulting electrostatic force. In inhomogeneous electrical fields and under suitable conditions, the resultant force causes the desired formation of particle chains between the field-forming electrodes (Pohl et al., 1978; Nocke et al., 2009). The aim of such an alignment is the creation of a conductive network between appropriate electrodes at a preferably low filler loading (here: 0.1 wt. % in relation to the polymer matrix), which is essential for a technological malleability of the polymer composite.

2.3   Characterization

Electrical measurements were performed using the Electrometer 617 with the testbox 8002A (manufacturer: Keithley), which is well suited for high ohmic samples. For temperature-dependent measurements, a climate chamber HC0020 (manufacturer: Voetsch) with additional humidity control was used. All measurements were performed at 0 % relative humidity to minimize humidity influence. Noise and responsivity characteristics were measured with the lock-in amplifier 7265 DSP (manufacturer: EG&G Instruments). In order to establish the sensor parameters of a microbolometer pixel, the frequency-dependent responsivity R_V was measured metrologically by detecting the voltage response of a modulated IR radiation source (manufacturer: DIAS infrared system CS 500 with chopper wheel) with the lock-in amplifier 7265 DSP. The measurement was performed in an evacuated measuring chamber at an ambient pressure lower than 10 Pa.

3   Results and discussion

The main quality parameters of a microbolometer are the (voltage) responsivity R_V, the detectivity D^* and the noise equivalent temperature difference (NETD) (Gerlach and Budzier, 2010). The responsivity of a sensor is defined by the change in its output in proportion to the change of the input parameter and therefore should preferably be high. For a microbolometer, these parameters are the output voltage U_O and the radiation flux Φ_S, respectively. According to the relation

$$R_V = \frac{dU_O}{d\Phi_S} = \frac{\alpha \alpha_R U_B}{G_{th}\sqrt{1 + \omega_s^2 \tau_{th}^2}} \tag{1}$$

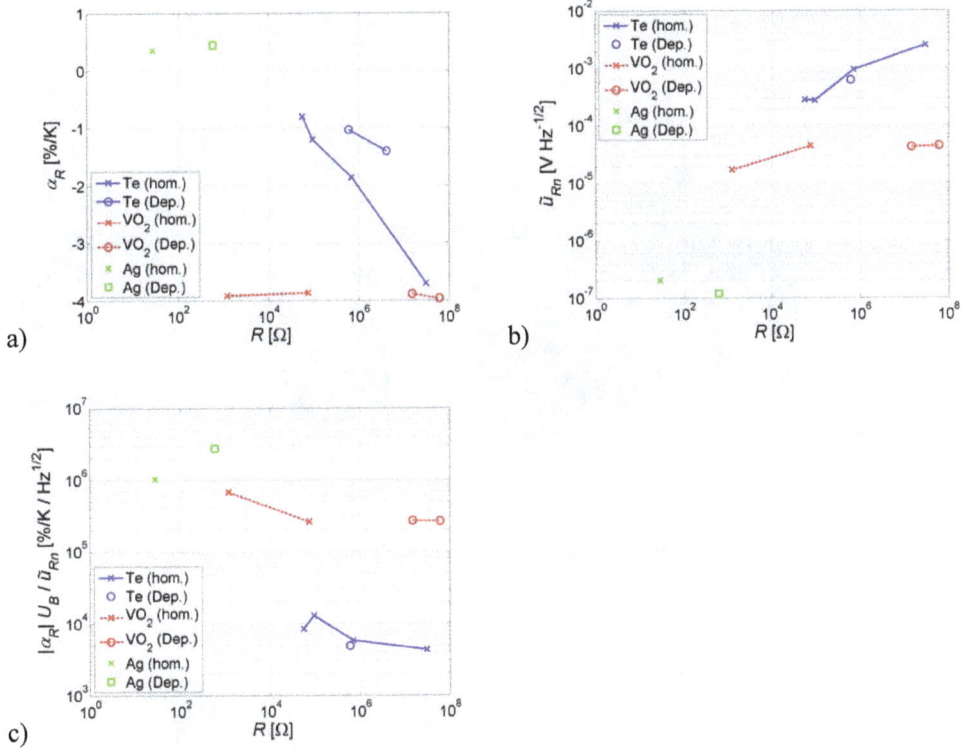

Figure 4. **(a)** Temperature coefficient α_R of resistance and **(b)** spectral noise voltage density $\tilde{u}_{Rn}$ (measured at 52 Hz) and **(c)** influence on detectivity $D^* \propto |\alpha_R| U_B / \tilde{u}_{Rn}$ of a microbolometer, calculated according to Eq. (3) in dependence on the resistance R of the examined polymer composites with homogenous particle distribution (hom.) and particles aligned by dielectrophoresis (Dep.).

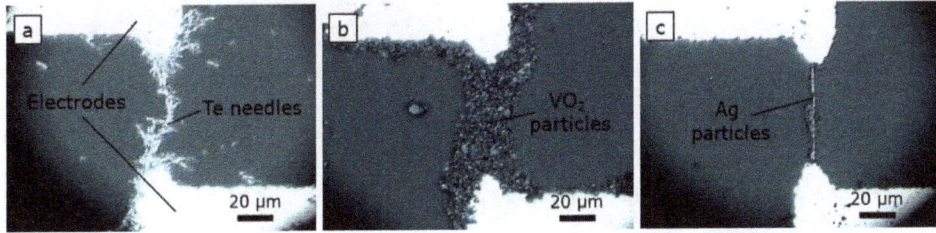

Figure 5. Optical microscopic pictures of dielectrophoretically aligned particles between electrodes: **(a)** tellurium needles, **(b)** vanadium dioxide, and **(c)** silver particles.

with

$$\tau_{th} = C_{th}/G_{th}, \tag{2}$$

the responsivity R_V of a microbolometer depends on the absorption coefficient α, on the temperature coefficient α_R of resistance, on the bias voltage U_B, on the thermal conductance G_{th} between the sensitive area and the surrounding media, on the heat capacity C_{th} of the microbolometer pixel, on the thermal time constant τ_{th}, and on the angular frequency ω_S. The detectivity D^* and the noise equivalent temperature difference (NETD) are sensor parameters typically used to characterize the overall sensor performance by taking into account its ratio of measurement and noise signal. The corresponding basic relation is given by

$$D^* = \frac{\sqrt{A_P} R_V}{\tilde{u}_{Rn}} \propto \frac{1}{NETD}, \tag{3}$$

where $\tilde{u}_{Rn}$ is the spectral noise voltage density and A_P is the pixel size.

Figure 4 shows the relevant electrical properties of characteristic polymer composites for homogeneously distributed filling particles as well as for composites with particles aligned between the electrodes by dielectrophoresis (Fig. 5).

Stemming from the temperature dependence of the resistance behavior (Fig. 4a) and additional measurements of the current-voltage characteristics (Nocke, 2011), the respective

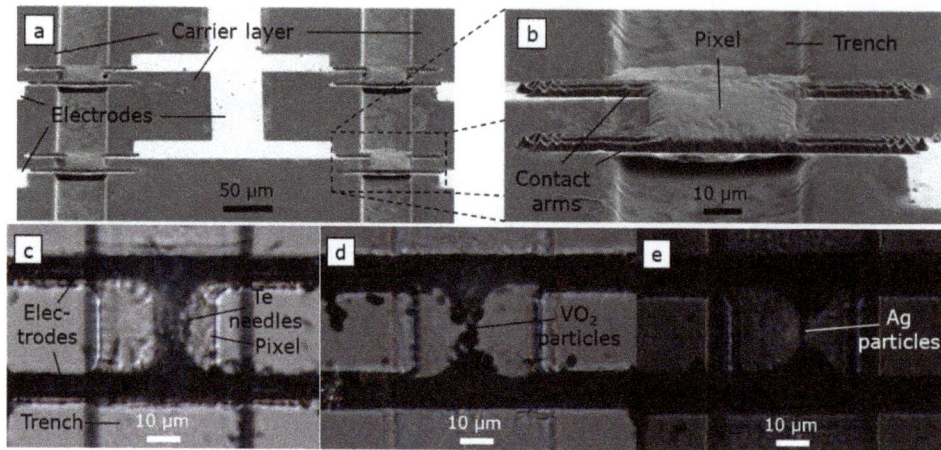

Figure 6. Microscopic pictures of the polymer composite based microbolometers realized: SEM-views of **(a)** the 2×2 array, and **(b)** a magnified individual pixel with a sensitive layer containing aligned silver particles, view of the structures at a tilting angle of 75° across the top view; optical microscopic pictures of individual pixels with aligned **(c)** tellurium needles, **(d)** vanadium dioxide particles and **(e)** silver particles in the sensitive polymer composite.

composites could be connected to individual dominant conduction mechanisms:

- In tellurium composites, the electric conductivity is considerably influenced by potential barriers between the particles. The related hopping conductivity mechanism exhibits a characteristic exponential relation of the resistance R and its temperature coefficient α_R (Mott and Davis, 1979).

- In vanadium oxide composites, the measured temperature coefficient α_R corresponds approximately with the value of the filling material, which is $\alpha_{R,VO_2} = -4.19\,\%\,K^{-1}$. Therefore, the total conductivity of these composites is determined largely by the semiconducting properties of the vanadium dioxide.

- The electrical properties of silver composites are characterized by the metallic conductivity mechanism of the particles, which exhibit a positive and, relative to amount, small temperature coefficient of resistance $\alpha_{R,Ag}$ of $0.41\,\%\,K^{-1}$.

The noise spectra of all composites display significant $1/f$ dependence. The noise levels of silver composites are lower by magnitudes compared to those of composites with semiconducting filling particles (Fig. 4b). The resulting influence on detectivity D^* (Fig. 4c) is greater than the differences in the temperature coefficient of resistance α_R. The detected electrical behavior of the polymer composites has proven largely independent of the variety of particle distribution.

Microscopic pictures of the produced 2×2 microbolometer arrays with the dielectrophoretically aligned filling materials tellurium needles, vanadium oxide and silver particles are shown in Fig. 6. They verify that the technological

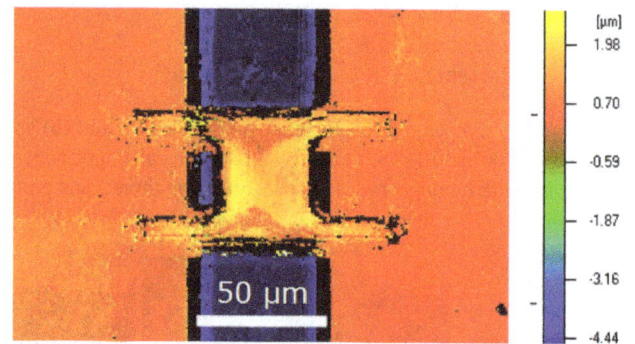

Figure 7. Confocal microscopic view of the surface topography of a microbolometer pixel with a sensitive layer containing aligned tellurium needles.

process is reproducible, independent of the filling material used. The square pixel surface approximately conforms to the specified measurements of $40 \times 40\,\mu m^2$. Figure 6a and b shows tilted scanning electron microscope (SEM) pictures of the microbolometer pixels filled with silver particles, proving the self-supporting nature of these structures. The roughness of the contact arms is caused by diffraction effects occurring during the photolithographic process. The characteristics of the surface topography are examined in the example of a pixel with aligned tellurium needles (Fig. 7). The pixel has the desired layer thickness of $2\,\mu m$ and runs above the trench without significant bending. The trench has a depth of ca. $4.5\,\mu m$.

Table 2 shows the application-relevant electrical and thermal sensor properties (see also Nocke, 2011). The relevant thermal parameters of the microbolometer arrays, thermal conductance G_{th} between the sensitive area, and the surrounding carrier layer and heat capacity C_{th} of the sensitive

Table 2. Electrical and thermal properties of realized polymer composite based microbolometers; calculation of sensor responsivities R_V and detectivities D^* according to the Eqs. (1) and (3); assumed absorption coefficient $\alpha = 1$.

Filling material	Te needles	VO$_2$ particles	Ag particles
Bolometer resistance R_B (293 K) [Ω]*	$7.5 \pm 1.6 \times 10^6$	$28 \pm 10 \times 10^7$	820 ± 200
Temperature coefficient α_R of resistance (293 K) [% K^{-1}]*	-1.4 ± 0.1	-3.9 ± 0.1	0.51 ± 0.02
Bias voltage U_B [V]*	9	9	0.5
Thermal conductance G_{th} [W K^{-1}]	4.6×10^{-6}	4.6×10^{-6}	4.6×10^{-6}
Heat capacity C_{th} [J K^{-1}]	6.1×10^{-9}	6.1×10^{-9}	6.1×10^{-9}
Thermal time constant τ_{th} [s]	1.3×10^{-3}	1.3×10^{-3}	$1.3 \times 10^{-3}/1.1 \times 10^{-4*}$
Responsivity R_V ($f = 50$ Hz) [V W^{-1}]	2.5×10^4	7.6×10^4	$4.8 \times 10^2/2 \times 10^{2*}$
Spectral noise voltage density $\tilde{u}_{Rn}$ ($f = 50$ Hz) [V Hz$^{-1/2}$]*	1.7×10^{-3}	8.1×10^{-5}	8.5×10^{-8}
Detectivity D^* ($f = 50$ Hz) [cm Hz$^{1/2}$ W^{-1}]	5.9×10^4	3.5×10^6	2.1×10^7
NETD [K] (Nocke, 2011)	2.4×10^3	41	6.7

* Measured values; statistical values refer to the four pixels of the respective array.

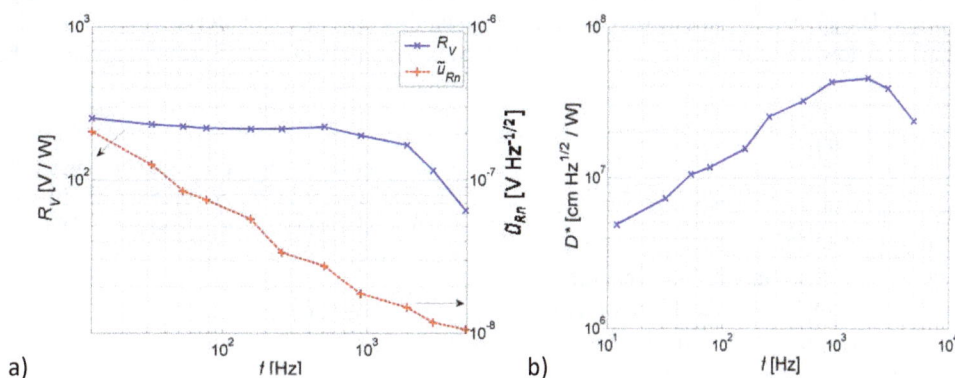

Figure 8. (a) Measured responsivity R_V and spectral noise voltage density $\tilde{u}_{Rn}$ and (b) calculational derived detectivity D^* for a microbolometer pixel with a sensitive polymer layer containing dielectrophoretically aligned silver particles; bias voltage $U_B = 0.5$ V, bolometer resistance $R_B = 820\,\Omega$.

layer are determined from the pixel dimensions and the respective material parameters. The calculated thermal parameters results in a thermal time constant of $\tau_{th} = 1.3$ ms. The responsivities R_V of the produced microbolometer arrays are calculated according to Eq. (1), using the electrical parameters of bolometer resistance R_B, temperature coefficient of resistance α_R, and bias voltage U_B determined in practice, as well as calculated thermal parameters. The responsivity values of the composites filled with tellurium and vanadium oxide filling materials are approximately two orders of magnitude above those of sensitive silver composites. This difference results from the smaller (according to amount) temperature coefficient of resistance α_R and the lower bias voltage U_B for the values with silver composite.

For the given similar geometric structures, the relation of the individual distinctive sensor parameters results from the electrical properties of the respective polymer composites shown in Fig. 4. Thus, the silver composites, due to their little noise behavior, have highest detectivity D^* and therefore the smallest NETD. The microbolometer arrays with vanadium

dioxide composites display the greatest responsivity R_V due to their high temperature coefficient of resistance α_R.

The sensor characteristics of a microbolometer pixel in terms of responsivity and spectral noise voltage density were measured exemplarily on a pixel structure filled with silver particles (Fig. 8a). At low frequencies, the detected responsivity is approximately frequency-independent with a value of ca. $R_V = 2 \times 10^2$ V W^{-1}. The distinct low-pass behavior sets in at a critical frequency of ca. 1.5 kHz, corresponding to a (thermal) time constant of $\tau_{th} = 0.11$ ms. The subsequent detectivity of the measured pixel was calculated according to Eq. (3). As can be seen in Fig. 8b, it shows a strong frequency-dependence affected by the $1/f$ noise characteristic and the dynamic behavior of the responsivity.

While the measured responsivity is in a range comparable to the previously calculated one, the measured time constant is approximately one order of magnitude smaller than the calculated one (Table 2). The deviations can be traced back to erroneous estimates of thermal influences or the presumption of an optimum wavelength-independent absorption coefficient.

In comparison to values from the relevant literature, which places conventional microbolometers' NETD values at around 30–100 mK (Gerlach and Budzier, 2010), fundamentally higher NETD values were observed. One essential reason for this lies in the higher thermal conductances of the contact arms of the design presented here. Furthermore, the noise levels of polymer composites with semiconducting filling particles are higher than those of semiconducting sensitive layers in conventional microbolometers. The same relation holds true with regards to the organic sensitive layer consisting of pyrolyzed parylene C (Liger, 2006). The NETD value comparison of microbolometers with metallic sensitive layers shows the smallest differences: NETD = 500 mK (microbolometer with sensitive titanium layer: Mansi et al., 2003) as opposed to NETD = 6.7 K (microbolometer pixel with silver composite). These metallic sensitive layers are distinguished by their very low noise levels.

4   Conclusions

The aim of this work was the basic suitability assessment of polymeric materials and related technological methods for the production of polymeric materials for the manufacture of microbolometer arrays with a sensitive layer of electrically conductive polymer composites. The all-polymer compatible technology chain is an innovative approach to the manufacture of polymer-based, self-supporting MEMS (microelectromechanical systems) structures and allows for a prospective economization potential as well as highly parallel processing suitability. Concerning their suitability for use as sensitive layers in a microbolometer pixel, additional metrological and physical observations were made regarding the electrical properties of polymer composites filled with either tellurium needles, vanadium dioxide particles or silver particles.

The best noise equivalent temperature difference NETD, which is similar to the temperature-dependent resolution limit of the measuring object, was detected for the microbolometer array with a sensitive silver composite layer, with its peak value at 6.7 K. This opens new applications for low-cost thermal imaging devices targeted at simple object detection.

In the future, the relevant sensor parameters have to be further improved in order to ensure a proliferation of possible applications. This may be attained particularly well by an enhancement of thermal insulation of the individual pixels and a reduction of noise levels in the composites with semiconducting filling particles. For a better thermal insulation of the individual pixels, the geometric properties of the contact arms have to be adapted to lower their thermal conductance. This requires longer contact arms with a smaller cross section, which mechanically destabilizes the self-supporting

pixel structures. Thus, future inquiries will have to aim at further optimization of the polymer-based manufacturing process for microbolometer arrays described herein. Concerning the reduction of noise levels in the composites with semiconducting filling particles, the number and energetic height of potential barriers within the conductive network have to be reduced, as they are major factors in the noise behavior between particles. One possible approach to this is to perform the chemical synthesis (of the sensitive semiconductors) in the vicinity of the microbolometer pixels themselves, leading to barrier-free conducting paths with reduced bolometer resistances. An alternative is offered by depositing the thermoresistive layer in a separate step below or above the self-supporting polymer layer. With such a constructional approach, the simple technology chain developed in this project could be used for the manufacture of self-supporting polymer layers. Thanks to the relevant photoresist's great temperature stability for organic materials, conventional deposition methods known from CMOS technology could be used, as applicable.

Acknowledgements. The author gratefully acknowledges financial support by the Saxon State Ministry of Science and Art and from German Research Foundation (DFG), Sonderforschungsbereich 287.

Edited by: B. Jakoby
Reviewed by: three anonymous referees

References

Aliev, A. E.: Bolometric detector on the basis of single-wall carbon nanotube/polymer composite, Infrared Phys. Techn., 51, 541–545, 2008.

Ambrosio, R., Moreno, M., Mireles Jr., J., Torres, and Kosarev, A.: An overview of uncooled infrared sensors technology based on amorphous silicon and silicon germanium alloys, Phys. Status Solidi C, 7, 1180–1183, 2010.

Ashcroft, N. W. and Mermin, N. D.: Festkörperphysik, Oldenbourg Verlag, München, 2001.

Berglund, C. N. and Guggenheim, H. J.: Electronic properties of VO_2 near semiconductor-metal transition, Phys. Rev., 1, 1022–1033, 1969.

Fukukawa, K.-I. and Ueda, M.: Recent progress of photosensitive polyimides, Polym. J., 40, 281–296, 2008.

Gerlach, G. and Budzier, H.: Thermische Infrarotsensoren: Grundlagen für Anwender, Wiley-VCH, Weinheim, 2010.

Kaufmann, J., Moss, M. G., Wang, Y., and Giedd, R. E.: Conductive polymer films for microbolometer applications, in: Infrared Technology and Applications XXII. SPIE 2744, Orlando, USA, 334–344, 1996.

Liger, M.: Uncooled carbon microbolometer imager, Dissertation (Ph.D.), California Institute of Technology, 2006.

Loferski, J. J.: Infrared optical properties of single crystals of tellurium, Phys. Rev., 93, 707–716, 1954.

Mansi, M. V., Brookfield, M., Porter, S. G., Edwards, I., Bold, B., Shannon, J., Lambkin, P., and Mathewson, A.: AUTHENTIC: a very low-cost infrared detector and camera system, in: Infrared Technology and Applications XXVIII. SPIE 4820, Seattle, USA, 227–238, 2003.

Mayers, B. and Xia, Y. N.: One-dimensional nanostructures of trigonal tellurium with various morphologies can be synthesized using a solution-phase approach, J. Mater. Chem., 12, 1875–1881, 2002.

Moonen, P. F., Yakimets, I., and Huskens, J.: Fabrication of Transistors on Flexible Substrates: from Mass-Printing to High-Resolution Alternative Lithography Strategies, Adv. Mater., 24, 5526–5541, 2012.

Mott, N. F. and Davis, E. A.: Electronic Processes in Non-Crystalline Materials, 2nd Edn., Clarendon Press, Oxford, 1979.

Nocke, A.: Mikrobolometer auf der Basis von Polymerkompositen, TUD Press, Dresden, 2011.

Nocke, A., Wolf, M., Budzier, H., Arndt, K.-F., and Gerlach, G.: Dielectrophoretic alignment of polymer compounds for thermal sensing, Sensor. Actuat. A-Phys., 156, 164–170, 2009.

Nussbaum, A.: Electrical properties of pure tellurium and tellurium-selenium alloys, Phys. Rev., 94, 337–342, 1954.

Pohl, H., Pollock, K., and Crane, J.: Dielectrophoretic force: A comparison of theory and experiment, J. Biol. Phys., 6, 133–160, 1978.

Roy, D., Basu, P. K., Raghunathan, P., and Eswaran, S. V.: DNQ-novolac photoresists revisited: H-1 and C-13 NMR evidence for a novel photoreaction mechanism, Magn. Reson. Chem., 41, 84–90, 2003.

Son, H. J., Kwon, I. W., and Lee, H. C.: Passivation Effect for the Reduction of 1/f Noise in Poly(3,4-ethylenedioxythiophene):Poly(styrene sulfonate) Thin Films Based on Uncooled Type Microbolometer Applications, Appl. Phys. Express, 2, 041501, 2009.

Zeng, Q., Wang, S., Yang, L., Wang, Z., Pei, T., Zhang, Z., Peng, L.-M., Zhou, W., Liu, J., Zhou, W., and Xie, S.: Carbon nanotube arrays based high-performance infrared photodetector, Optical Materials Express, 2, 839–848, 2012.

5

Overview on conductometric solid-state gas dosimeters

I. Marr, A. Groß, and R. Moos

Department of Functional Materials, University of Bayreuth, Bayreuth, Germany

Correspondence to: R. Moos (functional.materials@uni-bayreuth.de)

Abstract. The aim of this article is to introduce the operation principles of conductometric solid-state dosimeter-type gas sensors, which have found increased attention in the past few years, and to give a literature overview on promising materials for this purpose. Contrary to common gas sensors, gas dosimeters are suitable for directly detecting the dose (also called amount or cumulated or integrated exposure of analyte gases) rather than the actual analyte concentration. Therefore, gas dosimeters are especially suited for low level applications with the main interest on mean values. The applied materials are able to change their electrical properties by selective accumulation of analyte molecules in the sensitive layer. The accumulating or dosimeter-type sensing principle is a promising method for reliable, fast, and long-term detection of low analyte levels. In contrast to common gas sensors, few devices relying on the accumulation principle are described in the literature. Most of the dosimeter-type devices are optical, mass sensitive (quartz microbalance/QMB, surface acoustic wave/SAW), or field-effect transistors. The prevalent focus of this article is, however, on solid-state gas dosimeters that allow a direct readout by measuring the conductance or the impedance, which are both based on materials that change (selectively in ideal materials) their conductivity or dielectric properties with gas loading. This overview also includes different operation modes for the accumulative sensing principle and its unique features.

1 Introduction

In times of strict environmental requirements and an increasing demand for mobility and energy, a growing need for reliable, low concentration level gas-sensing devices comes along. Especially the detection of toxic, harmful gases, like SO_2, NO and/or NO_2 (= NO_x), H_2S, CO, or NH_3, which can affect human health, particularly in workplaces and urban environments, are important target gases for gas sensors. Inexpensive sensor devices based on semiconducting metal oxides manufactured in ceramic technology or in thin- or thick-film technology are installed in annual quantities of millions (Yamazoe, 2005). The measurand is the conductance of metal oxides, as it depends on the analyte concentration. For information on the physical background of such sensors, the reader is referred to reviews in the literature (Göpel, 1994; Williams, 1999; Kohl, 2001; Yamazoe et al., 2003; Barsan et al., 2007). Semiconducting solid-state gas sensors can be miniaturized and operated with low power consumption (Simon et al., 2001; Semancik et al., 2001). However,

most sensors show poor selectivity, nonlinear characteristics (Williams, 1999; Yamazoe et al., 2003), and according to Padilla et al. (2010), the main drawback is their lack of stability over time, leading to recalibration costs. Another issue is that increased sensitivity is often accompanied by a slow sensor kinetic. Even if the response time is fast enough for special applications, in many cases the recovery time is not. This does not really jeopardize low-cost applications for detecting dangerous conditions or alarm concentration levels. If, however, maximum allowable concentrations in emission or immission regulations are given in time-weighted average concentrations or cumulated amounts, like the immission standards for air quality monitoring as annual or hourly mean values, a mathematical integration of the linearized sensor signal is error-prone due to both slow sensor recovery times and baseline drifts.

For instance, NO_2 thresholds are given by the EU immission legislation Directive 2008 (Directive, 2008) or the German air quality standards (BImSchV, 2010) as an annual mean value of $40\,\mu g\,m^{-3}$, a one-hour value of $200\,\mu g\,m^{-3}$ that should not be exceeded more than 18 times per calendar year, and an alert limit of $400\,\mu g\,m^{-3}$ measured over three consecutive periods of one hour each. For comparison, the one hour limit of $200\,\mu g\,m^{-3}$ NO_2 corresponds to a concentration of 0.1 ppm. In order to detect such small integrated concentration values by conventional semiconducting gas sensors with the required accuracy, complex procedures like continuous switching between clean air and analyte are required (Schütze et al., 1995).

Solid-state dosimeter-type gas sensing principles comprise a promising method for overcoming the above-mentioned drawbacks. Like passive samplers, solid-state gas sensors working in the accumulating mode are adequate devices for the reliable detection of lowest pollution levels (Seethapathy et al., 2008) because the dosimeter principle eludes several problems of a conventional gas sensor (Yamazoe and Shimanoe, 2007).

There are various publications on sensing devices mentioning an "irreversible reaction" or a very slow recovery time, not recognizing the benefits of this sensor behavior that differs from conventional gas sensors. The intention of this overview is to show exemplarily research papers that observed accumulating, integrating or irreversible gas-sensing properties. Sensor responses that do not recover in the absence of analyte under sorption conditions are denoted as "irreversible" in the following, even if the samples can be regenerated under modified operation conditions. It is a further aim of this study to elucidate the sensing principle and some special operation modes that allow a direct readout of the concentration as well as an accumulated concentration value at the same time and a plausibility check during regeneration.

This overview lists the different dosimeter-type gas sensing principles using the change of the electrical properties of the gas-sensitive material as the measurand. The focus of this overview is on conductometric integrating gas sensors but there are also optical (mostly organic dye-based coloration reactions) (Dasgupta et al., 1998; Tanaka et al., 1998, 1999; Sasaki et al., 2001; Maruo, 2007; Maruo et al., 2009; Small et al., 2009; Bhalla et al., 2010), SAW (Martin et al., 1996; Nieuwenhuizen and Harteveld, 1997), field-effect transistors (FET) (Andringa et al., 2012; Klug et al., 2013), quartz microbalance (QMB) (Matsuguchi et al., 2005; Jung et al., 2009), and passive sampling (Roadman et al., 2003; Varshney and Singh, 2003; Seethapathy et al., 2008) devices that are able to detect very low gas concentrations in the ppb range.

This article reviews the basic working principles of conductometric solid-state gas dosimeters (Sect. 2). It considers the direct amount determination and the indirect concentration determination, influencing factors, and compares gas dosimeters with classical gas sensors. Certain basic examples for sensing devices using the dosimeter-type sensing principle are shown in Sect. 3. Section 4 describes modifications regarding the dosimeter setup and the operation modes. Fur-

thermore, suitable or already applied materials are listed in Sect. 5.

2 Basic working principle of a solid-state gas dosimeter

The basic working principle of gas dosimeters relies on the accumulation of analyte gas molecules in the sensitive layer of the detecting device and the concomitant accumulation level dependent change of the electrical properties. The irreversible, selective, and progressive analyte sorption resulting in analyte accumulation can be caused by strong analyte sorption or a chemical reaction between the analyte and the receptor. Hence, gas dosimeters working in the accumulation mode (Yamazoe and Shimanoe, 2007) are often also denoted as "integrating" or "accumulating" sensors (Tanaka et al., 1999; Maruo 2007; Shu et al., 2010; Geupel et al., 2010; Groß et al., 2012b). For the explanation of the accumulating sensing principle, a planar setup consisting of an insulating support with electrodes, e.g., interdigitated electrodes, covered with a sensitive layer, was chosen. This setup makes it possible to measure the conductance of the sensor layer.

2.1 Direct amount and indirect concentration detection

The benefit of the dosimeter-type sensing behavior is the direct determination of the dose of the analyte gas species or the mean concentration in a defined period of time without the need for a mathematical integration of the sensor signal. In the case of a constant gas flow rate, the analyte dose as the measurand can be given in a simplified manner as the analyte amount or cumulated exposure A_c, calculated according to Eq. (1) as the time integral of the analyte concentration $c(t)$ with the starting point of the sorption interval t_0:

$$A_c = \int_{t_0}^{t} c(t)dt. \tag{1}$$

A_c is often given in the unit ppm s, which can be converted to μL or μmol if accounting for the applied gas flow rate.

The determination of the amount A_c as the measurand of the gas dosimeter is in accordance with national or international environmental immission or emission regulations, given, for instance, as mean time values, like the annual or 1 h limit.

Similar to passive samplers, the operation of dosimeter-type gas sensors with an analyte sorbent as sensitive material can be divided into two alternating periods: the sensing or analyte sorption interval and the regeneration or analyte release step. The fundamental operation principle of an accumulating gas sensor is depicted schematically in Fig. 1. Under sorption conditions, the analyte gas is sorbed "irreversibly" by the sensitive sensor layer since the sorption rate is enhanced compared to the desorption rate, provoking the

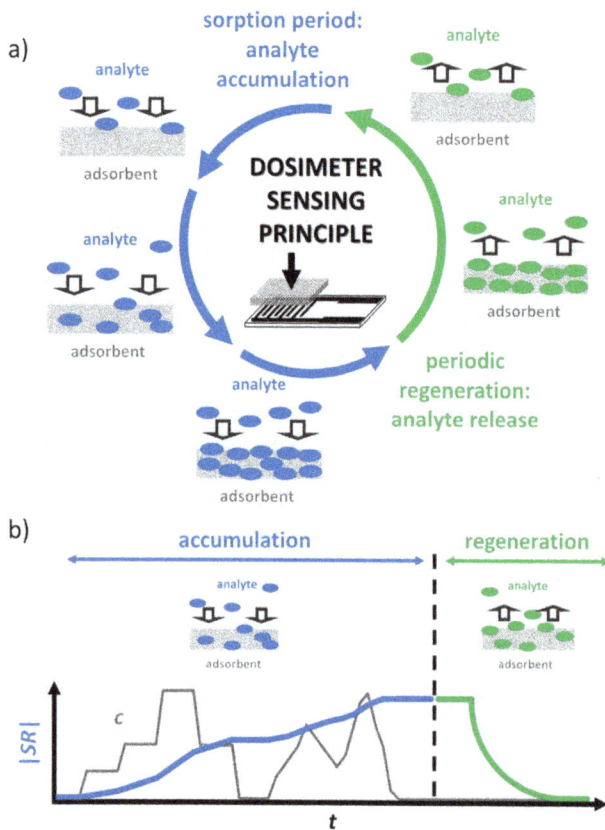

Figure 1. Scheme of the operation principle of a dosimeter-type gas sensing device with a sensitive layer as an adsorbent: **(a)** alternation of analyte accumulation in the sorption period (blue) and analyte release upon regeneration (green), **(b)** sensor response |SR| during the analyte accumulation (blue) and during the short-term regeneration period (green).

change of at least one material property due to analyte accumulation. In a short regeneration step under defined conditions, the formerly sorbed analyte molecules are released from the sorbent and the sorption capacity is recovered. Dependent on the interaction between the sorbent and the analyte, different regeneration procedures are effective to recover the sorption capacity as well as the sensor signal. They include thermal decomposition at elevated temperatures (Katayama et al., 2004; Helwig et al., 2008; Kubinski and Visser, 2008; Brunet et al., 2008; Shu et al., 2010; Brandenburg et al., 2013; Groß et al., 2013a), chemical reactions, e.g., by a changed gas atmosphere (Helwig et al., 2007, 2008; Groß et al., 2012a) or optical-induced desorption, e.g., by UV light (Li et al., 2003; Helwig et al., 2008).

As sketched in Fig. 1b, the absolute value of the accumulation-level dependent physical property that is monitored and evaluated as sensor response, |SR|, increases during the sorption period and recovers during regeneration. "… the sensor is designed to provide an output that intrinsically depends on the history of the input quantities" (Angelini et

al., 2009) and therefore the sensor response correlates with the cumulated exposure. Hence, the electrical detection of the analyte might occur either during the sorption or during the release period (Kubinski and Visser, 2008; Groß et al., 2013a). In the literature, this change in the accumulation-level is detected mainly optically (Tanaka et al., 1999; Sasaki et al., 2001; Maruo, 2007) or gravimetrically (Matsuguchi et al., 2005). Recently impedimetric approaches have emerged (Varghese et al., 2001; Mattoli et al., 2007; Helwig et al., 2008; Geupel et al., 2010; Shu et al., 2010; Hennemann et al., 2012a; Groß et al., 2012c). Regeneration has to be initiated if saturation effects limit the sorption rate of the sensitive layer or if the sensing characteristic is deteriorated.

The dosimeter-type sensing properties are often demonstrated in two different ways, but always at a constant gas flow: first, by looking at the signal change as a function of the exposure duration at a constant analyte concentration; or secondly, by plotting the concentration-related characteristic line of the measured data after specific exposure duration. However, more information on the accumulating sensing properties is obtained by a repeated exposure to pulses of analyte gas with a defined concentration and duration.

A more detailed analysis of the ideal sensor response of a gas dosimeter with a linear sensing characteristic during the analyte sorption at a constant gas velocity is shown in Fig. 2. Like the cumulated amount, A_c, being the time integral of the concentration according to Eq. (1), the absolute value of the sensor response, |SR|, in Fig. 2a increases in the presence of the analyte gas. Thereby, the slope of the time-dependent sensor response reflects the sorption rate and correlates with the actual analyte concentration $c(t)$ (Dasgupta et al., 1998; Sasaki et al., 2001; Reyes et al., 2006; Shu et al., 2010; Groß et al., 2012a; Mukherjee et al., 2012). In the absence of the analyte, i.e., if $c = 0$ ppm, no signal recovery occurs, the sensor response remains constant and the response change due to the analyte is irreversible (Nieuwenhuizen and Harteveld, 1997). The course of |SR| in Fig. 2a indicates progressive analyte accumulation in the sensitive layer. Ideally, the sensor response |SR| depends linearly on the analyte amount, A_c (Fig. 2b). When the characteristic line becomes nonlinear, the sensitive layer needs to be regenerated. The amount sensitivity S_A is defined as the slope of the characteristic line in the linear measurement range.

In consequence of the concentration-dependent slope of the sensor response, the course of the current analyte concentration over time can be determined by the time derivative of the sensor response |dSR/dt|, as shown on the timescale in Fig. 2c. The corresponding concentration-related characteristic line with the concentration sensitivity S_c is depicted in Fig. 2d. Instead of the equilibrated state, in the case of gas dosimeters, the sensor response change is evaluated.

Figure 2 clarifies that with gas dosimeters two pieces of information can be obtained from just one sensitive device: the sensor response correlates directly and without mathematical signal operations with the cumulated analyte exposure,

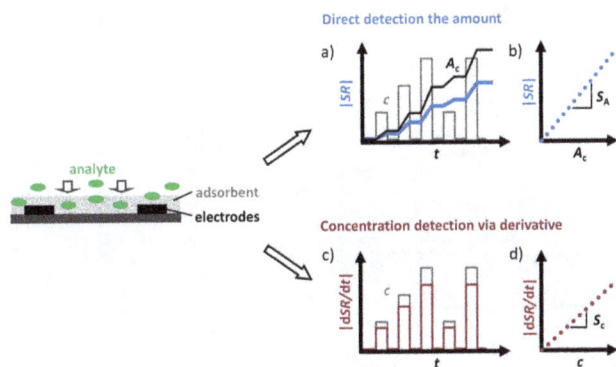

Figure 2. Dual-mode functionality of gas dosimeters: **(a)** correlation between the analyte amount A_c and the sensor response $|SR|$ and **(b)** associated amount-related characteristic line, **(c)** correlation between the time-dependent analyte concentration $c(t)$ and the time derivative of the signal $|dSR/dt|$ and **(d)** associated concentration-related characteristic line.

and the current course of the analyte concentration can be determined indirectly from the signal derivative. For an ideal dosimeter, the material acts as an integrator and the derivative reflects the actual concentration. This is in contrast to an ideal classical gas sensor, which determines the actual concentration but the dose has to be obtained by integration.

The accumulating sensing principle is well suited for the long-term detection of very low analyte concentrations. By the chemical accumulation of the analyte molecules in the sensitive sorbent, the dosimeter "counts" every gas molecule that reaches the sensitive layer and causes a change of the physical properties of the material. This chemical integration increases the accuracy of the amount determination compared to conventional concentration detecting gas sensors. There, the determination of the average concentration values or the cumulated exposure by mathematical integration of the concentration-related sensor response is error-prone in the case of very low gas concentrations due to slow sensor kinetics, zero-point drifts and nonlinearities in the measurement range (Corcoran and Shurmer, 1994).

As described above, periodic regeneration recovers the sorption capacity of the sensitive layer as well as the sensor response. Concerning the observed saturation of gas dosimeters, e.g., in Geupel et al. (2010), it needs to be distinguished whether those effects originate from the limited number of sorption sites in the analyte sensitive material or from a nonlinear correlation between the sensor response and the analyte loading level.

2.2   Influencing factors

Similar to passive samplers, different sensor setup parameters and measurement conditions influencing the dosimeter-type gas sensing properties are reported in the literature and will be summarized and discussed.

Like conventional sensors, gas dosimeters may show cross sensitivities to other gases if those interact with the sensitive layer. They might restrict or promote the analyte sorption process, which would affect the analyte sensitivity. Additionally, the sensor response might change reversibly (Varghese et al., 2001) or irreversibly due to other gas components dependent on the strength of interaction with the receptor. For instance, the dosimeter-type NO_x sensing properties of a carbonate-based sensitive layer were found to be affected by SO_2 in two ways: the competition between SO_2 and NO_x for the sorption sites lowers the NO_x sensitivity as well as the measurement range. Furthermore, the conductivity is irreversibly affected by sulfate formation, enabling SO_2 dosimetry (Groß et al., 2012d).

The analyte sorption properties of the sensitive material are also influenced by the operation temperature. Due to kinetic limitations, the sorption rate diminishes at lower temperatures; however, the strength of the analyte receptor interaction increases, resulting in accumulating properties due to a reduced desorption rate. With increasing temperature, analyte sorption occurs faster and the sensitivity is enlarged. However, since the sorption capacity is reduced and the strength of interaction is weakened, the sensitive layer tends to release formerly sorbed analyte molecules. Hence, higher temperatures increase the sensitivity of gas dosimeters but reduce the linear measurement range (Groß et al., 2012a).

While the equilibrated sensor response is monitored with conventional sensors as a measure of the analyte concentration, the progressive change of the material properties of gas dosimeters during analyte sorption serves as the measure of the cumulated amount. Hence, for the accumulating properties of the gas dosimeter it is very important that the chemical equilibrium is strongly shifted to the product side. Since the chemical equilibrium is strongly affected by temperature, it is possible that the same sensitive device can act like a gas dosimeter at low temperatures, while at higher temperatures it shows the typical concentration-related gas sensor behavior (Reyes et al., 2006; Shu et al., 2010; Andringa et al., 2012). Brandenburg et al., 2013 demonstrate this temperature-dependent behavior for a sensitive layer of a carbonate-based NO_x storage material. The sensor shown in Fig. 3 acts as an accumulating-type gas sensor up to 450 °C, whereas the enhanced NO_x desorption rate in the absence of analyte deteriorates the sensor response at 550 °C. At 650 °C, response and recovery of the sensor signal to pulses of NO_x are very fast and the sensor behaves like a concentration detecting device (Fig. 3c) (Brandenburg et al., 2013).

Another influencing factor is the thickness of the sensitive layer being related to the sorption capacity (Nieuwenhuizen and Harteveld, 1997; Shu et al., 2011). Groß et al. demonstrated the dependency of the dosimeter-type NO_x sensing properties on the thickness of the carbonate-based sorbent. Since nitrates form mainly at the surface of the sorbent, the sensitivity of the NO_x accumulating sensor decreases with a growing layer thickness as the linear measurement range

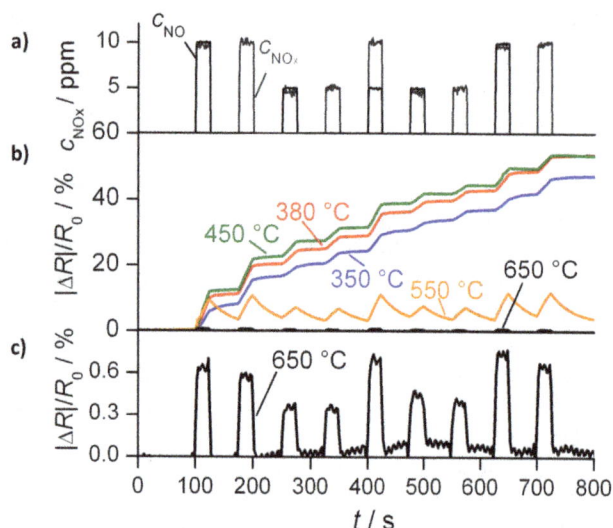

Figure 3. Temperature-dependence of the gas sensing behavior of a carbonate-based NO_x storage material for NO_x sensing purposes: **(a)** dosed NO_x (NO or NO_2) concentration, **(b)** sensor response $|\Delta R|/R_0 = |SR|$ for different temperatures and **(c)** sensor response at 650 °C. Please note the different scale of the y axes in **(b)** and **(c)**. Reprinted from Brandenburg et al. (2013), with permission of Elsevier.

increases (Groß et al., 2012b). Additionally, the sensor response time concerning the concentration detecting properties was found to increase with the layer thickness.

Similarly, it is expected that the analyte uptake can be influenced by the sensitive surface area and hence the porosity of the sorbent.

As will be discussed in detail in Sect. 4, the setup-dependent gas velocity across the sensitive material was found to affect the analyte sorption (Beulertz et al., 2011, 2012).

The main characteristics of the classical and the dosimeter-type sensing principle – particularly for low level detection – are briefly compared in Fig. 4.

Some further advantages of the dosimeter-type sensor principle should be mentioned here. The long-term zero drift, being a drawback of conventional gas sensors, is avoided by redefining the zero level at the end of each regeneration period.

While the sensors' response and recovery times, defined as the time span to reach the equilibrated states, are critical parameters for classical gas sensors, in the case of gas dosimeters the sensor signal change due to analyte accumulation is analyzed. Since the actual concentration value is given by the derivative, the sensor response is very fast.

It should also be mentioned that the measurement range and the sensitivity of the gas dosimeter can be adjusted during operation by varying the working temperature.

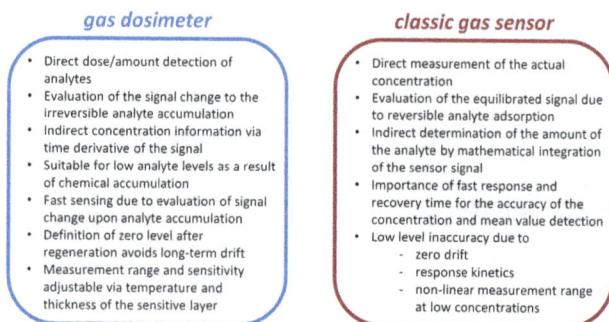

Figure 4. Comparison of the main features of a gas dosimeter for the long-term low-level analyte detection with the properties of a conventional gas sensor.

3 Examples for conductometric gas dosimeters

In the following, some work on conductometric or impedancemetric sensors working in the accumulation mode for different analytes and with various sensitive materials are discussed exemplarily.

Shu and colleagues presented a resistive NO_2 sensor based on iron (II) phthalocyanine that is operated at room temperature (Shu, 2010; Shu et al., 2010, 2011). Figure 5 depicts the course of the measured resistance R during cyclic exposure to NO_2 for 30 min in concentrations of 0.5, 1, and 2 ppm. The total flow rates were 0.95, 0.43, and 0.25 L min^{-1}, respectively. For all applied concentrations, the typical curve shape of a dosimeter-type gas sensor signal is observed. While dosing NO_2 in N_2 with a constant concentration, the resistance decreases with a steady slope and remains at its level during the intermediate N_2 phases, resulting in plateaus in the response course. Unfortunately, the effect of NO_2 is relatively low ($R/R_0 \approx 1$ to 0.88) and the characteristic line is not linear. The regeneration of this NO_2 chemo-dosimeter is achieved thermally. The influence of the temperature was investigated in the range from −46 °C to 71 °C. For the difference of ~ 50 °C the resistance decreases approximately one order of magnitude. With increased temperature the desorption effects increase. Regarding the long-term stability, an increase of the conductivity during storage in air due to the adsorption of H_2O and O_2, an effect which is known for metal containing phthalocyanines, can be observed (Shu, 2010).

Helwig et al. realized a dosimeter-like p-type conductometric gas sensor with a hydrogenated diamond (HD) as pH sensitive material for NO_2 and NH_3 as analytes (Beer et al., 2013; Helwig et al., 2013). Figure 6 compares the resistance changes of the HD-sample to those of a conventional metal oxide (MOX)-based device in dry synthetic air. Due to the electrolytic dissociation of the gases in the sensitive layer, the resistance at room temperature decreases stepwise in the presence of NO_2 pulses (Fig. 6a, red line) and increases in the NH_3-containing atmosphere (Fig. 6b, red line). For both analytes, the sensitive hydrogenated diamond device works

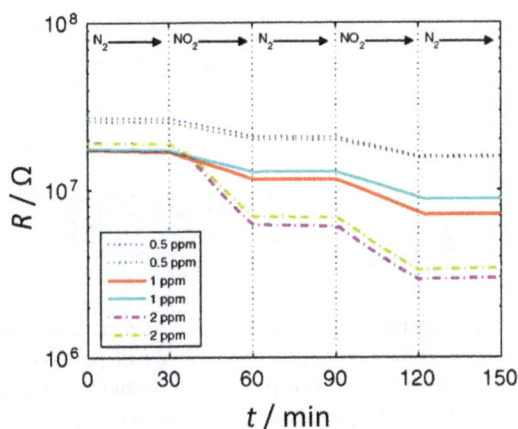

Figure 5. Course of the resistance R of iron(II)phthalocyanine in the presence of different NO_2 concentrations at room temperature. Adapted from Shu et al. (2010); reprinted with permission from Elsevier.

like an integrator-like gas sensor, whereas the signal of the metal oxide-based gas sensor (black line) shows the expected behavior of a conventional concentration related gas sensor. The recovery times of the dosimeter can be enhanced by a temperature increase, or by ozone or UV light. Figure 6c shows the reset of the sensor response after exposure to NH_3 in different concentrations by a short O_3 pulse after each ammonia phase, causing the oxidation of NH_3 at the sensor surface at room temperature (Helwig et al., 2008).

In another example, a lean NO_x-trap material known from automotive NO_x storage catalysts was applied as the NO_x sensitive layer (Geupel et al., 2010, 2011). At 380 °C, the carbonate-based material accumulates NO and NO_2 chemically by forming nitrates. Due to nitrate formation, the resistance decreases, enabling total NO_x dosimetry. Figure 7 demonstrates the suitability of the carbonate-based dosimeter for the long-term detection of NO_2 even in the sub-ppm range. The sensor response $|SR| = |\Delta R|/R_0$ in Fig. 7a follows the characteristic behavior of an accumulating gas sensor during cyclic exposure to 0.2 to 2 ppm NO_2 in steps of 75 s over 50 min. $|\Delta R|/R_0$ completely overlaps with the normalized time integral of the concentration, $\int c_{NO_2}(t)\mathrm{d}t$ (with $c_{NO_2}(t)$ calculated from mass flow controller output data), which is shown in green as a reference for the course of the cumulated amount. The corresponding characteristic line gives a linear correlation between the sensor response $|\Delta R|/R_0$ and the analyte amount A_{NO_2} (Fig. 7b). Due to this linearity, the time derivative of the resistance $|\mathrm{d}R/\mathrm{d}t|$ reflects the course of the NO_2 concentration, as shown in Fig. 7c. In accordance with the European air quality standards limit of 200 μg m^{-3} NO_2 (≈ 0.1 ppm) and the applied flow rate of 2 L min^{-1}, this hourly threshold of about 360 ppm s could be monitored for at least 3.7 h with the sensitivity of the NO_x dosimeter shown in Fig. 7b, whereas it is even 18.5 h for the annual value of 40 μg m^{-3}. Besides the sensitivity, the lin-

ear measurement range of the presented sensor was found to be affected by temperature (Geupel et al., 2011) as well as by the thickness (Groß et al., 2012b) of the sensitive layer. The linear range always ends at a sensor response change of about 30 to 40 %. Regeneration is achieved by increasing the temperature to 650 °C (Groß et al., 2013a) or switching to reducing gas atmospheres (Geupel et al., 2010).

Further examples for resistive-type gas dosimeters will be given in the next sections with the focus on modifications of the gas dosimeter setup and with special respect to the operation mode as well as to suitable sorbent materials.

4 Modifications of the dosimeter principle

Several conductometric sensors relying on the accumulating detection principle are reported. In this section, examples for modifications of the above-described dosimeter-type sensing principle are summarized. First of all, different configurations of the sensor setup and the associated influencing factors are described before discussing alternative operation modes.

4.1 Dosimeter setup configurations

The access of the analyte molecules to the sensitive layer, impacting the sensing properties particularly in the case of gas velocity fluctuations, can be varied by the gas dosimeter setup. Equation (1) defines the concentration-related amount of analyte A_c in the gas phase as the time integral of the concentration. However, for fluctuating flow rates, the analyte dose is better reflected by the volume-related amount A_V as the time integral of the concentration and the gas flow rate $\dot{V}(t)$ according to Eq. (2):

$$A_V = \int_{t_0}^{t} c(t) \cdot \dot{V}(t)\mathrm{d}t. \tag{2}$$

If $\dot{V}(t)$ denotes a volume flow ($[\dot{V}(t)] = \mathrm{m}^3\,\mathrm{s}^{-1}$), the unit of A_V is m^3 (or here μL). If, however, $\dot{V}(t)$ is a molar flow ($[\dot{V}(t)] = \mathrm{mol}\,\mathrm{s}^{-1}$), then the unit of A_V is mol, i.e., $[A_V] = \mathrm{mol}$. Hence, the sensor signal of real amount-integrating gas dosimeters is expected to depend on the gas velocity. Otherwise the sensitive device works as a concentration integrator, allowing for concentration information by analyzing the signal derivative (Sect. 2.1).

Beulertz et al. (2011) investigated the influence of the flow rate on the sensor response of a carbonate-based NO_x dosimeter. Figure 8a depicts the time course of the sensor response $|SR| = |\Delta R|/R_0$ of the NO_x dosimeter in the planar sensor setup at various flow rates. Exposed to the same analyte concentration profile, the sensor response of the planar sensor with a sensitive area of 42 mm^2 placed in a quartz tube with an inner diameter of 9.2 mm increases – but independently of the gas flow rate, ranging from 200

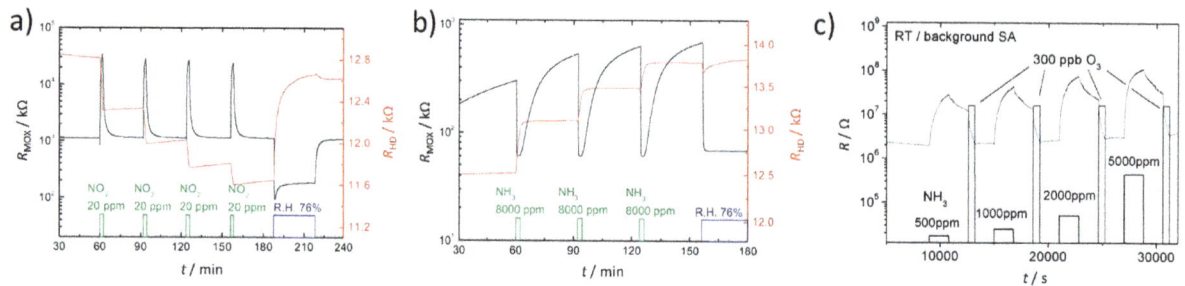

Figure 6. Dosimeter-like response of a hydrogenated diamond (HD) sensor at room temperature acting as pH sensor towards a sequence of **(a)** NO$_2$ and **(b)** NH$_3$ pulses in dry synthetic air and the effect of humidity. Adapted from Helwig et al. (2013); reprinted with permission from Elsevier. **(c)** Response of a hydrogenated diamond sample in synthetic air (80 % N$_2$ / 20 % O$_2$) towards pulses of NH$_3$ and the O$_3$-induced reset of the sensor surface by NH$_3$ oxidation. Modified after Helwig et al. (2008); reprinted with permission from Elsevier.

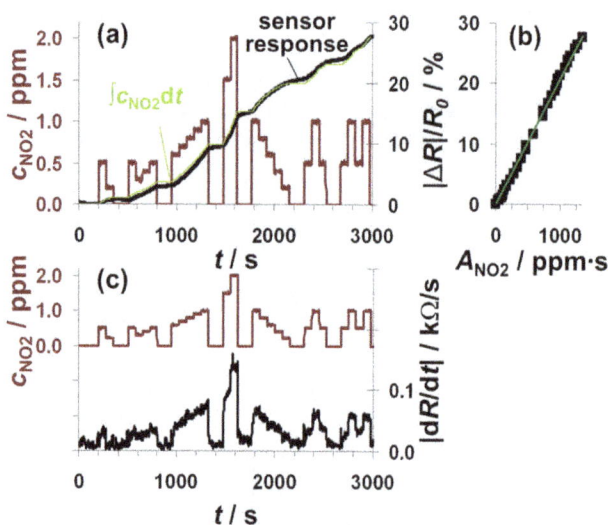

Figure 7. Suitability for the long-term low-level NO$_2$ detection of a planar sensor with a carbonate-based LNT-material in lean atmosphere at 350 °C: **(a)** sensor response $|SR| = |\Delta R|/R_0$ in 0.2 to 2 ppm NO$_2$ for 75 s each, **(b)** linear correlation of $|SR| = |\Delta R|/R_0$ and the total NO$_2$ amount A_{NO_2}, and **(c)** time derivative of the resistance $|dR/dt|$ follows the course of the NO$_2$ concentration. Adapted from Groß et al. (2012a), with permission of the authors.

centration near the sensitive layer is always constant and unaffected by the sorption capability of the sensitive material, resulting in same sensor responses independent of the gas velocity.

For further investigations on the NO$_x$ dosimeter, a flow-through channel-type sensor, as sketched in Fig. 9a, was developed in order to obtain full analyte storage (Beulertz et al., 2012; Moos et al., 2012). Due to the use of two sensitive layers with bigger sensitive areas of 126 mm^2 (threefold longer compared to planar setup) with a small spacing in between, the diffusion of the passing analyte molecules to the sorption sites should be enhanced. The applied gas flow rate was adapted to the new geometry in such a way that the gas velocities across the sensitive layers were the same as in the planar setup. Hence, the measurements were carried out at flow rates of 30 to 60 mL min^{-1}. All these modifications yielded a residence time of 151 ms, a value which is in the same order of magnitude as in the small channels of an automotive lean NO$_x$ catalyst, in which one can observe a full storage without NO$_x$ slip. A second channel-type sensor was located downstream of the first dosimeter to detect the NO$_x$ slip. The sensor responses $|SR|$ of the two channel-type sensors upon various NO concentrations and for the different flow rates are summarized in Fig. 8c. The response of the downstream sensor ("second sensor") remains zero for all runs. This indicates that for all applied flow rates, NO is completely sorbed by the sensitive layer of the first channel-type dosimeter. Comparing the sensor responses of the first sensor, a clear dependence of the sensor response on the gas flow rate for this channel-type setup appears. The higher the gas flow rate and, consequently, the higher the total exposed NO dose is, the higher is the sensor response. The characteristic lines demonstrate that in fact the same volume-related amount of analyte gas in the gas stream causes the same sensor response change. The sensitivity is independent of the gas velocity (in Fig. 8d, A_V denoted as A_{NO}). Hence, the channel-type sensor setup can be used to establish a real amount integrating gas sensor, whereas the simple planar setup serves as concentration integrator with the benefit of a correlation

to 400 mL min^{-1}. The corresponding characteristic lines in Fig. 8b, correlating the sensor response with the volume-related amount A_V (here denoted as A_{NO}), reveal that for the same total amount of NO in the gas stream, the sensor response $|SR| = |\Delta R|/R_0$ differs for the applied gas flow rates. This implies that a planar gas dosimeter measures the integral of the concentration A_c independently of the gas flow rate, according to Eq. (1), and not the real, gas velocity-dependent amount of the analyte gas, A_V, according to Eq. (2). This behavior can be explained by the fact that the gas velocity is very high and that only a small fraction of the gas stream passing the setup reaches the sensitive layer to be accumulated. In the investigated flow rate range, the analyte con-

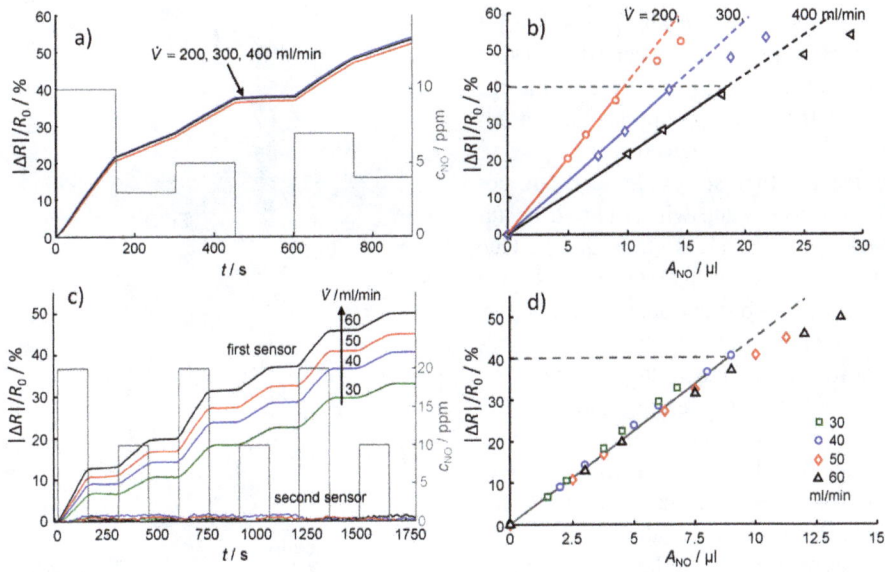

Figure 8. (a) and (b) Sensor response |SR| = |ΔR|/R_0 of a planar NO$_x$ accumulating sensor at various gas flow rates: (a) |SR| does not depend on the flow rates upon exposure towards 0 to 10 ppm NO, (b) corresponding characteristic lines related to the flow rate-dependent total amount A_V, here denoted as A_{NO}. (c) and (d) Sensor response |SR| = |ΔR|/R_0 of a channel-type NO$_x$ accumulating senor at various gas flow rates: (c) course of |SR| in 0 to 20 ppm NO for the different flow rates, (d) corresponding characteristic lines related to the flow rate-dependent total amount A_V, here denoted as A_{NO}. Slightly modified after Beulertz et al. (2012); reprinted with permission from Elsevier.

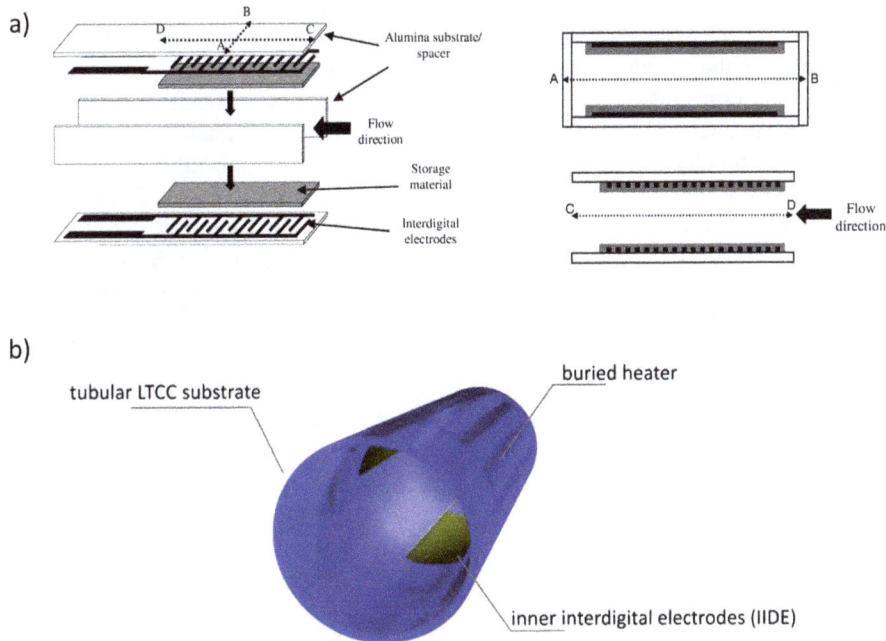

Figure 9. Scheme of setups of resistive gas dosimeters to obtain a flow rate-dependent analyte sorption. (a) Channel-type sensor setup with coated interdigital electrodes (IDE). Reprinted from Beulertz et al. (2011), with permission from Elsevier. (b) Tube-type LTCC transducer with a buried heater and coated inner interdigital electrodes (IIDE). Reprinted from Brandenburg et al. (2013), with permission from Elsevier.

between the sensor response derivative and the actual concentration. A typical application of a concentration integrating dosimeter might be in the field of health protection for working people or air quality monitoring where the sensing device must be independent of the gas flow rate, while the amount integrating dosimeter is an appropriate tool in exhaust lines where the gas flow rate changes rapidly.

In a further development, a tubular, low temperature cofired (LTCC)-based NO_x sensor setup with a buried heater structure was established (Fig. 9b) (Kita et al., 2012). Due to this rotational symmetric setup, the analyte can be sorbed uniformly by a sensitive layer deposited onto the inner surface of the tube. The enhanced interface between the gas phase and the sorbent also enables integrating sensing properties with linear characteristics concerning the volume-related amount. The buried heater structure allows a uniform and adjustable temperature distribution in the sensitive layer as well as periodic switching between the sorption and the regeneration temperature (Brandenburg et al., 2013).

Another design of a gas dosimeter, similar to the setup of passive sampling devices, was introduced by Mattoli et al. (2007) for a mercury sensor with a gold layer as sorbent. As shown in Fig. 10a, the sensitive layer is located inside a sampling chamber closed by a filter material functioning as diffusion barrier. To reach the sensitive sorbent layer, the gas needs to diffuse through the filter, resulting in a gradient of the analyte concentration (here Hg) between zero at the sorbent (acting as analyte sink) and the surrounding gas concentration (Fig. 10b). Hence, the uptake of the analyte in the sorbent is diffusion controlled, leading to an analyte current that depends linearly on the analyte concentration in the ambience. The independency of the analyte uptake on the velocity of the passing gas stream results in concentration integrating properties of the sensing device. Upon exposure to gaseous elemental Hg, the dosimeter changes its sensor signal irreversibly. With growing number of mercury injections, the total amount of adsorbed Hg is concomitantly increased, which is reflected by the sensor signal. With this sensor design, real-time monitoring of low levels of gaseous analyte is possible with a small size, low weight and low cost device applicable for personal safety.

Different options for the arrangement of gas dosimeters in gas pipes are possible to decrease the gas velocity across the sensitive layer, for instance to place the sensing device in a bypass with or without a gas pump to ensure a constant gas stream (Moos et al., 2010).

4.2 Operation modes of gas dosimeters

As discussed, the proportionality between sensor signal and cumulated analyte amount of linear gas dosimeters allows for time-continuous direct dose detection during the sorption period and, depending on the setup, for concentration information by analyzing the signal derivative. However, gas dosimeters might also be operated in different modes. Some

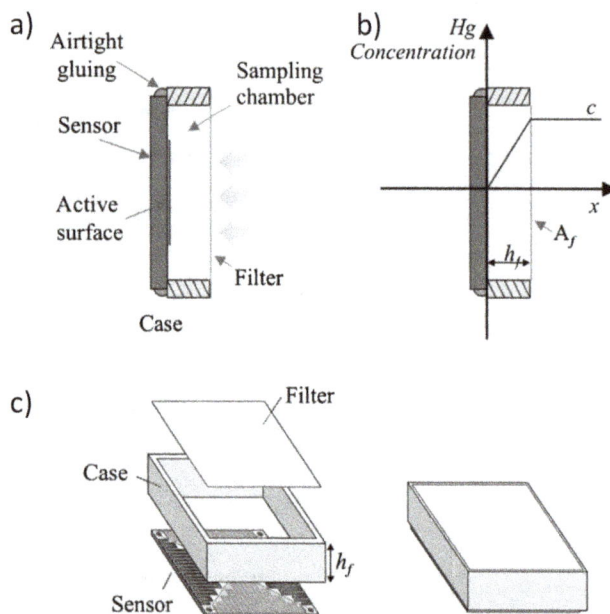

Figure 10. Gas dosimeter setup for a diffusion-controlled analyte sorption: **(a)** cross-section of the dosimeter sampling chamber with a filter acting as diffusion layer, **(b)** profile of the resulting Hg concentration, and **(c)** top view of the sampling chamber. Reprinted from Mattoli et al. (2007), with permission from Elsevier.

examples are described in the following. A summary concludes this section.

As described above and depicted in Fig. 1b, the analyte accumulation level of the sorbent is reflected by the electrical properties during the sorption period as well as during the subsequent analyte release in the regeneration interval. Hence, the analysis of the sensor signal change during the regeneration interval is another method for operating gas dosimeters. It was demonstrated on a potassium and manganese containing NO_x sensor that, after the sorption period under sorption conditions, the change of the electrical conductance during thermal decomposition of the formerly formed nitrates correlates with the preceding sorbed analyte amount (Groß et al., 2013a). A schematic drawing of this novel method is given in Fig. 11. Since it combines electrical characterization of the sorbent with temperature programmed desorption of the analyte, it is also denoted as eTPD. Due to the heating from the sorption to the desorption temperature, the thermal activated conductance G of the sensitive layer increases as well (Fig. 11b). Thereby, the conductance of the material in the NO_x loaded state, G, exceeds those of the unloaded sample, G_0. The thermal release of the formerly sorbed analyte is indicated by a desorption peak with the concentration $c_{released}$ (measurable by an NO_x analyzer downstream of the sensor). Upon this thermal release, the curve of the conductance of the sensitive layer in the analyte-loaded state, G, converges to those in the unloaded state, G_0 (Fig. 11a). The monitored desorption peak and the

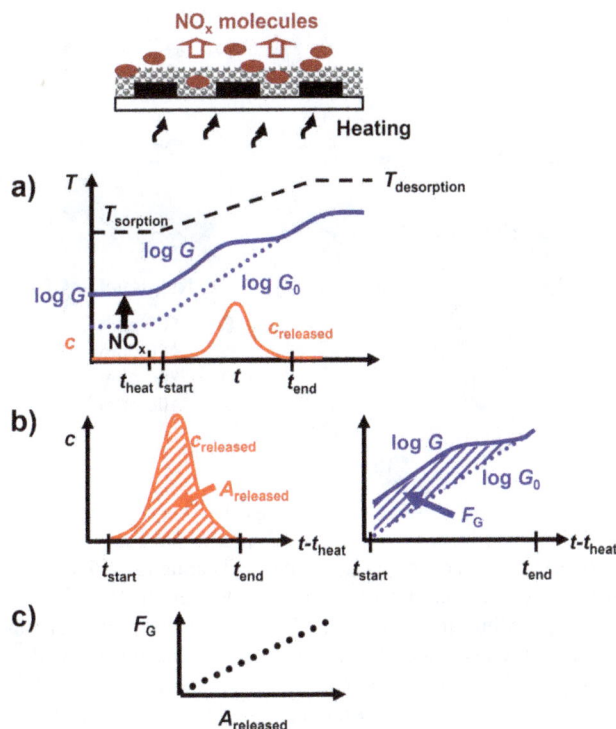

Figure 11. Scheme of the electrical evaluation during thermal regeneration (denoted as eTPD): **(a)** time dependence of the conductance G and the outlet analyte concentration $c_{released}$ upon heating from the sorption to the regeneration temperature, **(b)** determination of the released amount $A_{released}$ from the desorption peak and the electrical response F_G as deviation of G from the course of the unloaded state, and **(c)** analysis of F_G as a function of $A_{released}$. Reprinted from Groß et al. (2013a), with permission of the authors.

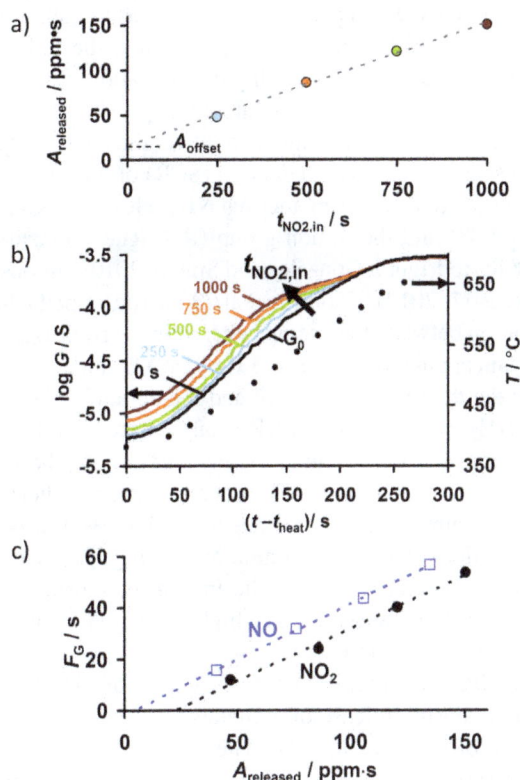

Figure 12. eTPD results of the potassium and manganese containing NO_x dosimeter after exposure to 8 ppm NO or NO_2 for different NO_2 exposure durations t_{NO_2} up to 1000 s: **(a)** linear increase of the desorbed amount $A_{released}$ with t_{NO_2}; **(b)** course of the conductance, G, in different loading states upon heating; and **(c)** linear correlation between $A_{released}$ and the electrical response, F_G (deviation of G from G_0), during desorption. Reprinted from Groß et al. (2013a), with permission of the authors.

course of the conductance can be analyzed, as sketched in Fig. 11b. The desorption peak gives information on the formerly sorbed analyte amount, being equal to the released amount $A_{released} = \int c_{released} \, dt$ if the regeneration is complete. The deviation of the thermally activated conductance after NO_x loading, G, from the course in the unloaded state, G_0 (Fig. 11b), was evaluated as sensor signal during the regeneration interval (Fig. 11c) and is defined as F_G according to Eq. (3). Figure 11c depicts the dependency of the resulting sensor signal, F_G, and the released analyte amount, $A_{released}$.

$$F_G = \int_{t_{start}}^{t_{end}} [\log G(t) - \log G_0(t)] \, dt. \qquad (3)$$

Figure 12 summarizes the results of an eTPD analysis of the potassium and manganese containing NO_x dosimeter operated at a sorption temperature of 380 °C and heated to 650 °C for thermal regeneration. During both intervals, the sensor was applied to a lean gas mixture of 2 L min^{-1}. To achieve different NO_x loading levels, the sensor sample was exposed to 8 ppm NO or 8 ppm NO_2 for a duration $t_{NO_x,in}$ between 0 s and 1000 s. The NO_x loading level, evaluated as the

released amount, $A_{released}$ (measured by a chemiluminescence detector), increases proportionally to the NO_x exposure time, $t_{NO_2,in}$, at the fixed NO_x concentration of 8 ppm, indicating a time-constant sorption rate (Fig. 12a). In Fig. 12b, the curves of $\log G$ during heating after different NO_2 exposure intervals from 250 to 1000 s are compared to those without NO_2 dosing ($\log G_0$, after 0 s NO_2). The temperature at which all the curves meet, here about 620 °C, represents the end of the nitrate decomposition. Figure 12c clarifies that for NO or NO_2 dosage, the electrical response, F_G (Eq. 3), during thermal regeneration serves as an electrical measure for the former NO_x loading level, since F_G correlates with $A_{released}$. Figure 12 reveals that, besides of the real-time and time-continuous detection of the analyte dose during the sorption period, the NO_x loading state as well as the cumulated NO_x exposure can also be determined electrically during the short-term thermal release. In this operation mode, the gas dosimeter can be seen as passive sampler with an internal sensor function to determine the sampled analyte dose electrically instead of applying an external subsequent gas analysis in the laboratory. By combining these two operation modes (measurement during sorption and desorption), it is expected that the redundant information enables a plausibility consideration of the time-resolved sensor signal (Groß et al., 2013a).

Similarly, the NH_3 loading level of zeolites being applied for the selective catalytic reduction (SCR) of NO_x was electrically investigated during thermal NH_3 release by Kubinski et al. (2008) and the Simon group (Rodríguez-González et al., 2008; Rodríguez-González and Simon, 2010; Simons and Simon, 2011, 2012). Kubinski et al. (2008) described a linear correlation between the average alternating current $I_{Average}$ at a constant applied AC voltage (5 V_{P-P}) at a frequency of 4 Hz during the thermal NH_3 release and the duration of the previous NH_3 exposure of an SCR catalyst-based conductivity sensor (Fig. 13). For reaching a lowly loaded state, the sensor was exposed to a constant NH_3 concentration, as indicated in a base gas atmosphere of 5 % O_2 and $\sim$ 1 % H_2O in N_2 for time durations from 0 to 40 min. Measuring $I_{Average}$ during thermal NH_3 release enables the in situ monitoring of the former NH_3 loading level with a higher sensitivity compared to those in the sorption mode.

A similar example for the investigation of the electrical properties during release of a chemisorbed gas species was the work of Rossé et al, 1984. They found that a CdSe thin film irreversibly sorbs oxygen and can be thermally regenerated, causing a measurable change in the resistance.

The different operation modes of gas dosimeters are summarized in Fig. 14. As discussed in detail, the cumulated analyte exposure as well as the analyte loading level of the sensitive material can be electrically detected either during sorption (including concentration information from the derivative) (Fig. 14a) or regeneration (Fig. 14b). Due to the correlation between the analyte amount and the sensor signal, it is possible to determine the mean analyte concentration also by the time Δt_{spec} that is needed to reach a defined loading state of the sensor, indicated by a specific signal change $|SR|_{spec}$ (Fig. 14c). For instance, the frequency of regeneration might serve as a measure for the analyte concentration. Due to a limited linear measurement range, the regeneration needs to be initiated at a predefined sensor signal before the signal saturates. Further, the time to reach a percolation threshold, t_{perc}, indicated by a steep increase or decrease in the conductivity of the percolation dominated dosimeter, may serve as the measure for the mean concentration (Fig. 14d). For both time-measuring modes, no instantaneous information on the cumulated exposure or the actual concentration is obtained.

4.3   Gas dosimeters based on percolation effects

Percolation effects (Ulrich et al., 2004) may modify the sensing characteristics of gas dosimeters, as investigated and described in detail for soot sensors being similar to gas sensors. Several soot sensors for diesel particulate filters consisting of an insulating substrate with electrodes on top have been reported (Ochs et al., 2010; Weigl et al., 2010; Hagen et al.,

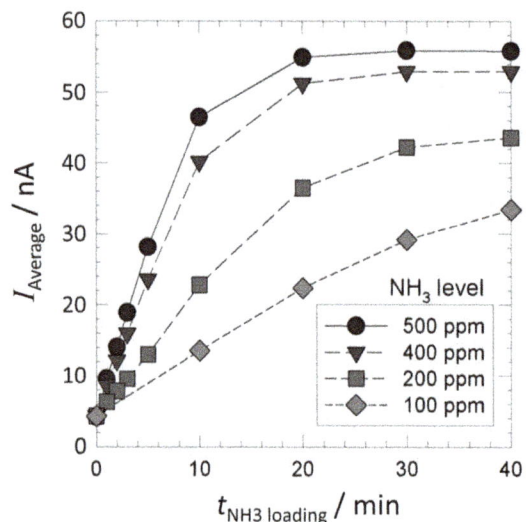

Figure 13. Average current $I_{Average}$ during heating up to 520 °C for thermal regeneration of a zeolite-based NH_3 sensor depending on the NH_3 loading time, $t_{NH_3\ loading}$. The SCR catalyst material was loaded at a temperature of $\sim$ 267 °C in a gas stream composed of 5 % O_2 and $\sim$ 1 % H_2O in N_2 containing NH_3, as indicated for 0 to 40 min. Adapted from Kubinski and Visser (2008); reprinted with permission from Elsevier.

2010; Kondo et al., 2011; Husted et al., 2012; Bartscherer and Moos, 2013). The resistance between these electrodes is analyzed. It serves as a measure for accumulated soot particles on an insulating support. Figure 15 schematically depicts the formation of conductive pathways for the soot sensing device in its different states of soot loading. The soot deposition is driven by electrophoresis due to the applied voltage, U. Soot paths start to grow at the electrodes. From Fig. 15a to d, the amount of deposited conductive soot particles on the substrate surface between the electrodes increases. For the unloaded and lowly soot loaded state, almost no current, I, flows and the resistance between the electrodes, R, is almost infinite ($R \rightarrow \infty$). The sensor remains blind until the percolation threshold is reached after an exposure time of t_{perc}. Then, the formed conducting paths cause a fast decrease of the resistance ($R < \infty$). Further soot accumulation leads to an additional current increase or resistance decrease ($R \downarrow$). An additional low-conducting layer on top of the electrodes made of a resistive paste may reduce the sensor blind time. Owing to the low conductivity of the added layer, the soot amount can be detected in the accumulation mode without percolation effects (Bartscherer and Moos, 2013). The actual soot concentration can be determined from the percolation time t_{perc} (time to reach a certain value of I) (Ochs et al., 2010) or from the slope dI / dt (Weigl et al., 2010). As soon as a predefined current is reached, the sensor device is heated, the soot burns off, and after cooling to operation temperature, the regeneration is completed and a new accumulation period starts.

Different operation modes of the dosimeter principle

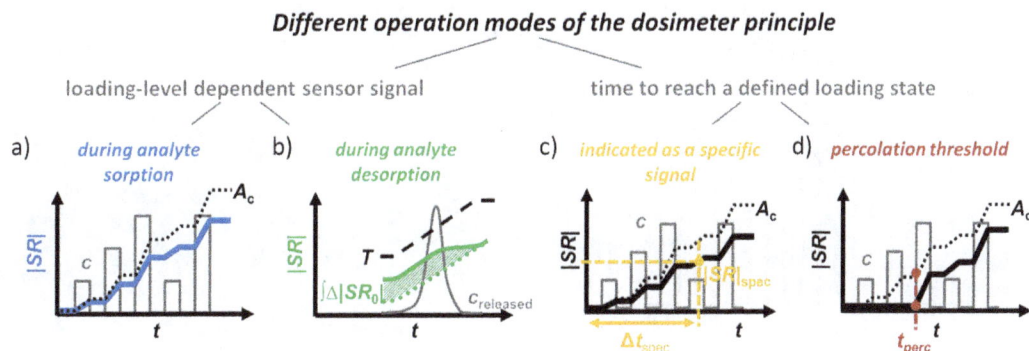

Figure 14. Scheme of the different operation modes of gas dosimeters: consideration of the loading level-dependent sensor signal during **(a)** analyte sorption and **(b)** analyte desorption and of the period of time to reach a defined loading state indicated by **(c)** a specific sensor signal like the end of the linear measurement range, or, e.g., **(d)** the percolation threshold.

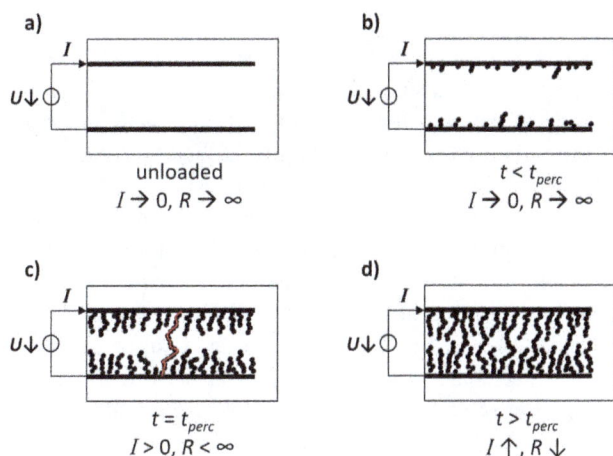

Figure 15. Scheme of percolation-dominated resistive soot detection with planar electrodes on an insulating substrate: **(a)** regenerated soot sensor with an infinite resistance R, **(b)** sensor covered by a few soot particles below the percolation threshold, **(c)** sensor at point in time when the percolation threshold t_{perc} is reached due to a sufficient number of deposited particles forming a conductive pathway with a specific resistance, and **(d)** more soot particles at the surface, causing a further decrease of R.

Another example for dosimeter type gas sensors using percolation effects is given by Hennemann et al. (2012b, c) and Sauerwald et al. (2013). Applying CuO nanofibers as H_2S sensitive materials, the change of conductance of the fibers upon pulsed H_2S dosing is measured. When adsorbing H_2S, the conductance of the fibers increases due to their transition from CuO to CuS. However, as shown in Fig. 16a, the conductance G does not change during the first H_2S pulses, but after reaching the percolation threshold, a conductive pathway of CuS in the CuO layer is formed and the conductance increases promptly. Beyond the percolation threshold, the sensor behaves like a H_2S integrating gas sensor. Nevertheless, saturation effects deteriorate the sensor signal soon after reaching the percolation threshold. If one plots the re-

ciprocal value of the percolation time t_{perc} (defined as the time when the percolation threshold is reached, denoted as "switching time" in the work of Hennemann et al., 2012a) as a function of the H_2S concentration, Fig. 16b implies a linear dependency, demonstrating clearly that t_{perc} decreases with increased H_2S concentration. Hence, a direct measurement of the actual concentration and dose of H_2S is not possible; however, by the evaluation of the switching time the average concentration and the total amount of H_2S in a certain period of time can be determined indirectly (Hennemann et al., 2012a).

Similarly, the electrical properties of a Ni-based catalyst pellets were examined by impedance spectroscopy during the exposure to H_2S. Before sulfur poisoning, the conductance is very low due to the high dispersion of the Ni particles in the catalyst, i.e., the impedance is dominated by the permittivity of the pellet material. With the addition of H_2S, nickel sulfide forms. Due to the enhanced volume of the nickel sulfides compared to the Ni particles, the distance between the conductive particles diminishes. Reaching the percolation threshold, the conductance increases immediately due to the formation of conductive pathways of nickel sulfides, allowing also dosimeter-type H_2S detection (Fremerey et al., 2012).

It should be clear, however, that accumulating percolation-type gas sensors cannot display the instant value directly. Since they act like switches if a certain accumulation level has been reached, however, the time span between two percolation events can be seen as an indicator for the average concentration between the two events.

5 Suitable materials for accumulating gas sensors

To find adequate materials in gas dosimeter application, it is necessary to define the required properties of these materials. In general, the following five main criteria were found to be crucial for the proper selection of materials as sensitive

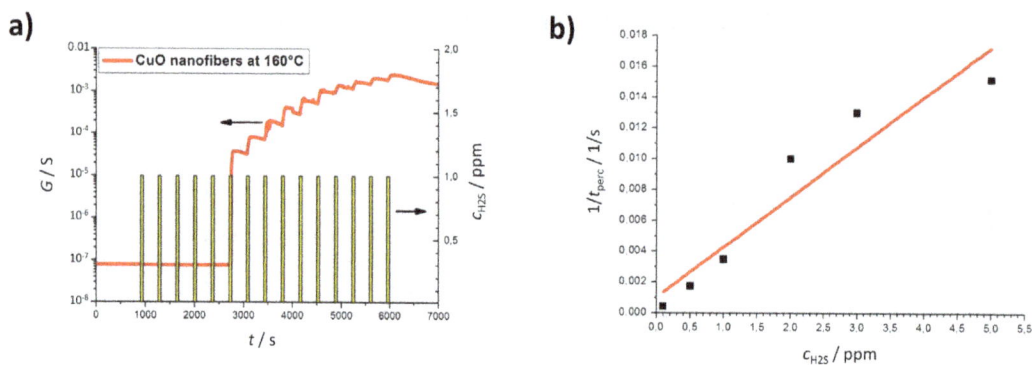

Figure 16. Percolation-dominated H_2S detection with CuO nanofibers at 160 °C: **(a)** conductance during consecutive 1 ppm H_2S pulses for 1 min each with a sudden increase of the conductance as the percolation threshold is reached, denoted as switching time, **(b)** correlation between the reciprocal switching time and the H_2S concentration. Modified after Hennemann et al. (2012a); reprinted with permission from John Wiley and Sons.

components of gas dosimeters with a linear sensing characteristic (Groß et al., 2012a).

1. Selective sorption of the analyte: only one distinct gas species should accumulate in the sensitive layer, e.g., by chemisorption, including the chemical transformation of the material. To obtain linear sensing characteristics, the sorption rate should be proportional to the current analyte concentration in the gas phase.

2. Strength of sorption: the analyte sorption rate has to exceed largely the desorption rate to ensure accumulation of analyte molecules in the sensitive material without releasing formerly sorbed molecules in the absence of the analyte in the gas atmosphere (i.e., holding capability).

3. At least one measurable electrical property must change with analyte accumulation, e.g., the impedance or the conductivity. The relation between the analyte loading level and the change of the electrical measurand has to be linear.

4. The sorption rate must be independent of the present analyte loading level of the sensitive layer so that no deterioration of the sorption efficiency with time occurs.

5. Ability to actively initiate a regeneration: the sensitive material must be able to release the formerly sorbed analyte molecules from the storage sites under defined conditions that are different from the sorption conditions (e.g., different temperature or gas composition). If there is no possibility to regenerate the storage material, the sensor is a single-use device.

Conductometric gas sensors using the accumulating sensing principle have been reported by several working groups for various analytes and sensitive materials (Sect. 3). The applied materials range from organic semiconductors to metal oxides, zeolites and carbon nanotubes. In the following an overview on the materials is given.

Suitable materials for the accumulation sensing mode are materials that have been successfully applied as storage components, for instance in filters or catalysts, e.g., lean NO_x traps or selective catalytic reduction (SCR) materials storing NH_3. Geupel et al. (2011) applied NO_x storage catalysts based on alkaline (earth) carbonates as coatings for NO_x dosimeter operated at temperatures in the range of 300 to 400 °C. However, $BaCO_3$ serves also as a sulfur oxide adsorbing layer on a gas sensor to protect it from sulfur poisoning (Rettig et al., 2003). The irreversible interaction of SO_2 with the NO_x storage catalyst material causes a cross-sensitivity of the NO_x dosimeter but enables SO_2 dosimetry (Groß et al., 2012d). Zeolites, being a common class of materials in SCR catalysis due to their NH_3 storage capability, are also viable materials as sensitive layers working in the accumulation mode. The gas sorption capacity of zeolites is very high and, due to their special framework structure, they hold a certain shape-selectivity. Upon gas sorption, zeolite materials are able to change their conductivity (Stamires, 1962; Eigenmann et al., 2000; Sahner et al., 2008), making them potential candidates for dosimeter-type sensors. An example for the dosimeter-type conductivity change is proton conducting H-ZSM-5 adsorbing NH_3 (Simon et al., 1998; Franke et al., 2003; Kubinski and Visser, 2008). Similar to those microporous zeolites, mesoporous materials might also be of interest, as they have been successfully applied as preconcentrators in gas sensors to enhance the sensitivity by analyte accumulation in the porous material via sorption (Wagner et al., 2013).

Metal oxides also show potential as sensitive layers of gas dosimeters. Hennemann et al. (2012a) showed that CuO is an appropriate material for the accumulative detection of H_2S being dominated by percolation effects. The conductivity of NiO shows a similar behavior in H_2S containing gas atmosphere (Fremerey et al., 2012). Other (transition) metal oxides, yet to be applied as sensitive material in conventional gas sensors, should be investigated concerning their suitability as gas dosimeter under modified sensing conditions that allow for irreversible analyte sorption, e.g., WO_3 and Al-activated WO_3 as H_2S dosimeter at room temperature (Reyes et al., 2006) and SnO_2 as O_2, H_2O, and H_2 dosimeter (Yamazoe et al., 1979). Mathieu et al., 2013, for example, give a broad overview on the sorption of SO_2 by various oxides, e.g., CaO, TiO_2, CeO_2, MnO_2 and mixed oxides, meaning metal oxides in association with, e.g., Al_2O_3. They also describe the regeneration of the metal oxides by thermal decomposition of the metal sulfates (Mathieu et al., 2013). With regard to the accumulating sensor principle, these materials hold a certain storage capacity, but it has to be clarified whether the SO_2 accumulation causes measurably changing electrical properties.

Another class of materials, providing a gas storage capacity that can be monitored by the resistance, are metal oxide-coated carbon nanotubes (Mangu et al., 2010). Varghese et al. (2001) reported impedimetric gas sensors based on SiO_2 coated multi-wall carbon nanotubes as sensitive layer, showing the behavior of a conventional gas sensor towards H_2O (Varghese et al., 2001), O_2 and CO_2 (Ong et al., 2002), but acting as a dosimeter in presence of NH_3 (Varghese et al., 2001; Ong et al., 2002). Single-walled carbon nanotubes with SnO_2 coatings were applied by Mubeen et al. (2013) as resistive gas sensors. Their tests with several gases reveal the classical gas sensor response; however the behavior towards NO_2 is dosimeter-like.

As already shown, hydrogenated diamond is a well-suited material for the accumulating sensing method to detect NO_2 and NH_3 due to a pH change upon electrolytic analyte dissociation in the surface electrolyte layer (Beer et al., 2013; Helwig et al., 2013). The hydrogenated diamond-based sensor can be regenerated thermally by O_3 or UV light (Helwig et al., 2007, 2008).

Conductive polymers have already been applied for gas sensing applications. Bai and Shi show an overview on different polymers for gas detection. They described that some of the conductive polymers show an irreversible reaction to certain gases, making them a notable class of materials for gas dosimeters. However, it has to be kept in mind that polymers often show a poor long-term stability (Bai and Shi, 2007; Liu et al., 2012). The conductive polymeric system PEDOT:PSS was applied as a gas sensitive layer by Liu et al. (2012), acting as a classical gas sensor in the presence of NO_x. However, initial tests of Marr et al. (2013) have shown that the conductive copolymer PEDOT:PSS is also a promising material for an accumulating gas sensing device. Their first measurements at room temperature with NO_x as test gas show a dosimeter-type change of the resistance in presence of NO_x.

Organic compounds also might be suitable materials for the integrating sensing principle. A resistive low-level NO_2 dosimeter based on iron (II) phthalocyanine was presented by Shu et al. (2010). Padma et al. (2009) and Brunet et al.

(2001, 2008) reported a NO_2 sensitivity of copper phthalocyanine. The Saltzmann reagent was applied in optical sensors for the detection of O_3 and NO_2 (Tanaka et al., 1998, 1999; Maruo, 2007; Maruo et al., 2009), morpholines for NO_2 sensing (Matsuguchi et al., 2005), eosin Y for NH_3 detection via the slope of the drain current of a based OFET (Klug et al., 2013), and naphthyl polyenes to detect organic vapors electrically (Sircar et al., 1983). Their suitability for resistive gas dosimeters should be investigated.

Even metals can be applied as sensitive materials for gas dosimeters: Mattoli et al. (2007) realized a sensing device based on a Au layer for the resistive, cumulative detection of gaseous Hg. Ag was found to be a suitable material for the detection of H_2S in the accumulation mode by Angelini et al. (2009) and Chen et al. (2013).

Additionally, Mukherjee et al. (2012) summarizes gas sensor materials interacting reversibly or partly irreversibly with various gases.

As already mentioned in Sect. 2, the temperature strongly influences the gas sensing behavior (Reyes et al., 2006; Andringa et al., 2012; Shu et al., 2012; Brandenburg et al., 2013; Groß et al., 2013a). Since conventional gas sensors are mostly based on metal oxides as sensitive layer and these materials are known to be able to adsorb gases, e.g., SO_2 (Mathieu et al., 2013), they might be suitable for applications in a gas dosimeter under modified operation conditions. The desorption of the analyte gas predominates at elevated temperatures if the analyte concentration in the gas phase decreases. The sensor signal changes reversibly, according to the established chemical equilibrium. Hence, the operation of the sensitive materials of conventional gas sensors at lower temperatures might reveal dosimeter-type sensing properties with analyte accumulation due to a reduced desorption rate. Similarly, $BaCO_3$ acts as NO_x dosimeter at low temperatures (Groß et al., 2013b) but the recovery time after NO_x sorption can be reduced by increasing the temperature, resulting in a reversible sensing characteristic (Tamaki et al., 1995).

6 Conclusion and outlook

The accumulating sensing principle of solid-state gas dosimeters is a promising method in order to elude certain drawbacks of classical metal oxide gas sensors. Dosimeter-type gas sensors are well suited for the detection of harmful, toxic gases. This is in accordance with the growing demand for sensing devices that are able to detect lowest analyte concentrations reliably, fast and over a long period of time, particularly in the field of air quality monitoring in workplaces and urban environments. Acting similar to passive samplers, gas dosimeters are able to detect very low concentration levels of gaseous pollutants. However, the dosimeter offers a real-time detection of the analyte dose, according to mean time values in emission and immission legislations. Due to the chemical accumulation of the analyte in the sensing material, the error-prone mathematical integration of the concentration-related sensor signal, as is necessary in cases of conventional gas sensors to measure doses, is not needed.

Being affected by several influencing factors, the accumulating sensing principle might be adapted to common gas sensing materials under modified measurement conditions.

This overview shows a number of works applying and describing accumulating sensing behavior, but further investigations regarding the long-term stability of gas dosimeter have to be conducted, especially when the regeneration step is a thermal one. A comparison between classical gas sensors and gas dosimeters lists the advantages and disadvantages of both sensing principle. Nevertheless, different principles have to be tested under same conditions to compare them directly.

Acknowledgements. Ralf Moos thanks the German Research Foundation (DFG) for supporting this work under grant number MO 1060/15-1.

Edited by: M. Penza
Reviewed by: two anonymous referees

References

Andringa, A.-M., Vlietstra, N., Smits, E. C. P., Spijkman, M.-J., Gomes, H. L., Klootwijk, J. H., Blom, P. W. M., and de Leeuw, D. M.: Dynamics of charge carrier trapping in NO_2 sensors based on ZnO field-effect transistors, Sensor. Actuator. B: Chemical, 171–172, 1172–1179, doi:10.1016/j.snb.2012.06.062, 2012.

Angelini, E., Grassini, S., Neri, A., Parvis, M., and Perrone, G.: Plastic Optic Fiber Sensor for Cumulative Measurements, in: Proceedings of the International Instrumentation and Measurement Technology Conference – I2MTC 2009, Singapore, 5–7 May 2009, 1666–1670, doi:10.1109/IMTC.2009.5168723, 2009.

Bai, H. and Shi, G.: Gas Sensors Based on Conducting Polymers, Sensors, 7, 267–307, doi:10.3390/s7030267, 2007.

Barsan, N., Koziej, D., and Weimar, U.: Metal oxide-based gas sensor research: How to?, Sensor. Actuator. B: Chemical, 121, 18–35, doi:10.1016/j.snb.2006.09.047, 2007.

Bartscherer, P. and Moos, R.: Improvement of the sensitivity of a conductometric soot sensor by adding a conductive cover layer, Journal of Sensors and Sensor Systems, 2, 95–102, doi:10.5194/jsss-2-95-2013, 2013.

Beer, S., Helwig, A., Müller, G., Garrido, J., and Stutzmann, M.: Water adsorbate mediated accumulation gas sensing at hydrogenated diamond surfaces, Sensor. Actuator. B: Chemical, 181, 894–903, doi:10.1016/j.snb.2013.02.072, 2013.

Beulertz, G., Geupel, A., Moos, R., Kubinski, D. J., and Visser, J. H.: Accumulating gas sensor principle – how to come from concentration integration to real amount measurements, Procedia Engineering, 25, 1109–1112, doi:10.1016/j.proeng.2011.12.273, 2011.

Beulertz, G., Groß, A., Moos, R., Kubinski, D. J., and Visser, J. H.: Determining the total amount of NO_x in a gas stream – Advances

in the accumulating gas sensor principle, Sensor. Actuator. B: Chemical, 175, 157–162, doi:10.1016/j.snb.2012.02.017, 2012.

Bhalla, V., Singh, H., and Kumar, M.: Facile Cyclization of Terphenyl to Triphenylene: A New Chemodosimeter for Fluoride Ions, Organic Lett., 12, 628–631, doi:10.1021/ol902861b, 2010.

BImSchV 2010, Neununddreißigste Verordnung zur Durchführung des Bundes-Immissionsschutzgesetzes (Verordnung über Luftqualitätsstandards und Emissionshöchstmengen-39. BImSchV). Bundesgesetzblatt Jahrgang 2010, ausgegeben zu Bonn am 5 August 2010, Nr. 40, 1065, 2010.

Brandenburg, A., Kita, J., Groß, A., and Moos, R.: Novel tube-type LTCC transducers with buried heaters and inner interdigitated electrodes as a platform for gas sensing at various high temperatures, Sensor. Actuator. B: Chemical, 189, 80–88, doi:10.1016/j.snb.2012.12.119, 2013.

Brunet, J., Talazac, L., Battut, V., Pauly, A., Blanc, J. P., Germain, J. P., Pellier, S., and Soulier, C.: Evaluation of atmospheric pollution by two semiconductor gas sensors, Thin Solid Films, 391, 308–313, doi:10.1016/S0040-6090(01)01001-X, 2001.

Brunet, J., Garcia Parra, V., Pauly, A., Varenne, C., and Lauron, B.: An optimized gas sensor microsystem for accurate and real-time measurement of nitrogen dioxide at ppb level, Sensor. Actuator. B: Chemical, 134, 632–639, doi:10.1016/j.snb.2008.06.010, 2008.

Chen, R., Morris, H. R., and Whitmore, P. M.: Fast Detection of Hydrogen Sulfide Gas in the ppmv Range with Silver Nanoparticles Films at Ambient Conditions, Sensor. Actuator. B: Chemical, 186, 431–438, doi:10.1016/j.snb.2013.05.075, 2013.

Corcoran, P. and Shurmer, H. V.: An intelligent gas sensor, Sensor. Actuator. A: Physical, 41–42, 192–197, doi:10.1016/0924-4247(94)80110-X, 1994.

Dasgupta, P. K., Genfa, Z., Poruthoor, S. K., Caldwell, S., Dong, S., and Liu, S.-Y.: High-Sensitive Gas Sensors Based on Gas-Permeable Liquid Core Waveguides and Long-Path Absorbance Detection, Anal. Chem., 70, 4661–4669, doi:10.1021/ac980803t, 1998.

Directive 2008/50/EC of the European Parliament and of the Council of 21 May 2008 on Ambient Air Quality and Cleaner Air for Europe, Off. J. EU 2008, L152/1, 2008.

Eigenmann, F., Maciejewski, and Baiker, A.: Gas adsorption studied by pulse thermal analysis, Thermochimica Acta, 359, 131–141, doi:10.1016/S0040-6031(00)00516-5, 2000.

Franke, M. E., Simon, U., Moos, R., Knezevic, A., Müller, R., and Plog, C.: Development and working principle of an ammonia gas sensor based on a refined model for solvate supported proton transport in zeolites, Phys. Chem. Chem. Phy., 5, 5195–5198, doi:10.1039/B307502H, 2003.

Fremerey, P., Jess, A., and Moos, R.: Direct in-situ detection of sulfur loading on fixed bed catalysts, in: Proceedings of the 14th International Meeting on Chemical Sensors – IMCS 2012, Nuremberg, Germany, 20–23 May 2012, 76–79, doi:10.5162/IMCS2012/1.1.5, 2012.

Geupel, A., Schönauer, D., Röder-Roith, U., Kubinski, D. J., Mulla, S., Ballinger, T. H., Chen, H.-Y., Visser, J. H., and Moos, R.: Integrating nitrogen oxide sensor: A novel concept for measuring low concentrations in the exhaust gas, Sensor. Actuator. B: Chemical, 145, 756–761, doi:10.1016/j.snb.2010.01.036, 2010.

Geupel, A., Kubinski, D. J., Mulla, S., Ballinger, T. H., Chen, H.-Y., Visser, J. H., and Moos, R.: Integrating NO_x Sensor for Automotive Exhausts – A Novel Concept, Sensor Letters, 9, 311–315, doi:10.1166/sl.2011.1471, 2011.

Göpel, W.: New materials and transducers for chemical sensors, Sensor. Actuator. B: Chemical, 18–19, 1–21, doi:10.1016/0925-4005(94)87049-7, 1994.

Groß, A., Beulertz, G., Marr, I., Kubinski, D. J., Visser, J. H., and Moos, R.: Dual Mode NO_x Sensor: Measuring Both the Accumulated Amount and Instantaneous Level at Low Concentrations, Sensors, 12, 2831–2850, doi:10.3390/s120302831, 2012a.

Groß, A., Richter, M., Kubinski, D. J., Visser, J. H., and Moos, R.: The Effect of the Thickness of the Sensitive Layer on the Performance of the Accumulating NO_x Sensor, Sensors, 12, 12329–12346, doi:10.3390/s120912329, 2012b.

Groß, A., Bishop, S. R., Yang, D. J., Tuller, H. L., and Moos, R.: The electrical properties of NO_x-storing carbonates during NO_x exposure, Solid State Ionics, 225, 317–323, doi:10.1016/j.ssi.2012.05.009, 2012c.

Groß, A., Hanft, D., Beulertz, G., Marr, I., Kubinski, D. J., Visser, J. H., and Moos, R.: The effect of SO_2 on the sensitive layer of a NO_x dosimeter, Sensor. Actuator. B: Chemical, 187, 153–161, doi:10.1016/j.snb.2012.10.039, 2012d.

Groß, A., Kremling, M., Marr, I., Kubinski, D. J., Visser, J. H., Tuller, H. L., and Moos, R.: Dosimeter-Type NO_x Sensing Properties of $KMnO_4$ and Its Electrical Conductivity during Temperature Programmed Desorption, Sensors, 13, 4428–4449, doi:10.3390/s130404428, 2013a.

Groß, A., Weller, T., Tuller, H. L., and Moos, R.: Electrical conductivity study on NO_x trap materials $BaCO_3$ and $K_2CO_3/La-Al_2O_3$ during NO_x exposure, Sensor. Actuator. B: Chemical, 187, 461–470, doi:10.1016/j.snb.2013.01.083, 2013b.

Hagen, G., Feistkorn, C., Wiegärtner, S., Heinrich, A., Brüggemann, D., and Moos, R.: Conductometric Soot Sensor for Automotive Exhausts: Initial Studies, Sensors, 10, 1589–1598, doi:10.3390/s100301589, 2010.

Helwig, A., Müller, G., Weidemann, O., Härtl, A., Garrido, J. A., and Eickhoff, M.: Gas Sensing Interactions at Hydrogenated Diamond Surfaces, IEEE Sensors Journal, 7, 1349–1353, doi:10.1109/JSEN.2007.905019, 2007.

Helwig, A., Müller, G., Garrido, J. A., and Eickhoff, M.: Gas sensing properties of hydrogen-terminated diamond, Sensor. Actuator. B: Chemical, 133, 156–165, doi:10.1016/j.snb.2008.02.007, 2008.

Helwig, A., Beer, S., and Müller, G.: Breathing mode gas detection, Sensor. Actuator. B: Chemical, 179, 131–139, doi:10.1016/j.snb.2012.07.088, 2013.

Hennemann, J., Sauerwald, T., Kohl, C.-D., Wagner, T., Bognitzki, M., and Greiner, A.: Electrospun copper oxide nanofibers for H_2S dosimetry, Phys. Status Solidi A, 209, 911–916, doi:10.1002/pssa.201100588, 2012a.

Hennemann, J., Sauerwald, T., Wagner, T., Kohl, C.-D., Dräger, J., and Russ, S.: Electrospun Copper(II)oxide Fibers as Highly Sensitive and Selective Sensor for Hydrogen Sulfide Utilizing Percolation Effects, in: Proceedings of the 14th International Meeting on Chemical Sensors – IMCS 2012, Nuremberg, Germany, 20–23 May 2012, 197–200, doi:10.5162/IMCS2012/2.3.4, 2012b.

Hennemann, J., Sauerwald, T., Wagner, T., Kohl, C.-D., Dräger, J., and Russ, S.: Gassensoren für Schwefelwasserstoff mit integrierender Funktion, in: Proceedings of the XXVI. Messtechnisches Symposium des Arbeitskreises der Hochschullehrer für

Messtechnik, Aachen, Germany, 20–22 September 2012, 5–16, 2012c.

Husted, H., Roth, G., Nelson, S., Hocken, L., Fulks, G., and Racine, D.: Sensing of Particulate Matter for On-Board Diagnosis of Particulate Filters, SAE paper 2012-01-0372, doi:10.4271/2012-01-0372, 2012.

Jung, W., Sahner, K., Leung, A., and Tuller, H. L.: Acoustic wave-based NO_2 sensor: Ink-jet printed active layer, Sensor. Actuator. B: Chemical, 141, 485–490, doi:10.1016/j.snb.2009.07.010, 2009.

Katayama, S., Yamada, N., and Awano, M.: Preparation of alkaline-earth metal silicates from gels and their NO_x-adsorption behavior, Journal of the European Ceramic Society, 24, 421–425, doi:10.1016/S0955-2219(03)00211-5, 2004.

Kita, J., Brandenburg, A., Groß, A., and Moos, R.: Novel tube-type LTTC transducers with buried heaters and inner interdigitated electrodes for high-temperatures gas sensors, Proc. Eng., 47, 60–63, doi:10.1016/j.proeng.2012.09.084, 2012.

Klug, A., Denk, M., Bauer, T., Sandholzer, M., Scherf, U., Slugovc, C., and List, E. J. W.: Organic field-effect transistor based sensors with sensitive gate dielectrics used for low-concentration ammonia detection, Organic Electronics, 14, 500–504, doi:10.1016/j.orgel.2012.11.030, 2013.

Kohl, D.: Function and applications of gas sensors, J. Phys. D: Applied Physics, 34, R125–R149, doi:10.1088/0022-3727/34/19/201, 2001.

Kondo, A., Yokoi, S., Sakurai, T., Nishikawa, S., Egami, T., Tokuda, M., and Sakuma, T.: New Particulate Matter Sensor for On Board Diagnosis, SAE paper 2011-01-0302, doi:10.4271/2011-01-0302, 2011.

Kubinski, D. J. and Visser J. H.: Sensor and method for determining the ammonia loading of a zeolite SCR catalyst, Sensor. Actuator. B: Chemical, 130, 425–429, doi:10.1016/j.snb.2007.09.007, 2008.

Li, J., Lu, Y., Ye, Q., Cinke, M., Han, J., and Meyyappan, M.: Carbon Nanotube Sensors for Gas and Organic Vapor Detection, Nano Letters, 3, 929–933, doi:10.1021/nl034220x, 2003.

Liu, X., Cheng, S., Liu, H., Hu, S., Zhang, D., and Ning, H.: A Survey on Gas Sensing Technology, Sensors, 12, 9635–9665, doi:10.3390/s120709635, 2012.

Mangu, R., Rajaputra, S., Clore, P., Qian, D., Andrews, R., and Singh, V. P.: Ammonia sensing properties of multi-walled carbon nanotubes embedded in porous alumina templates, Materials Science and Engineering B, 174, 2–8, doi:10.1016/j.mseb.2010.03.003, 2010.

Marr, I., Stöcker, T., and Moos, R.: Resistives Gasdosimeter auf Basis von PEDOT:PSS zur Detektion von NO und NO_2, in: Proceedings of the 11th Dresdner Sensor-Symposium, Dresden, Germany, 9–11 December 2013, 317–320, doi:10.5162/11dss2013/F3, 2013.

Martin, S. J., Frye, G. C., Spates, J. J., and Butler, M. A.: Gas Sensing with Acoustic Devices, in: Proceedings of the IEEE Ultrasonics Symposium 1996, 1, San Antonio, Texas/USA, 3–6 November 1996, 423–434, doi:10.1109/ULTSYM.1996.584005, 1996.

Maruo, Y. Y.: Measurement of ambient ozone using newly developed porous glass sensor, Sensor. Actuator. B: Chemical, 126, 485–491, doi:10.1016/j.snb.2007.03.041, 2007.

Maruo, Y. Y., Kunioka, T., Akaoka, K., and Nakamura, J.: Development and evaluation of ozone detection paper, Sensor. Actuator. B: Chemical, 135, 575–580, doi:10.1016/j.snb.2008.09.016, 2009.

Mathieu, Y., Tzanis, L., Soulard, M., Patarin, J., Vierling, M., and Molière, M.: Adsorption of SO_x by oxide materials: A review, Fuel Proc. Technol., 114, 81–100, doi:10.1016/j.fuproc.2013.03.019, 2013.

Matsuguchi, M., Kadowaki, Y., and Tanaka, M.: A QCM-based NO_2 gas detector using morpholine-functional cross-linked copolymer coatings, Sensor. Actuator. B: Chemical, 108, 572–575, doi:10.1016/j.snb.2004.11.044, 2005.

Mattoli, V., Mazzolai, B., Raffa, V., Mondini, A., and Dario, P.: Design of a new real-time dosimeter to monitor personal exposure to elemental gaseous mercury, Sensor. Actuator. B: Chemical, 123, 158–167, doi:10.1016/j.snb.2006.08.004, 2007.

Moos, R., Geupel, A., Visser, J. H., and Kubinski, D. J.: Vorrichtung und Verfahren zur Detektion einer Menge einer Gaskomponente, German Patent Application, DE 10 2010 023 523 A1, 2010.

Moos, R., Beulertz, G., Geupel, A., Visser, J. H., and Kubinski, D. J.: Vorrichtung und Verfahren zur Detektion der Menge und der Konzentration einer Gaskomponente, German Patent Application, DE 10 2012 206 788 A1, 2012.

Mubeen, S., Lai, M., Zhang, T., Lim, J.-H., Mulchandani, A., Deshusses, M. A., and Myung, N. V.: Hybrid tin oxide-SWNT nanostructures based gas sensor, Electrochimica Acta, 92, 484–490, doi:10.1016/j.electacta.2013.01.029, 2013.

Mukherjee, K., Gaur, A. P. S., and Majumder, S. B.: Investigations on irreversible- and reversible-type gas sensing for ZnO and $Mg_{0.5}Zn_{0.5}Fe_2O_4$ chemi-resistive sensors, J. Phys. D: Applied Physics, 45, 505306, doi:10.1088/0022-3727/45/50/505306, 2012.

Nieuwenhuizen, M. S. and Harteveld, J. L. N.: Studies on a surface acoustic wave (SAW) dosimeter sensor for organophosphorous nerve agents, Sensor. Actuator. B: Chemical, 40, 167–173, doi:10.1016/S0925-4005(97)80257-2, 1997.

Ochs, T., Schittenhelm, H., Genssle, A., and Kamp, B.: Particulate matter sensor for on board diagnostics (OBD) of diesel particulate filters (DPF), SAE paper 2010-01-0307, doi:10.4271/2010-01-0307, 2010.

Ong, K. G., Zeng, K, and Grimes, C. A.: A Wireless, Passive Carbon Nanotube-Based Gas Sensor, IEEE Sensors Journal, 2, 82–88, doi:10.1109/JSEN.2002.1000247, 2002.

Padilla, M., Perera, A., Montoliu, I., Chaudry, A., Persaud, K., and Marco, S.: Drift compensation of gas sensor array data by Orthogonal Signal Correction, Chemometrics and Intelligent Laboratory Systems, 100, 28–35, doi:10.1016/j.chemolab.2009.10.002, 2010.

Padma, N., Joshi, A., Singh, A., Deshpande, S. K., Aswal, D. K., Gupta, S. K., and Yakhmi, J. V.: NO_2 sensors with room temperature operation and long term stability using copper phthalocyanine thin films, Sensor. Actuator. B: Chemical, 143, 246–252, doi:10.1016/j.snb.2009.07.044, 2009.

Rettig, F., Moos, R., and Plog, C.: Sulfur adsorber for thick-film exhaust gas sensors, Sensor. Actuator. B: Chemical, 93, 36–42, doi:10.1016/S0925-4005(03)00334-4, 2003.

Reyes, L. F., Hoel, A., Saukko, S., Heszler, P., Lantto, V., and Granqvist, C. G.: Gas sensor response of pure and activated WO_3 nanoparticle films made by advanced reactive gas deposition, Sensor. Actuator. B: Chemical, 117, 128–134, doi:10.1016/j.snb.2005.11.008, 2006.

Roadman, M. J., Scudlark, J. R., Meisinger, J. J., and Ullman, W. J.: Validation of Ogawa passive samplers for the determination of gaseous ammonia concentrations in agricultural settings, Atmos. Environ., 37, 2317–2325, doi:10.1016/S1352-2310(03)00163-8, 2003.

Rodríguez-González, L. and Simon, U.: NH_3-TPD measurements using a zeolite-based sensor, Measurement Science and Technology, 21, 027003, doi:10.1088/0957-0233/21/2/027003, 2010.

Rodríguez-González, L., Rodríguez-Castellón, E., Jiménez-López, A., and Simon, U.: Correlation of TPD and impedance measurements on the desorption of NH_3 from zeolite H-ZSM-5, Solid State Ionics, 179, 1968–1973, doi:10.1016/j.ssi.2008.06.007, 2008.

Rossé, G., Raoult, F., and Fortin, B.: Regeneration of CdSe Thin Films After Oxygen Chemisorption, Thin Solid Films, 111, 175–181, doi:10.1016/0040-6090(84)90485-1, 1984.

Sahner, K., Hagen, G., Schönauer, D., Reiß, S., and Moos, R.: Zeolites – Versatile materials for gas sensors, Solid State Ionics, 179, 2416–2423, doi:10.1016/j.ssi.2008.08.012, 2008.

Sasaki, D. Y., Singh, S., Cox, J. D, and Pohl, P. I.: Fluorescence detection of nitrogen dioxide with perylene/PMMA thin films, Sensor. Actuator. B: Chemical, 72, 51–55, doi:10.1016/S0925-4005(00)00632-8, 2001.

Sauerwald, T., Hennemann, J., Kohl, C.-D., Wagner, T., and Russ, S.: H_2S detection utilizing percolation effects in copper oxide, in: Proceedings of Sensor 2013, Nuremberg, Germany, 14–16 May 2013, E6.4, doi:10.5162/sensor2013/E6.4, 2013.

Schütze, A., Pieper, N., and Zacheja, J.: Quantitative ozone measurement using a phthalocyanine thin-film sensor and dynamic signal evaluation, Sensor. Actuator. B: Chemical, 23, 215–217, doi:10.1016/0925-4005(94)01281-L, 1995.

Seethapathy, S., Górecki, T., and Li, X.: Passive sampling in environmental analysis, J. Chromatogr. A, 1184, 234–253, doi:10.1016/j.chroma.2007.07.070, 2008.

Semancik, S., Cavicchi, R. E., Wheeler, M. C., Tiffany, J. E., Poirier, G. E., Walton, R. M., Suehle, J. S., Panchapakesan, B., and DeVoe, D. L.: Microhotplate platforms for chemical sensor research, Sensor. Actuator. B: Chemical, 77, 579–591, doi:10.1016/S0925-4005(01)00695-5, 2001.

Shu, J. H.: Passive Chemiresistor Sensor Based on Iron (II) Phthalocyanine Thin Films for Monitoring of Nitrogen Dioxide, Dissertation, Auburn University, Auburn, Alabama/USA, 13 December 2010.

Shu, J. H., Wikle, H. C., and Chin, B. A.: Passive chemiresistor sensor based on iron (II) phthalocyanine thin films for monitoring of nitrogen dioxide, Sensor. Actuator. B: Chemical, 148, 498–503, doi:10.1016/j.snb.2010.05.017, 2010.

Shu, J. H., Wikle, H. C., and Chin, B. A.: Passive Detection of Nitrogen Dioxide Gas by Relative Resistance Monitoring of Iron (II) Phthalocyanine Thin Films, IEEE Sensors Journal, 11, 56–61, doi:10.1109/JSEN.2010.2051024, 2011.

Simon, I., Bârsan, N., Bauer, M., and Weimar, U.: Micromachined metal oxide gas sensors: opportunities to improve sensor performance, Sensor. Actuator. B: Chemical, 73, 1–26, doi:10.1016/S0925-4005(00)00639-0, 2001.

Simon, U., Flesch, U., Maunz, W., Müller, R., and Plog, C.: The effect of NH_3 on the ionic conductivity of dehydrated zeolites Na beta and H beta, Microporous and Mesoporous Materials, 21, 111–116, doi:10.1016/S1387-1811(97)00056-5, 1998.

Simons, T. and Simon, U.: Zeolite H-ZSM-5: A Microporous Proton Conductor for the in situ Monitoring of $DeNO_x$-SCR, in: Proceedings of the Materials Research Society Spring Meeting 2011, 1330, San Francisco, California/USA, 25–29 April 2011, mrss11-1330-j01-03-k03-03, doi:10.1557/opl.2011.1337, 2011.

Simons, T. and Simon, U.: Zeolites as nanoporous, gas-sensitive materials for in situ monitoring of $DeNO_x$-SCR, Beilstein Journal of Nanotechnology, 3, 667–673, doi:10.3762/bjnano.3.76, 2012.

Sircar, A., Mallik, B., and Misra, T. N.: Effect of Adsorption of Vapours on the Electrical Conductivity of a Series of Some Naphthyl Polyenes: Adsorption and Desorption Kinetics, J. Phys. Chem. Solids, 44, 401–405, doi:10.1016/0022-3697(83)90067-7, 1983.

Small IV, W., Maitland, D. J., Wilson, T. S., Bearinger, J. P., Letts, S. A., and Trebes, J. E.: Development of a prototype optical hydrogen gas sensor using a getter-doped polymer transducer for monitoring cumulative exposure: Preliminary results, Sensor. Actuator. B: Chemical, 139, 375–379, doi:10.1016/j.snb.2009.03.020, 2009.

Stamires, D. N.: Effect of Adsorbed Phases on the Electrical Conductivity of Synthetic Crystalline Zeolites, J. Chem. Phys., 36, 3174–3181, doi:10.1063/1.1732446, 1962.

Tamaki, J., Fujimori, K., Miura, N., and Yamazoe, N.: Sensing characteristics of semiconductor barium carbonate sensor to nitrogen oxides at elevated temperature, in: Proceedings of the 2nd East Asian Conference on Chemical Sensors, Xian, China, 5–8 October 1995, 1995.

Tanaka, T., Ohyama, T., Maruo, Y. Y., and Hayashi, T.: Coloration reactions between NO_2 and organic compounds in porous glass for cumulative gas sensor, Sensor. Actuator. B: Chemical, 47, 65–69, doi:10.1016/S0925-4005(98)00051-3, 1998.

Tanaka, T., Guilleux, A., Ohyama, T., Maruo, Y. Y., and Hayashi, T.: A ppb-level NO_2 gas sensor using coloration reactions in porous glass, Sensor. Actuator. B: Chemical, 56, 247–253, doi:10.1016/S0925-4005(99)00185-9, 1999.

Ulrich, M., Bunde, A., and Kohl, C.-D.: Percolation and gas sensitivity in nanocrystalline metal oxide films, Appl. Phys. Lett., 85, 242–244, doi:10.1063/1.1769071, 2004.

Varghese, O. K., Kichambre, P. D., Gong, D., Ong, K. G., Dickey, E. C., and Grimes, C. A.: Gas sensing characteristics of multiwall carbon nanotubes, Sensor. Actuator. B: Chemical, 81, 32–41, doi:10.1016/S0925-4005(01)00923-6, 2001.

Varshney, C. K. and Singh, A. P.: Passive Samplers for NO_x Monitoring: A Critical Review, The Environmentalist, 23, 127–136, doi:10.1023/A:1024883620408, 2003.

Wagner, T., Haffer, S., Weinberger, C., Klaus, D., and Tiemann, M.: Mesoporous materials as gas sensors, Chem. Soc. Rev., 42, 4036–4053, doi:10.1039/c2cs35379b, 2013.

Weigl, M., Roduner, C., and Lauer, T.: Particle-Filter-Onboard-Diagnosis by Means of a Soot-Sensor Downstream of the Particle-Filter, in: Proceedings of the 6th International Exhaust Gas and Particulate Emissions Forum, Ludwigsburg, Germany, 9–10 March 2010, 62–69, 2010.

Williams, D. E.: Semiconducting oxides as gas-sensitive resistors, Sensor. Actuator. B: Chemical, 57, 1–16, doi:10.1016/S0925-4005(99)00133-1, 1999.

Yamazoe, N.: Toward innovations of gas sensor technology, Sensor. Actuator. B: Chemical, 108, 2–14, doi:10.1016/j.snb.2004.12.075, 2005.

Yamazoe, N. and Shimanoe, K.: Overview of Gas Sensor Technology, in: Science and Technology of Chemiresistor Gas Sensors, edited by: Aswal, D. K. and Gupta, S. K., Nova Science Publishers Inc., New York, 2007.

Yamazoe, N., Fuchigami, J., Kishikawa, M., and Seiyama, T.: Interactions of Tin Oxide Surface with O_2, H_2O and H_2, Surface Science, 86, 335–344, doi:10.1016/0039-6028(79)90411-4, 1979.

Yamazoe, N., Sakai, G., and Shimanoe, K.: Oxide Semiconductor Gas Sensors, Catalysis Surveys from Asia, 7, 63–75, doi:10.1023/A:1023436725457, 2003.

6

New ferromagnetic core shapes for induction sensors

C. Coillot[1], J. Moutoussamy[2], M. Boda[3], and P. Leroy[2]

[1]Laboratoire Charles Coulomb, BioNanoNMRI group, University Montpellier II, Place Eugene Bataillon, 34090 Montpellier, France
[2]LPP/CNRS/UPMC/Ecole Polytechnique, Route de Saclay, 91128 Palaiseau, France
[3]SubSeaStem, 25 rue des Ondes, 12000 Rodez, France

Correspondence to: C. Coillot (christophe.coillot@univ-montp2.fr)

Abstract. Induction sensors are used in a wide range of scientific and industrial applications. One way to improve these is rigorous modelling of the sensor combined with a low voltage and current input noise preamplifier aiming to optimize the whole induction magnetometer. In this paper, we explore another way, which consists in the use of original ferromagnetic core shapes of induction sensors, which bring substantial improvements. These new configurations are the cubic, orthogonal and coiled-core induction sensors. For each of them we give modelling elements and discuss their benefits and drawbacks with respect to a given noise-equivalent magnetic induction goal. Our discussion is supported by experimental results for the cubic and orthogonal configurations, while the coiled-core configuration remains open to experimental validation. The transposition of these induction sensor configurations to other magnetic sensors (fluxgate and giant magneto-impedance) is an exciting prospect of this work.

1 Introduction

The function of induction magnetometers is to measure extremely weak magnetic fields. Their field of application is very large and covers soil characterization for agriculture (Sudduth et al., 2001), earthquake survey or magnetotelluric waves observation, and natural electromagnetic waves near the surface of the Earth (lightning observations (Ozaki et al., 2012), whistlers (Lichtenberger et al., 2008)) or in space (Roux et al., 2007). In these applications the induction sensors must cover a wide frequency range from millihertz (mHz), for magnetotelluric observations, up to megahertz (MHz), for plasma waves observation. In order to remove the resonance of the induction sensor, they are combined either with a feedback flux (Seran and Fergeau, 2005) or a current amplifier (Prance et al., 2000). The extension of their frequency range is made possible using two windings on the same ferromagnetic core separated by a magnetic mutual reducer (Coillot et al., 2010), while a relevant solution, called the dual-resonant search coil, permits combination of the two windings into a single one (Ozaki et al., 2013). The design of such an instrument requires obtaining a reliable modelling tool to match the measurement requirements. The main specification of the measurement is usually given in terms of noise-equivalent magnetic induction (NEMI in $T/\sqrt{Hz}$) either at a given frequency or by its spectrum over a frequency range.

In previous works authors have focused on physical modelling of the induction sensor performances in terms of NEMI (Seran and Fergeau, 2005; Korepanov and Pronenko, 2010) or low-noise amplifier design (Rhouni, 2012; Shimin et al., 2013).

The appropriate core and coil parameters can be found by reformulating the problem as a mathematical optimization problem (Coillot et al., 2007; Yan, 2013). Some analytical formulae are proposed in order to dimension the system in very limited cases. In Grosz and Paperno (2012), the authors present the design of a low-frequency induction magnetometer, their assumption being that the impedance of the coil is simply equal to its resistance.

In this work we explore new tracks of improvement guided by modifications in the shape of the ferromagnetic core. The first solution is the cubic induction sensor inspired by Dupuis (2003), where the core consists of 12 rods assembled in a cubic configuration (which could be extended to an array). For a given measurement direction the coil is distributed on the four edges of the cube. In the following, we will present the orthogonal induction sensor, which consists in an helical ferromagnetic core which canalizes the magnetic flux. The magnetic flux through the core turns is then measured by a coil wound orthogonally to the direction of the external magnetic field. This allows for significant reduction of the number of coil turns and reduction of the resistance of the coil for a given flux. The third induction sensor is the coiled ferromagnetic wire core, where the core is assumed to be made with a coilable ferromagnetic wire.

2 Induction sensor basics

2.1 Elements of the electrical model

Induction sensors (Ripka, 2000; Tumanski, 2007) are classically built with an N-turns coil. According to Lenz's law, when the coil is immersed in a magnetic field, a voltage e is induced.

The resistance of the coil (R) can be approximately computed using the following formula:

$$R = 4\rho N \frac{\left(d + N(d_w + 2t)^2 / L_w\right)}{d_w^2},$$ (1)

where ρ is the material resistivity (copper or aluminium are usually preferred), d is the internal diameter of the coil, d_w is the wire diameter, L_w is the length over which the winding is distributed and t is the thickness of the wire insulator.

The voltage difference between turns and layers is associated with electrostatic energy storage. This is usually represented by a capacitance C on the electrokinetic model. The computation of this capacitance depends on the winding strategy; one can notice that discontinuous winding should be preferred as to avoid parasitic resonances (Coillot and Leroy, 2012).

Optionally, the wire is coiled around a ferromagnetic core, taking advantage of its magnetic gain (Bozorth and Chapin, 1942), known as apparent permeability (μ_{app}) and given in Eq. (2):

$$\mu_{app} = \frac{\mu_r}{1 + N_z(\mu_r - 1)},$$ (2)

where μ_r is the relative permeability and N_z is the magnetometric demagnetizing coefficient in the z direction. For a long cylinder core (i.e. length-to-diameter ratio: $m = L_c/d \gg 1$), the approximation of ellipsoid demagnetizing coefficient given in Osborn (1945) is valid:

$$N_z(m) = \frac{1}{m^2}(\ln(2m) - 1).$$ (3)

When a "diabolo" core is used (Coillot et al., 2007) the apparent permeability is increased thanks to the magnetic flux

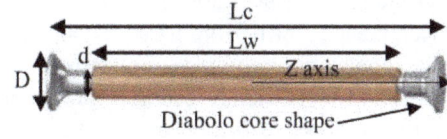

Figure 1. Diagram of an induction sensor with a diabolo core shape.

concentrators (shown in Fig. 1). In this work the various ferromagnetic core shapes will be linked to the one of the diabolo cores. In the case of diabolo core shape the apparent permeability equation becomes

$$\mu_{app\text{-}diab} = \frac{\mu_r}{1 + N_z(m')\frac{d^2}{D^2}(\mu_r - 1)},$$ (4)

where $N_z(m')$ is the magnetometric demagnetizing coefficient for a cylinder of length-to-diameter ratio $m' = L_c/D$, while d^2/D^2 represents the surface ratio between the centre and the end surfaces of the core.

In the case of ferromagnetic core induction sensor, the inductance equation given in Tumanski (2007) is recalled here:

$$Ł = \lambda N^2 \mu_0 \frac{\mu_{app} S}{L_c},$$ (5)

where S is the ferromagnetic core section, μ_0 is the vacuum permeability and $\lambda = (L_c/L_w)^{2/5}$ is a correction factor proposed in Lukoschus (1979).

2.2 Transfer function of the induction sensor

Using a ferromagnetic core exhibiting an apparent permeability μ_{app}, the induced voltage is expressed (in harmonic regime at the pulsation ω) as

$$e = -j\omega NS\mu_{app} B,$$ (6)

where j represents the unit imaginary number $j^2 = -1$ and B is the magnetic flux density. The electrokinetic modelling assumes that the induced voltage is in series with the resistance and the inductance, while the accessible voltage (V) is measured at the capacitance terminals. Thus, the transmittance ($T(j\omega)$) is given by the following equation:

$$T(j\omega) = \frac{V}{B} = \frac{-j\omega NS\mu_{app}}{1 - LC\omega^2 + jRC\omega}.$$ (7)

This transmittance exhibits a resonance at pulsation $\omega_0 = 1/\sqrt{(LC)}$. Beyond the resonance the induced voltage will decrease. The very high value of the induced voltage at the resonance frequency can be useful for some applications, while it should be removed and the transfer function must be flattened in induction magnetometer applications requiring wide-band measurements. This is typically true for applications where natural electromagnetic waves are measured, such as earthquake measurements, whistler observations and space plasmas. In such applications two kinds of electronic conditioners are classically implemented: feedback flux amplifiers or current amplifiers. In both cases, the transfer function will be flattened over about 3 to ~ 6 decades.

2.3 Noise-equivalent magnetic induction

Noise-equivalent magnetic induction (NEMI), expressed in $\frac{T}{\sqrt{(Hz)}}$, is the relevant quantity to determine the ability of the magnetometer to measure weak magnetic fields. The NEMI is defined as the square root of the total power spectrum density of the input reffered noise (PSD_{INPUT}) related to the transfer function modulus ($T(j\omega)$):

$$NEMI = \sqrt{\frac{PSD_{INPUT}}{|T(j\omega)^2|}}, \tag{8}$$

where

$$PSD_{INPUT} = 4kTR + e_{PA}^2 + (Zi_{PA})^2, \tag{9}$$

where k is the Boltzmann constant, T is the temperature, Z is the impedance and the electronic amplifier noise parameters are $e_{PA} = 4\,nV/\sqrt{(Hz)}$ and $i_{PA} < 20\,fA/\sqrt{(Hz)}$. Due to the low-frequency context, the current noise contribution (i.e. $(Zi_{PA})^2 \ll e_{PA}$) will be neglected.

3 Cubic induction sensor

3.1 Description of the sensor configuration

The cubic magnetometer proposed in Dupuis (2003) combines multiple induction sensors to form a cubic array (a diagram and a prototype are shown in Fig. 2). The advantages claimed by the author are the increase in sensitivity and the reduction of the self-inductance since the required turn number can be dispatched between the different edges. The cores are implemented in such a way that they are not coupled on the magnetic point of view but are connected in series on the electrical point of view. The sensitivity benefit is related to an increase in the apparent permeability. Our current goal is to confirm the behaviour of this original induction sensor configuration, briefly presented in Coillot and Leroy (2012), through experimental measurements with a prototype and the validation of the apparent permeability equation with the aim of comparison to classical induction sensors.

The edges of the cubic induction sensor are constituted by cylinder ferromagnetic cores of length L_c and diameter d. Due to the cubic shape, the demagnetizing coefficients are the same in the three directions:

$$N_x = N_y = N_z = \frac{1}{3}. \tag{10}$$

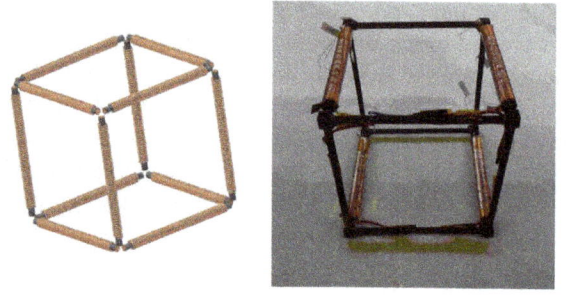

Figure 2. Cubic induction sensor diagram (left picture) and 85 cm × 85 cm × 85 cm prototype (right picture).

The flux caught by the square area face is distributed between the four ferromagnetic cores with a ratio corresponding to the surface ratio. The equation of the apparent permeability of the cubic sensor ($\mu_{app\text{-}cub}$) is then obtained as a special case of the diabolo core apparent permeability formula (given by Eq. 4) where the demagnetizing coefficient is equal to $\frac{1}{3}$ and the surface ratio is $L_c^2/(4 \times \pi d^2/4)$; this leads to

$$\mu_{app\text{-}cub_{x,y,z}} = \frac{\mu_r}{1 + N_{x,y,z}(\mu_r - 1)\frac{\pi d^2}{L_c^2}}. \tag{11}$$

For sufficiently high relative permeability (i.e. $\mu_r \gg 1$ and $N\mu_r \gg 1$), we can simply write

$$\mu_{app\text{-}cub_{x,y,z}} \simeq \frac{3L_c^2}{\pi d^2}. \tag{12}$$

The previous formula is different from the one presented in Coillot and Leroy (2012), which is only valid for rods with square sections (despite what is claimed in the article).

3.2 Experimental results

A prototype cubic induction sensor has been built to evaluate the modelling of the apparent permeability and the impact on the resonance frequency. The design equations which have been used are directly derived from the one of the classical induction sensors presented above. The design parameters (turns number, core diameter, copper wire diameter) have been computed to allow the prototype to reach a NEMI value close to $0.7\,pT/\sqrt{Hz}$ at 10 Hz. The design parameters of the prototype are summarized in Table 1.

The cubic induction sensor prototype (left picture in Fig. 2) is a 85 cm × 85 cm × 85 cm cube. A ferrite cube has been mounted at each corner of the cube to ensure a closed ferromagnetic path. The ferromagnetic material used for the ferromagnetic parts is B1 ferrite material whose initial relative permeability is typically about 2500. The ferromagnetic core has been wound in a single direction. In this direction, each of the four edges were wound with 8000 turns of 70 μm diameter copper coil. Each coil has been connected in serial from the electrical point of view but in opposition from the

Table 1. Design parameters of the cubic induction sensor prototype for a NEMI goal: $0.7 \, \text{pT}/\sqrt{(\text{Hz})}$ at 10 Hz, assuming $e_{\text{PA}} = 4 \, \text{nV}/\sqrt{(\text{Hz})}$ and $i_{\text{PA}} = 20 \, \text{fA}/\sqrt{(\text{Hz})}$.

Cylinder core length (L_{c} in mm)	85
Core diameter (d in mm)	4
Copper wire turns (N per core)	8000
Wire diameter (d_{w} in mm)	0.07
Layer number	$n_{\text{l}} = 4$
Winding length (L_{w} in mm)	80
Resistance (R in Ω)	1300
Apparent permeability ($\mu_{\text{app-cub}}$)	354

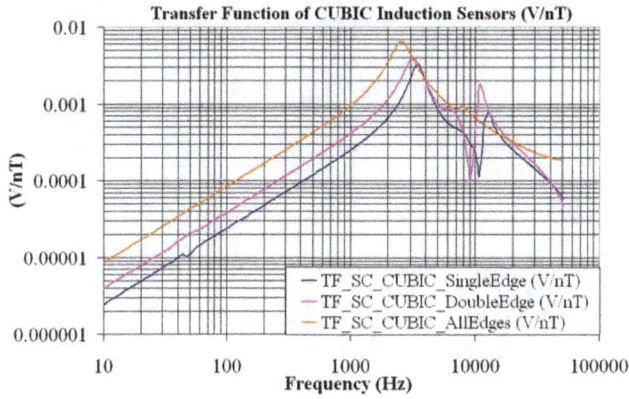

Figure 3. Transfer function of the cubic induction sensor: one edge (dark blue), two edges (pink) and four edges (orange curve).

magnetic field point of view in order to cancel the mutual induction between edges. It results that from a single edge to four edges, an increase in inductance by a ratio of 4 is expected (instead of 16 when inductances are magnetically coupled). The transfer function has been measured in three cases: single edge, double edges and four edges. This shows that the increase in gain (before the resonance) is proportional to the edge number and is multiplied by almost 4 from a single edge to four edges. The reason why the ratio between four edges and single edges is not precisely equal to 4 could be explained by a unbalanced magnetic path (namely the fluxes seen by each edge are not exactly identical). The resonance frequency varies weakly (from 3400 Hz for the single edge to 2700 Hz for four edges). This could be related to the fact that when the edges are connected together, the total inductance increases, while the total capacitance decreases; consequently the resonance frequency does not vary significantly. This property is of great interest for the design of wide-band and compact induction sensors. However, the multiple resonance next to the main resonance could make the use of feedback flux or current amplifiers difficult.

The apparent permeability deduced from the measurement (cf. Fig. 3) is $\mu_{\text{app-cub-meas}} = 354$, while the numerical application of Eq. (11) gives a rather close approximation ($\mu_{\text{app-cub}} = 368$), which validates our modelling attempt.

3.3 Discussion

In order to evaluate the benefit of the cubic induction sensor, we will compare its apparent permeability and inductance to the one of a cylinder core induction sensor of the same length (L_{c}) and diameter (d). Under the ellipsoid shape approximation, its demagnetizing coefficient is given by Eq. (3). Thus the long cylinder's apparent permeability (Eq. 2) can be written (assuming $N_z(m)\mu_{\text{r}} \gg 1$)

$$\mu_{\text{app}} = \frac{1}{N_z(m)} = \frac{L_{\text{c}}^2}{d^2} \frac{1}{\ln(2\frac{L_{\text{c}}}{d}) - 1}. \tag{13}$$

The comparison between the cubic sensor and the usual induction sensor can be done through the ratio between their apparent permeability (from Eqs. 12 and 13), which leads to

$$\frac{\mu_{\text{app-cub}}}{\mu_{\text{app}}} = \frac{3}{\pi}(\ln(2\frac{L_{\text{c}}}{d}) - 1). \tag{14}$$

This comparison suggests that the cubic induction sensor has a higher apparent permeability than the single rod since $\frac{L_{\text{c}}}{d} \gg 1$. For example, for a L_{c}/d ratio equal to 10, the cubic sensor apparent permeability will be about two times higher than the one of the single rod.

4 Orthogonal induction sensor

4.1 Description of the sensor configuration

In an orthogonal search coil, a helical core (shown in Fig. 4) is used to enhance the flux catched by each turn. Because of the two ferromagnetic discs (diameter D) mounted at the ends of the helicore (whose total length is L_{c}), and assuming a high relative permeability (μ_{r}), the external magnetic flux is canalized by the ends of the core and driven through each ferromagnetic core turn. The ferromagnetic turns are assumed to be square with side d.

To compute the induced voltage, we have to consider the angle between the normal vector of the coil turn section and the direction of the magnetic field inside the helicore; this angle (φ), called the helix angle, is defined as

$$\varphi = \arctan\left(\frac{b}{a}\right). \tag{15}$$

where the pitch of the helicore ($2\pi b$) is determined as

$$2\pi b = \frac{L_{\text{w}}}{n}, \tag{16}$$

where L_{w} is the length of the core on which the copper wire is wound and $2a$ is the average diameter of the ferromagnetic core helix (cf. Fig. 4).

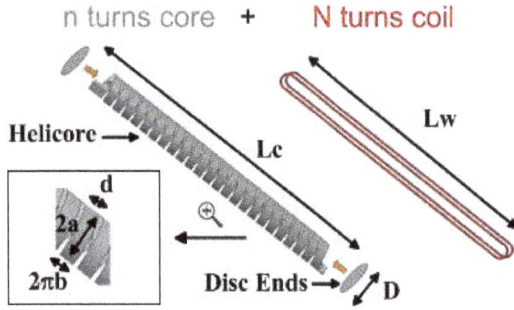

Figure 4. Diagram of the orthogonal induction sensor.

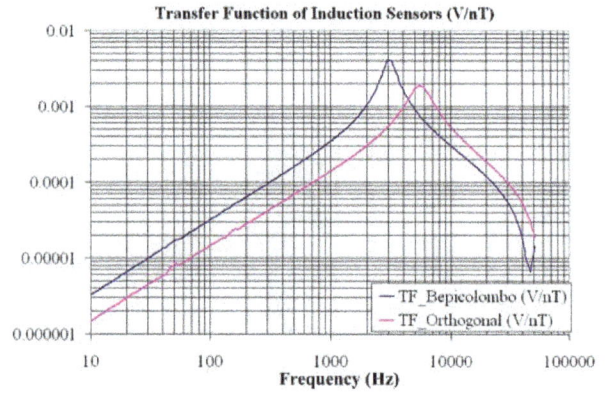

Figure 5. Picture of the orthogonal induction sensor prototype.

For N turns of coil surrounding the n core section, the induced voltage (e_{hc}) becomes

$$e_{hc} = -j\omega N n \mu_{app\text{-}hc} B S \cos(\varphi), \qquad (17)$$

where S is the section of the core turns (equal to d^2) and $\mu_{app\text{-}hc}$ is the apparent permeability of the helical ferromagnetic core.

This apparent permeability is directly derived from the formula of the diabolo core, given in Eq. (4). In that case $N_z(m')$ is the demagnetizing coefficient for the cylinder of length-to-diameter ratio $m' = L/D$ and $\frac{d^2}{\pi D^2/4}$ is the surface ratio between the square section of the core turns and the end discs' section of the core.

Finally, the sensitivity of the induction sensor, assuming low-frequency operation ($\omega \ll \omega_0$), is obtained:

$$| T(j\omega) | = | \frac{e_{hc}}{B} | = \omega N n \mu_{app\text{-}hc} S \cos(\varphi). \qquad (18)$$

The resistance of the coil (R_{hc}) is a derivation of the Eq. (1), while the inductance formula is intuitively obtained:

$$Ł_{hc} = \lambda (nN)^2 \mu_0 \frac{\mu_{app\text{-}hc} S \cos(\varphi)}{l}. \qquad (19)$$

The NEMI of the orthogonal induction sensor can be estimated using Eq. (20).

$$\text{NEMI}_{hc} = \frac{\sqrt{4kTR_{hc} + e_{PA}^2}}{\omega N n \mu_{app\text{-}hc} S \cos(\varphi)}. \qquad (20)$$

4.2 Experimental results

By using the previous set of equations we have determined the number of turns of an orthogonal sensor to get the same NEMI as the diabolo core sensor designed within the context of the BepiColombo space mission (namely $2\,\text{pT}/\sqrt{(\text{Hz})}$ at 10 Hz from Coillot et al., 2010). The following set of design parameters was chosen: $d_w = 140\,\mu\text{m}$, $L_c = 100\,\text{mm}$, $n = 28$, $d = 3\,\text{mm}$, $D_O = 20\,\text{mm}$, $a = 3.5\,\text{mm}$ and $L_w = 90\,\text{mm}$. The design result led to a 400-turn coil on a single layer. The

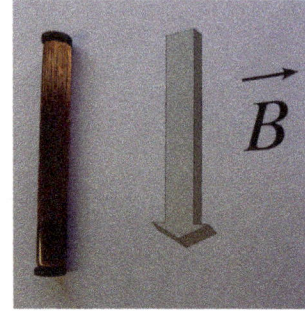

Figure 6. Transfer function of the orthogonal induction sensor (pink curve) versus that of the BepiColombo sensor (blue curve).

expected sensitivity at 10 Hz is about 2300 V/T, while the expected resistance should be four times lower than the one of the BepiColombo sensor (cf. Table 2). The orthogonal induction sensor prototype is shown in Fig. 5. It can be noticed that the coil normal direction is orthogonal to the direction of the magnetic field.

The advantage of the helicore is that the sensitivity criterion can be met with a small number of turns; the drawback is the difficulty in winding the core.

The transfer function reported in Fig. 6 demonstrates that the orthogonal induction sensor is able to efficiently measure magnetic fields. Its transfer function is compared to the BepiColombo induction sensor (blue curve), which uses many more turns (14 000 turns for the BepiColombo one versus 400 turns for the orthogonal one). The sensitivities at 10 Hz for the BepiColombo and the orthogonal induction sensors are 3200 and 1500 V/T respectively. The sensitivity of the orthogonal induction sensor prototype is 35 % lower than the expected one (cf. summary of performances in Table 2). A gap between the ferromagnetic end discs and the helicore part or a small crack in the core is suspected to explain the difference. Since the resistance of the orthogonal induction sensor prototype is four times lower than that of the BepiColombo one, it implies that the signal-to-noise ratio (SNR) of the two sensors is comparable. An advantage of

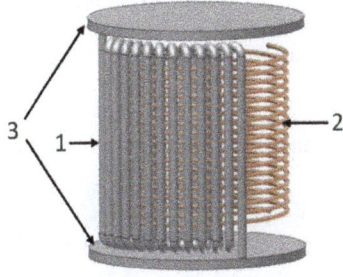

Figure 7. Diagram of the coiled-core induction sensor. (1) shows the ferromagnetic coil, (2) the conductive coil and (3) the ferromagnetic discs.

the orthogonal induction sensor is the higher resonance frequency, which permits extension of the frequency range of the measurement. The helical angle could reduce the performance, and thinner cores (on the helicore part) would reduce this angle and consequently increase the sensitivity. Lastly, the resonance frequency is lower than expected, which indicates that some leakage flux between core turns occurs. The use of a ferromagnetic wire to design an orthogonal induction sensor could solve the main problems encountered with the ferrite core.

5 Coiled-core induction sensor

5.1 Description of the sensor configuration

The coiled-core induction sensor, presented in Fig. 7, consists of an n-turns ferromagnetic wire (part 1) coiled around an n_{coil}-turns winding (part 2) made of conductive material like copper or aluminium. Similarly to the classical insulated conductive wire used to make the classical winding, the ferromagnetic wire should be insulated to leave a space between turns as to avoid a short-circuited magnetic path.

A ferromagnetic disc (part 3), acting as magnetic concentrators, is mounted at each end of the ferromagnetic wire in order to canalize the magnetic field. As a result, the magnetic field "flows" through the ferromagnetic wire and each conductive coil turn "sees" n times the derivative of the flux of the ferromagnetic coil turns. The advantage of such a sensor is quite obvious, but we propose a modelling attempt to convince the reader of the potential interest of this theoretical sensor.

5.2 Modelling of the coiled-core induction sensor

Let us consider a copper winding with N turns wound on a diameter D_{coil} made with a conductive wire of resistivity ρ and diameter d_w on a length L_w. Similarly to the resistance formula of the induction sensor (cf. Eq. 1), the winding re-

sistance of the coiled-core (R_{cc}) sensor is expressed as

$$R_{cc} = 4\rho N \frac{\left(D_{coil} + N(d_w + 2t)^2/L_w\right)}{d_w^2}. \tag{21}$$

On the one hand, the available area to coil the ferromagnetic wire is given by

$$S_{coil} = \frac{\pi D_{coil}^2}{4}. \tag{22}$$

On the other hand the core section for a single wire is given by

$$S_{core} = \frac{\pi d^2}{4}. \tag{23}$$

For a given core-coil filling factor k_f (a value of $\simeq 0.9$ is considered for the design example), the coil-core number (n) is deduced:

$$n = \frac{k_f S_{coil}}{S_{core}}. \tag{24}$$

In this relation we assume that the cored coil could be distributed on many layers inside the coil winding area. Let us now consider the sensor length L_c and the diameter of the ferromagnetic discs D_O, the length-to-diameter ratio being $m'' = L_c/D_O$. The apparent permeability, derived from Eq. (4), can be expressed as

$$\mu_{app-cc} = \frac{\mu_r}{1 + N_z(m'')\frac{d^2}{D_O^2}(\mu_r - 1)}. \tag{25}$$

The induced voltage modulus and the sensitivity of the induction sensor(assuming low-frequency operation ($\omega \ll \omega_0$)) are given by Eqs. (26) and (27) respectively:

$$e_{cc} = NnS_{core}\mu_{app-cc}\omega B, \tag{26}$$

$$|T(j\omega)| = \frac{e_{cc}}{B} = NnS_{core}\mu_{app-cc}\omega. \tag{27}$$

The NEMI equation for the coiled-core sensor is

$$NEMI_{cc} = \frac{\sqrt{4kTR_{cc} + e_{PA}^2}}{NnS_{core}\mu_{app-cc}\omega}. \tag{28}$$

Similarly to the orthogonal induction sensor, a design attempt is performed (we choose the following set of design parameters: $d_w = 70\,\mu m$, $L = 20\,mm$, $d = 1\,mm$, $D_O = 20\,mm$ and $L_w = 18\,mm$). The electronic amplifier noise parameters (e_{PA} and i_{PA}) remain identical to previous cases. Because of the compactness of the coiled-core sensor ($L = D_O \Rightarrow m'' \approx 1$), a demagnetizing factor $N_z(m'') = 1/3$ was considered (similarly to the value usually considered for sphere or cube). The resolution of the induction sensor design problem (i.e. number of turns verifying NEMI $= 2\,pT/\sqrt{(Hz)}$ at 10 Hz, using Eq. 28) lead to the parameters given in Table 2.

Table 2. Design parameters for diabolo, orthogonal and coiled-core induction sensors for an equal NEMI goal: $2\,\mathrm{pT}/\sqrt{(\mathrm{Hz})}$ at $10\,\mathrm{Hz}$, assuming $e_{\mathrm{PA}} = 4\,\mathrm{nV}/\sqrt{(\mathrm{Hz})}$ and $i_{\mathrm{PA}} = 20\,\mathrm{fA}/\sqrt{(\mathrm{Hz})}$.

Parameters	Diabolo	Ortho.	Coiled core
Sensor length (L_c in mm)	100	100	20
Winding length (L_w in mm)	90	90	18
Core diameter/side (d in mm)	4	3	1
Pitch ($2\pi b$ in mm)	N/A	3.5	N/A
Helix radius (a in mm)	N/A	5	N/A
End diameter (D_O in mm)	14	20	20
Coil diameter (D_{coil} in mm)	N/A	N/A	10
Copper wire turns (N)	15600	400	465
Core wire turns (n)	N/A	28	90
Wire diameter (d_w in µm)	70	140	70
Resistance (R in Ω)	1260	377	200
Apparent permeability	295	377	1070
Sensitivity (V/T)	3260	2360	2200

This solution, which remains theoretical, could permit for strong reduction in the size of the sensor for a given sensor sensitivity. For the considered design, it suggests that a coiled-core induction sensor could be five times smaller than a classical induction sensor. The availability of windable ferromagnetic wire is the weakness of this conceptual sensor.

6 Conclusions

The three induction sensors reported in this work offer new possibilities for improvements. We believe that the coiled-core induction sensor is the most promising one even if it remains theoretical as the prototype has not been built. Its manufacturing is strongly dependent on the availability of coilable and insulated ferromagnetic wire. Nevertheless, the orthogonal induction sensor and the coiled-core one are similar. The orthogonal induction sensor prototype has allowed for confirming the predicted performance, which provides good confidence concerning the real wound-core sensor performance. These three induction sensors could be adapted to enhance other magnetic sensors, especially fluxgates and giant magneto-impedance (GMI). For instance, in the case of the GMI, the excitation current could flow through the ferromagnetic wire, while the conductive winding could be used as a pick-up coil. The closeness of the ferromagnetic wire could enhance the skin effect by means of the proximity effect, which could permit having high GMI ratio even at low frequency.

Acknowledgements. The authors would like to thank CNES (Centre National d'Etudes Spatiales), which has funded the prototype manufacturing within the context of a spacecraft mission study.

Edited by: B. Jakoby
Reviewed by: two anonymous referees

References

Bozorth. R. M. and Chapin, D.: Demagnetizing factors of rods, J. Appl. Phys., 13, 320–327, 1942.
Coillot, C. and Leroy, P.: Induction Magnetometers: Principle, Modeling and Ways of Improvement, Magnetic Sensors – Principles and Applications, edited by: Kuang, K., ISBN: 978-953-51-0232-8, InTech, 2012.
Coillot, C., Moutoussamy, J., Leroy, P., Chanteur, G., and Roux, A.: Improvements on the design of search coil magnetometer for space experiments, Sens. Lett., 5, 167–170 2007.
Coillot, C., Moutoussamy, J., Lebourgeois, R., Ruocco, S., and Chanteur, G.: Principle and performance of a dual-band search coil magnetometer: A new instrument to investigate fluctuating magnetic fields in space, IEEE Sens. J., 10, 255–260, 2010.
Dupuis, J. C.: Optimization of a 3-AXIS Induction Magnetometer for Airbone Geophysical Exploration, Master of Science in Egineering, University of New Brunswick, 2003.
Grosz, A. and Paperno, E.: Analytical Optimization of Low-Frequency Search Coil Magnetometers, IEEE Sens. J., 12, 2719–2723, 2012.
Grosz, A., Paperno, E., Amrusi, S., and Liverts, E.: Integration of the electronics and batteries inside the hollow core of a search coil, J. App. Phys., 107, 09E703–E709E703-3, 2010.
Korepanov, V. and Pronenko, V.: Induction Magnetometers – Design Peculiarities, Sensors and Transducers Journal, 120, 92–106, 2010.
Lichtenberger, J., Ferencz, C., Bodnar, L., Hamar, D., and Steinbach, P.: Automatic whistler detector and analyzer system, J. Geophys. Res., 113, 2156–2202, 2008.
Lukoschus, D.: Optimization theory for induction-coil magnetometers at higher frequencies, IEEE T. Geosci. Elect., GE-17, 56–63, 1979.
Osborn, J. A.: Demagnetizing factors of the general ellipsoids, 67, 351–357, 1945.
Ozaki, M., Yagitani, S., Takahashi, K., and Nagano, I.: Development of a new portable Lightning Location System, IEICE Trans. Commun., E95-B, No. 1, January 2012.
Ozaki, M., Yagitani, S., Takahashi, K., and Nagano, I.: Dual-Resonant Search Coil for Natural Electromagnetic Waves in the Near-Earth Environment, IEEE Sens. J., 13, 644–650, 2013.
Prance, R. J., Clarck, T. D., and Prance, H.: Ultra low noise induction magnetometer for variable temperature operation, Sensor. Actuator., 85, 361–364, 2000.
Rhouni, A., Sou, G., Leroy, P., and Coillot, C.: A Very Low 1/f Noise and Radiation-Hardened CMOS Preamplifier for High Sensitivity Search Coil Magnetometers, IEEE Sens. J., 13, 159–166, 2012.
Ripka, P.: Magnetic sensors and magnetometers, Ed. Artech House, 2000.
Roux, A., Le Contel, O., Coillot, C., Bouabdellah, A., de la Porte, B., Alison, D., Ruocco, R., and Vassal, M. C.: The search coil

magnetometer for THEMIS, Space Sci. Rev., 141, 265–275, 2008.

Seran, H. C. and Fergeau, P.: An optimized low frequency three axis search coil for space research, Rev. Sci. Instrum., 76, 044502–0044502-9, 2005.

Shimin, F., Suihua, Z., and Zhiyi, C.: A very low noise preamplifier for extremely low frequency magnetic antenna, Journal of Semiconductors, 34, 075003–075003-5, 2013.

Sudduth, K. A., Drummond, S. T., and Kitchen, N. R.: Accuracy issues in electromagnetic induction sensing of soil electrical conductivity for precision agriculture, Comput. Electron. Agr., 31, 239–264, 2001.

Tumanski, S.: Induction coil sensors – A review, Meas Sci. Technol., 18, R31–R46, 2007.

Yan, B.: An Optimization Method for Induction Magnetometer of 0.1 mHz to 1 kHz, IEEE T. Magn., 49, 5294–5300, 2013.

7

Validation and application of a cryogenic vacuum extraction system for soil and plant water extraction for isotope analysis

N. Orlowski[1], H.-G. Frede[1], N. Brüggemann[2], and L. Breuer[1]

[1]Research Centre for BioSystems, Land Use and Nutrition (IFZ), Institute for Landscape Ecology and Resources Management (ILR), Justus-Liebig-University Giessen (JLU), Giessen, Germany
[2]Forschungszentrum Jülich GmbH, Institute of Bio- and Geosciences – Agrosphere (IBG-3), Jülich, Germany

Correspondence to: N. Orlowski (natalie.orlowski@umwelt.uni-giessen.de)

Abstract. Stable isotopic analysis of water in plant, soil, and hydrological studies often requires the extraction of water from plant or soil samples. Cryogenic vacuum extraction is one of the most widely used and accurate extraction methods to obtain such water samples. Here, we present a new design of a cryogenic vacuum extraction system with 18 extraction slots and an innovative mechanism to aerate the vacuum system after extraction. This mobile and extendable multi-port extraction system overcomes the bottleneck of time required for capturing unfractionated extracted water samples by providing the possibility to extract a larger number of samples per day simultaneously. The aeration system prevents the loss or mixture of water vapor during defrosting by purging every sample with high-purity nitrogen gas. A set of system functionality tests revealed that the extraction device guarantees stable extraction conditions with no changes in the isotopic composition of the extracted water samples. Surprisingly, extractions of dried and rehydrated soils showed significant differences of the isotopic composition of the added water and the extracts. This observation challenges the assumption that cryogenic extraction systems to fully extract soil water. Furthermore, in a plant water uptake study different results for hydrogen and oxygen isotope data were obtained, raising problems in the definition from which depths plants really take up water. Results query whether the well-established and widely used cryogenic vacuum distillation method can be used in a standard unified method of fixed extraction times as it is often done.

1 Introduction

During the past decades, stable water isotopes as natural tracers have become a common tool in plant ecological and pedological research. Isotopic fractionation occurring during evaporation and condensation of water leads to observable variations of deuterium and oxygen isotopic composition in natural waters and its use as natural tracers (Araguás-Araguás et al., 1995; Unkovich et al., 2001).

In plant ecology stable water isotopes provide a powerful method for determining seasonal changes in plant water uptake (Corbin et al., 2005; Eggemeyer et al., 2009; Butt et al., 2010; Liu et al., 2010), intra- and interspecific resource competition of plants (Williams and Ehleringer, 2000; Yang et al.,

2011), partitioning evaporation and transpiration (Wang and Yakir, 2000; Phillips and Gregg, 2003; Rothfuss et al., 2010, 2012), partitioning of water resources between plants (Stratton et al., 2000; Rossatto et al., 2012), and community water-use patterns or the zones of root activity in soils (Ehleringer and Dawson, 1992; Thorburn and Ehleringer, 1995; Dawson and Pate, 1996; Liu et al., 2011). Plant water uptake is considered as a non-fractionating process (Wershaw et al., 1966; Dawson and Ehleringer, 1991; Walker and Richardson, 1991; Thorburn et al., 1993; Dawson and Ehleringer, 1993) for non-saline conditions (Lin and Sternberg, 1993), implying that the isotopic signature of the source water remains the same during soil water uptake and water transport through plants (White et al., 1985). Thus, it can be utilized to trace back the origin (i.e. soil depth) of water in non-transpiring plant tissues.

In soil isotopic research, potential applications are the examination of soil water movements (Barnes and Allison, 1988; Gazis and Feng, 2004; Song et al., 2009; Brooks et al., 2009) or the quantification of phreatic evaporation (Brunner et al., 2008). To determine the isotopic signature of water, it is necessary to separate or extract the water from other components of the sample material (plant roots, stems and leaves, soil). In recent decades, several extraction methods have been developed: azeotropic distillation with various toxic substances such as toluene, hexane, and kerosene (Revesz and Woods, 1990; Thorburn et al., 1993), mechanical squeezing (Wershaw et al., 1966; White et al., 1985), cryogenic vacuum extraction (Dalton, 1988; Dawson and Ehleringer, 1993; Sala et al., 2000; West et al., 2006; Goebel and Lascano, 2012), the batch-method for stem water extraction (Vendramini and Sternberg, 2007), the modified vacuum extraction technique of Koeniger et al. (2011), centrifugation with or without immiscible heavy liquids (Mubarak and Olsen, 1976; Batley and Giles, 1979; Barrow and Whelan, 1980; Peters and Yakir, 2008) as well as different equilibrium techniques especially for soil samples (Scrimgeour, 1995; Hsieh et al., 1998; McConville et al., 1999; Koehler et al., 2000; Wassenaar et al., 2008). Out of these, cryogenic vacuum extraction is the most widely utilized method (Ingraham and Shadel, 1992; West et al., 2006; Vendramini and Sternberg, 2007; Koeniger et al., 2011).

During cryogenic vacuum extraction the plant or soil material is heated in a tube under a defined vacuum. The sample water evaporates and the evolved vapor is frozen in a cryogenic (liquid nitrogen) trap (Ingraham and Shadel, 1992). After defrosting the sample, its water isotopic signature can be analyzed. Previous studies comparing diverse extraction methods have shown that cryogenic vacuum extraction provides similar, consistent, and high precision results – except for dry soils with a high proportion of heavily bound water and soils containing hydrated salts, such as gypsum (Ingraham and Shadel, 1992; Walker et al., 1994; Araguás-Araguás et al., 1995; Kendall and McDonnell, 1998; Vendramini and Sternberg, 2007). However, cryogenic extraction requires a complex vacuum system and the duration of the extraction process is much longer than for azeotropic distillation or centrifugation methods (Vendramini and Sternberg, 2007).

The aim of this study is to present a vacuum-tight, reliable, and user-friendly cryogenic vacuum extraction system, mainly consisting of standard off-the-shelf material with a modular, extendable design enabling a high sample throughput. The device is equipped with a new mechanism to aerate the vacuum system after water extraction. Due to its flexible setup, the system is easily transportable and can also be used at field sites with power supply. After a description of the technical setup, we show the results of a set of functionality tests that prove the system's reliability, reproducibility, and stability of extraction conditions. Finally, we applied the cryogenic vacuum extraction method to soil and plant samples, which raised critical questions for further research.

2 Materials and methods

2.1 Technical description and extraction methodology

The distillation system utilized in this study consists of a vacuum manifold with six independent extraction lines, each comprising three extraction-collection units, resulting in a total of 18 extraction slots (Figs. 1 and 2). A detailed parts list is provided in Table A1.

These independent extraction lines are mainly composed of different types of Swagelok® fittings (Swagelok Company, Solon, OH, US), flanges, and flexible hoses (Rettberg®, Rettberg Inc., Göttingen, DE) (Fig. 2). The vacuum is generated by a two-stage rotary vane pump (Edwards®, RV5, Edwards Inc., Kirchheim, DE) and monitored by a PIRANI® vacuum gauge (VAP-5, Vacuubrand Inc., Wertheim, DE) at the end of the manifold. Additional interchangeable vacuum gauges are attached to two out of six extraction lines (DCP 3000 + VSK 3000, Vacuubrand Inc., Wertheim, DE) (Fig. 2). The vacuum can be separately applied or shut off via diaphragm valves (Fig. 2), which enables independent application of the units. A high-purity nitrogen purging system is realized by attaching additional diaphragm vales to each extraction line and joining them to a dry N_2 gas source (Fig. 2).

To extract water from soils or plants, the vacuum manifold is pre-evacuated to draw out possible atmospheric water contamination. The extraction tubes are filled with frozen sample material, and connected to the extraction lines together with the U-tubes (collection units). To fix the sample material in the tubes, avoiding a spread of sample material throughout the extraction system, fleece (Fackelmann Inc, Hersbruck, DE) is packed on top of each sample. During sample fixing care is required to avoid getting filaments of fleece between the connections disturbing vacuum-tightness. A previous test showed that fleece is suitable to fix the sample material and, moreover, does not hold residual water contaminating the sample. Afterwards, the entire system is evacuated to a pressure of 0.3 Pa. Subsequently, the diaphragm valves isolating the extraction lines from the vacuum manifold are closed. During the entire extraction process, the samples are heated to 90 °C, leading to the evaporation of water, which is successively trapped in the frozen U-tubes. In order to avoid isotope fractionation, the extraction process has to be conducted until completion, as fractionation effects appear for the more strongly bound water towards the end of the extraction period (Raleigh distillation) (Kendall and McDonnell, 1998).

At the end of the water extraction, residual water in the angular fittings connecting the extraction tubes with the U-tubes is evaporated by heating the fittings with a heat gun to ~ 90 °C. The extracted water is purged by high-purity nitrogen gas and every tube is removed from the system. During

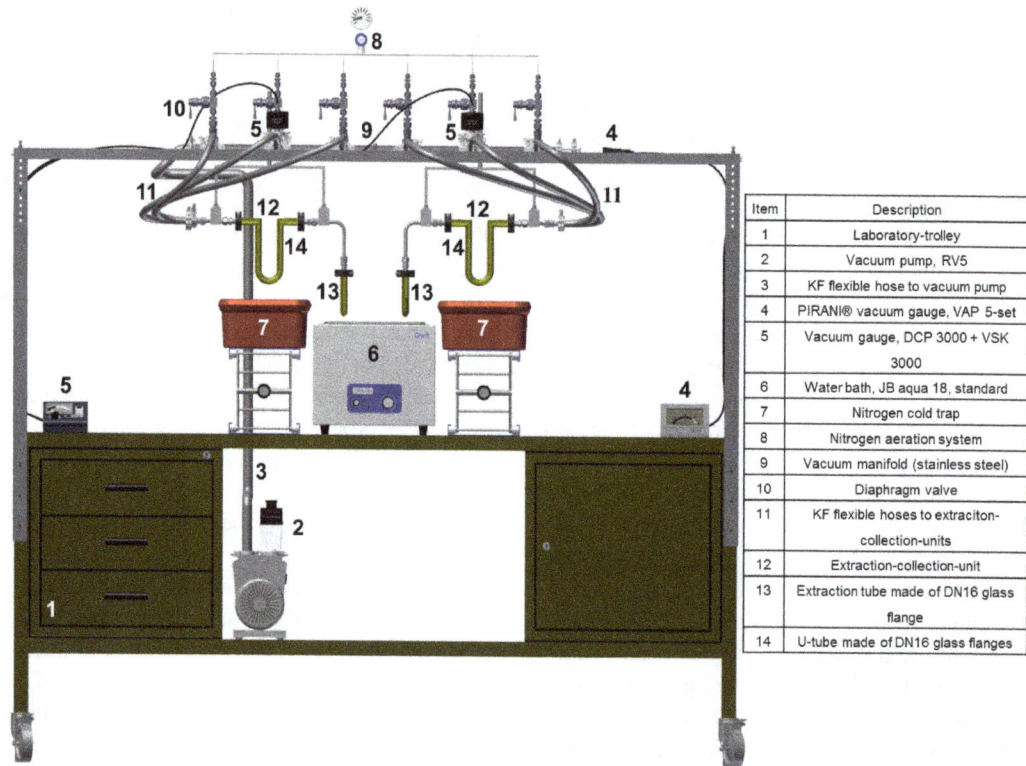

Figure 1. Schematic of the cryogenic vacuum extraction system fixed with a stainless steel frame to a laboratory trolley.

Item	Description
1	Laboratory-trolley
2	Vacuum pump, RV5
3	KF flexible hose to vacuum pump
4	PIRANI® vacuum gauge, VAP 5-set
5	Vacuum gauge, DCP 3000 + VSK 3000
6	Water bath, JB aqua 18, standard
7	Nitrogen cold trap
8	Nitrogen aeration system
9	Vacuum manifold (stainless steel)
10	Diaphragm valve
11	KF flexible hoses to extraciton-collection-units
12	Extraction-collection-unit
13	Extraction tube made of DN16 glass flange
14	U-tube made of DN16 glass flanges

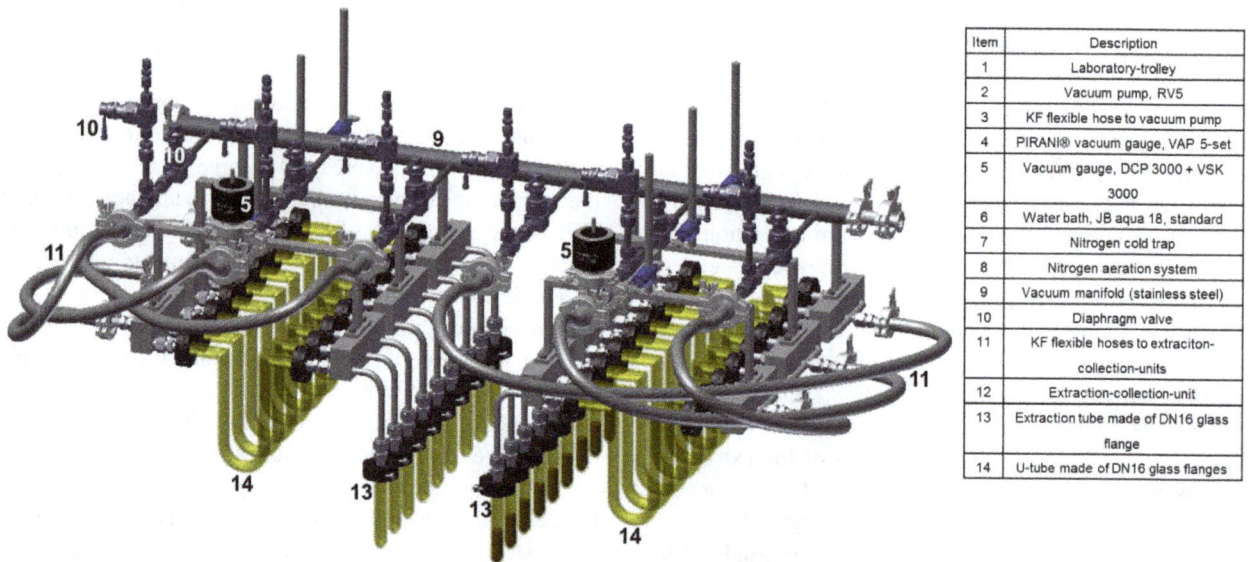

Figure 2. Detail view of the extraction manifold with six extraction lines and 18 extraction-collection units. Parts list is provided in Table A1.

thawing the U-tubes are sealed with silicon plugs. Finally, the extracted water is pipetted from the U-tubes into glass vials (2 mL) for isotopic analysis. All water and soil samples utilized in the following experiments were sealed with Parafilm® and stored light-excluded in vials or amber glass tubes at 7 °C.

2.2 Functionality tests of extraction system

To prove the system's reliability, reproducibility, and stability of extraction conditions, functionality tests were conducted (Table 1), which are supposed to serve as a basis for future extraction systems validations.

For a simple implementation and proper comparability of the results, three isotopically different types of pure water were chosen as testing materials for experiment #1 to #3: water from the Schwingbach creek (δ^2H: $-56.10\,\text{‰}$, δ^{18}O: $-8.46\,\text{‰}$), local precipitation (δ^2H: $-1.49\,\text{‰}$, δ^{18}O: $-0.99\,\text{‰}$), and tap water (δ^2H: $-56.74\,\text{‰}$, δ^{18}O: $-9.28\,\text{‰}$). Precision of analyses was $\pm0.60\,\text{‰}$ for δ^2H and $\pm0.20\,\text{‰}$ for δ^{18}O (LGR, 2013) (see Sect. 2.5). All waters were collected on 28 August 2009. Each extraction tube was filled with 10 mL test water. The extraction was complete when all test water had been transferred into the collection tubes.

Experiment #1 was conducted to test a potential effect of the applied cryogenic vacuum extraction procedure itself. Therefore, a water extraction with pure local tap water was performed and the isotopic signatures of untreated ($N = 5$) vs. extracted local tap water ($N = 6$) were statistically tested (Table 1).

For testing the effect of high-purity nitrogen purging on the extracted water isotopic signature, local tap water (experiment #2a) and Schwingbach creek water (experiment #2b) was extracted and purged with nitrogen gas (Table 1). Again, extracted water samples ($N = 6$) were compared with untreated, unpurged water samples ($N = 5$).

For experiment #3 local precipitation ($N = 3$) and water from the Schwingbach creek ($N = 3$), which strongly differ isotopically, were filled alternately into six extraction tubes to test if cross-contamination between the extraction lines occurred (Table 1).

2.3 Extraction process experiments

Testing the system's capability to recover water of known isotopic composition, experiments listed in Table 1 were performed. Each type of soil was sieved (2 mm), homogenized, oven-dried (105 °C, 24 h), and rehydrated with local tap water (δ^2H: $-59.49\pm0.79\,\text{‰}$, δ^{18}O: $-8.56\pm0.22\,\text{‰}$). Beside the Luvisol sample (highly clayey silt, pH: 7.0 ± 0.0), standard soils from the state research institute for agriculture (LUFA, Speyer, DE) were chosen as testing materials. The soils from LUFA 1 to 4 represent a gradient from sandy to loamy texture with increasing pH-values (LUFA 1: silty sand, pH: 5.1 ± 0.3; LUFA 2: loamy sand, pH: 5.5 ± 0.2; LUFA 3: silty sand, pH: 6.7 ± 0.3; and LUFA 4: clayey loam, pH: 7.1 ± 0.2). After rehydrating, soils were equilibrated for 24 h to ensure uniform water contents. For experiment #5b the extraction process was interrupted after 15 and 60 min. U-tubes were demounted, slowly defrosted, and an intermediate isotope sample (2 mL) was taken. Afterwards, the U-tubes with the remaining water were re-attached to the system and the water extraction was continued.

The isotopic signatures of added untreated tap water were statistically compared with extracted water isotopic signatures or groups of different extraction durations and approaches. Before rehydration, a 48 h oven-drying (105 °C) was conducted to check if soil samples had been dried to con-

stant mass containing no residual water contaminating the added isotopic signature. To check if the added water was entirely extracted, all samples were weighed (PM200, Mettler-Toledo Inc., Giessen, DE; precision 0.001 g) before and after water extraction, and after an additional oven-drying (24 h, 105 °C) of the extracted samples.

2.4 Applying cryogenic extraction to investigate crop water uptake

For herbaceous species like forbs and grasses, Barnard et al. (2006) identified the root crown tissue as the most suitable part of the plant to analyze the isotopic signature of the absorbed water. The root tissue itself is not a reliable identifier to quantify the actual source of the water as the rooting depth and the source of the water may differ (Thorburn and Ehleringer, 1995). Previous studies also analyzed stem or culm material exploring the water use by vegetation (Corbin et al., 2005; Gat et al., 2007; Rossatto et al., 2013). For this reason, a simultaneous probing of the root crown, stem tissue, and the actual isotopic signature of the soil water at different depths seemed to be the most reliable method to identify the source of water uptake. We conducted a pot experiment with two common field crops, i.e. barley (*Hordeum vulgare* L. cv. Barke) and wheat (*Triticum aestivum*, Xenos) under controlled-environment conditions to test if there is an impact of:

1. harvest time on the isotopic signature of sampled plant tissue (stem and root crown) water of two crop species (barley and wheat);

2. harvest time on water uptake zones of these crop species;

3. analyzed isotopes (either hydrogen or oxygen) on investigating plants' source water.

Plants were sown as monocultures on 3 December 2010, in free-draining pots ($16 \times 16 \times 16$ cm) filled with 4 L conventional potting soil (Fruhstorfer Erde, Hawita Group Inc., Vechta, DE). For every crop type 54 replicates were planted, each in a single pot, resulting in a sum of 108 pots. Plants were grown in the greenhouse with a 14 h photoperiod, day and night temperatures of 20 and 14 °C, respectively. Irrigation with local tap water was carried out three times per week until saturation, while excess water could freely drain. Irrigation water was sampled each time for isotopic analysis (200 times in total). The isotopic signature of irrigation water did not differ significantly between first and second harvest (means and standard deviations for the first ($N = 100$) and second harvest ($N = 100$), respectively, δ^2H: $-60.29\pm1.26\,\text{‰}$ and $-60.16\pm1.27\,\text{‰}$; δ^{18}O: $-8.85\pm0.35\,\text{‰}$ and $-8.92\pm0.27\,\text{‰}$).

On 2 and 3 March 2011, after 91 days, 27 individuals per species were harvested. The second harvest was carried out

Table 1. Description of the functionality tests #1 to #3 and the extraction process experiments #4 to #6.

Functionality tests

Experiment	Description	Testing material	N		
#1	Effect of extraction process	Local tap water	Untreated: 5, extracted: 6		
#2a	Effect of high-purity nitrogen purging	Local tap water ·	Untreated: 5, extracted and purged: 6		
#2b	Effect of high-purity nitrogen purging	Schwingbach creek water	Untreated: 5, extracted and purged: 6		
#3	Cross-contamination	Local precipitation, Schwingbach creek water	Untreated: 5, extracted: 3 per type of water		

Extraction process experiments

Experiment	Description	Testing material	Water content [%]	Extraction duration [min]	N
#4	Water recovery of rehydrated soils	Ah-horizon of Luvisol	10	180	18
#5a	Effect of extraction time	Ah-horizon of Luvisol	20	15, 30, 45, 60, 120, 180	3 per duration
#5b	Effect of extraction time	Ah-horizon of Luvisol	20	180; interruption after 15 and 60	18
#6	Effect of soil type	Ah-horizon of Luvisol, LUFA soil 1 to 4	20	15, 30, 45, 60, 120, 180	3 per duration

on 6 and 7 April 2011, after 126 days. At every harvest the following samples were taken: the lower 10 cm of the stem (after removal of the outer sheath), root crown, and an aliquot of the upper 8 cm and the lower 8 cm of the soil column (after removal of the roots). Plant and soil samples were immediately frozen until water extraction. Soils were extracted for 180 min and plant tissues for 90 min.

2.5 Isotopic analysis

Isotopes were analyzed at the Institute for Landscape Ecology and Resources Management (JLU Giessen, DE) according to the IAEA standard procedure (Newman et al., 2012) using a Los Gatos Research DLT-100-Liquid Water Isotope Analyzer (Los Gatos Research Inc., Mountain View, CA, USA). The DLT-100 is based on off-axis integrated cavity output spectroscopy (OA-ICOS), which enables a simultaneously quantification of $\delta^{18}O$ and $\delta^{2}H$ isotopic signatures. OA-ICOS measures with the same or even better precision than conventional stable isotope-ratio mass spectrometers (Lis et al., 2008; Penna et al., 2010). Isotopic ratios are reported in per mil (‰) relative to a standard, i.e. the Vienna Standard Mean Ocean Water (VSMOW) (Craig, 1961):

$$\delta^{2}H \text{ or } \delta^{18}O = \left(\frac{R_{\text{sample}}}{R_{\text{standard}}} - 1 \right) \times 1000. \qquad (1)$$

Here, R_{sample} and R_{standard} are $^{2}H/^{1}H$ or $^{18}O/^{16}O$ ratios of the sample and standard, respectively. Precision of analyses was ± 0.60‰ for $\delta^{2}H$ and ± 0.20‰ for $\delta^{18}O$ (LGR, 2013).

Leaf water extracts typically contain a high fraction of organic contaminations (West et al., 2010), which might lead to spectral interferences when using isotope ratio infrared absorption spectroscopy (Leen et al., 2012), causing erroneous isotope values (Schultz et al., 2011). Therefore, isotopic data of plant water extracts were checked for spectral interferences using the Spectral Contamination Identifier (LWIA-SCI) post-processing software (Los Gatos Research Inc.). No sample was found to be contaminated with organics.

2.6 Statistical analysis

All statistical analyses were performed utilizing IBM SPSS Statistics (Version 19.0, SPSS Inc., Chicago, IL, US). First, the data were tested for normal distribution. Subsequently, t tests were performed for experiments #1 and #2, whereas ANOVAs with Tukey-HSD tests were conducted for experiment #3 to #6 ($p \leq 0.05$). For the plant water uptake study, Multivariate Analyses of Variances (MANOVAs) were done and Tukey-HSD tests were run to determine significant differences between groups ($p \leq 0.05$).

Table 2. Means and standard errors for δ^2H and $\delta^{18}O$ values of system functionality tests #1 to #3 (superscripts indicate significant differences, $p < 0.05$).

	δ^2H [‰]	$\delta^{18}O$ [‰]
Experiment #1: Effect of extraction process		
Extracted local tap water	-57.49 ± 0.58^a	-9.40 ± 0.12^a
Untreated local tap water	-56.74 ± 0.36^a	-9.28 ± 0.11^a
Experiment #2a: Effect of high-purity nitrogen purging		
Local tap water without nitrogen purging	-57.66 ± 0.60^a	-9.38 ± 0.12^a
Local tap water with nitrogen purging	-58.51 ± 0.20^a	-9.33 ± 0.03^a
Experiment #2b: Effect of high-purity nitrogen purging		
Untreated Schwingbach water	-56.10 ± 0.72^a	-8.46 ± 0.14^a
Schwingbach water with nitrogen purging	-55.52 ± 0.28^a	-8.76 ± 0.11^a
Experiment #3: Cross-contamination		
Untreated local precipitation	-1.83 ± 0.21^a	-1.24 ± 0.19^a
Extracted local precipitation	-1.53 ± 0.10^a	-1.26 ± 0.03^a
Untreated Schwingbach water	-56.10 ± 0.72^b	-8.46 ± 0.14^b
Extracted Schwingbach water	-56.69 ± 0.43^b	-8.90 ± 0.12^b

3 Results and discussion

3.1 Functionality tests of extraction system

The functionality and extraction process tests #1 to #6 should serve as guidance for future rigorous validations of extraction systems. In the past, the performance of extraction systems was in most cases merely verified through water extractions from soils of various grain size, water contents, or extraction duration (Ingraham and Shadel, 1992; Walker et al., 1994; Araguás-Araguás et al., 1995; West et al., 2006; Koeniger et al., 2011), separately tested for the system's extraction ports (Goebel and Lascano, 2012). However, none of the studies tested the functionality of a vacuum extraction system in the same detail as performed here.

Experiment #1 demonstrated that the water extraction procedure did not lead to significant differences in isotopic signatures between untreated and extracted tap water samples (δ^2H: $p = 0.32$, $\delta^{18}O$: $p = 0.52$) (Table 2).

Purging with high-purity nitrogen (experiments #2a and #2b) to prevent a possible isotopic exchange of extracted water with air did also not result in significant differences in isotopic composition, neither for local tap water (δ^2H: $p = 0.22$, $\delta^{18}O$: $p = 0.69$) nor for Schwingbach creek water (δ^2H: $p = 0.48$, $\delta^{18}O$: $p = 0.13$) (Table 2).

In experiment #3, no significant differences were observed between untreated and extracted samples, neither for precipitation water (δ^2H: $p = 0.96$, $\delta^{18}O$: $p = 1.00$) nor for Schwingbach creek water (δ^2H: $p = 0.75$, $\delta^{18}O$: $p = 0.11$). Statistically highly significant differences between precipitation and Schwingbach water even after extraction ($p = 0.00$) were

found, demonstrating that there was no cross-contamination between the six extraction lines (Table 2).

3.2 Extraction process experiments

3.2.1 Experiment 4: water recovery of rehydrated soils

A simple, yet informative experiment to test the feasibility of the extraction system was to recover water of known stable isotopic composition that had been introduced to previously dried soil. Surprisingly, the isotopic signature of the added water could be recovered neither for hydrogen nor for oxygen (Fig. 3), despite a long extraction time of 180 min and a vacuum of 0.3 Pa.

Previous studies (Ingraham and Shadel, 1992; Walker et al., 1994; Araguás-Araguás et al., 1995; Kendall and McDonnell, 1998) indicated that extracting water from clay soils could be problematic due to interactions between pore water and weakly bound water in the clay matrix. Therefore, an additional 48 h of oven-drying of the untreated Luvisol sample was conducted before rehydration to check for residual water. Comparing the water content after 24 h drying (24.2 %) with that after 48 h drying (24.1 %) ($N = 3$) resulted in an additional water loss of 0.11 % on average. This small amount of tightly bound residual water could not have affected the isotopic signature in a way as we observed it. The fraction of residual water in the untreated soil becomes more significant as the quantity of introduced water in terms of water content is small (Walker et al., 1994) – which was not the case in our study with 10 or 20 % water content – or the isotopic signature of the residual water is significantly different from the introduced water. Thus, a memory effect in the soil

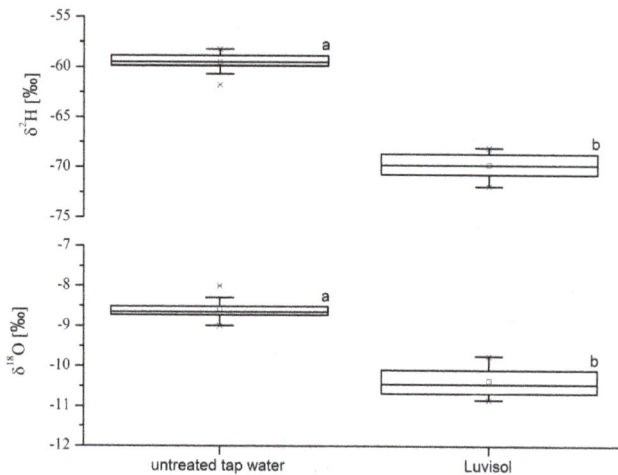

Figure 3. Comparison of isotopic ratios of untreated tap water with water extracted from an oven-dried Luvisol (180 min) rehydrated with the same tap water. Different letters indicate significant differences ($p < 0.05$).

as found by Koeniger et al. (2011) can be excluded, as hardly any residual water (0.11 %) was detected in the untreated soil samples. Walker et al. (1994) already stated a temperature dependency of the isotopic signature of the extracted water. Furthermore, even at extraction temperatures $> 100\,°C$ the introduced water in their experiments could not be fully recovered, neither by vacuum extraction nor by other extraction methods (azeotropic and microdistillation). The differences in extraction procedures, i.e. the applied vacuum or the extraction duration, remained unknown in their comparison. In our study, high-temperature extractions could not easily be tested without major rearrangement of the technical setup of the extraction system, for instance in changing the heating element and replacing it with a heated sand bath that facilitates heating of soils to several $100\,°C$. Further research is needed for fully explaining the observed incomplete recovery of soil water and its effect on soil water isotope studies.

3.2.2 Experiment 5: effect of extraction time

For specifying the effect of extraction time on isotopic ratios of extracted water, we conducted two experiments (#5a and #5b, see Table 1). For experiment #5a, the only significant differences in δ^2H values were found between soil extracted for 15 min and soil extracted for 60 and 120 min, respectively (Fig. 4a). For δ^{18}O, four statistical homogenous groups were identified (Fig. 4b). Samples extracted for 60 to 180 min exhibited somewhat lower mean δ^{18}O signatures around $-9\,‰$ as compared to the applied water with a δ^{18}O of $-8.5\,‰$. However, mean values were not significantly different, in contrast to results obtained for δ^2H.

Generally, isotopic values after 15 min extraction time were significantly depleted compared to the original tap water and thus indicated a large fractionation effect. Because

of Rayleigh fractionation, we found progressively less negative values towards the end of the extraction process for both isotopes (exception: δ^{18}O after 45 min extraction time). In agreement with the results of West et al. (2006) and Goebel and Lascano (2012), we observed no statistically significant changes in the isotopic signatures for extraction times longer than 30 min for hydrogen isotopes and 60 min for oxygen isotopes for this specific soil type. Nevertheless, as already observed in experiment #4 neither δ^2H nor δ^{18}O of extracted water matched the isotopic signature of the applied water.

In experiment #5b, no significant differences between the soil samples extracted for 15 and 60 min were found for both isotopes. However, they differed significantly from the soil samples extracted for 180 min (Fig. 4c and d). Again, isotopic composition became less depleted with longer extraction times. Most of the water had been extracted during the first 15 min, and decreasing amounts of water were extracted with increasing extraction time. The fraction of light isotopes is known to be extracted first (Kendall and McDonnell, 1998). After 60 min extraction time most of the water had already been removed, resulting in more positive isotopic signatures of the remaining water due to isotopic enrichment.

To test whether the isotopic ratios of extracted water at the same extraction times differed between the extraction experiments #5a and #5b, soil samples extracted for 15, 60, and 180 min were compared (Fig. 4, capital letters). Hydrogen isotopic ratios after 60 min extraction time differed significantly between the approaches, with more negative values in experiment #5b. However, no significant differences in δ^{18}O values were found between 15, 60, and 180 min extraction time in both experiments. Unless all soil samples were homogenized, sample heterogeneity could be a potential source of error, which becomes more important for smaller sample sizes. The greater variance of values in experiment #5a is most likely associated with a smaller sample size ($N = 3$). However, both approaches showed the same dynamic: isotopic signatures of extracted water approached the tap water signature with increasing extraction time. Therefore, in the following experiments examining the effect of soil type on isotopic signatures of the extracted water, the first approach (experiment #5a) was implemented due to its higher feasibility and better comparability with other studies.

3.2.3 Experiment 6: effect of soil type

To test the hypothesis that particle size distribution affects the tightness of water bound in the soil, five different dried–rehydrated soil types were extracted for 15 to 180 min (approach analogous to experiment #5a).

The δ^2H values of Luvisol showed no statistical differences between 15 to 45, and 180 min, just as for 30 to 180 min extraction time (Fig. 5). The δ^2H of water added for rehydration in soil water extracts could not be recovered, in contrast to δ^{18}O after 60 to 180 min extraction time (Fig. 6). For LUFA soil 1 no significant changes in δ^2H and δ^{18}O

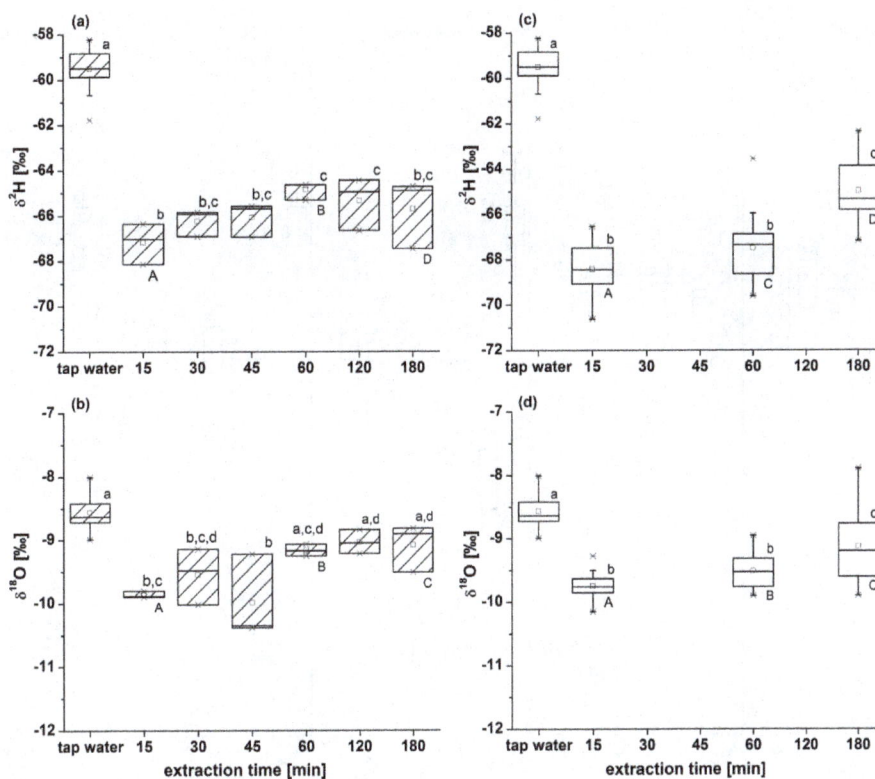

Figure 4. Comparison of isotopic ratios of untreated tap water with water recovered from dried Luvisol rehydrated with the same tap water and extracted increasingly long time intervals. Left panels: experiment #5a; right panels: experiment #5b. Different letters indicate significant differences ($p < 0.05$). Small letters indicate intra-experimental comparisons between different extraction times for the same approach; capital letters indicate inter-experimental comparisons between same extraction times for different experimental approaches (#5a vs. #5b).

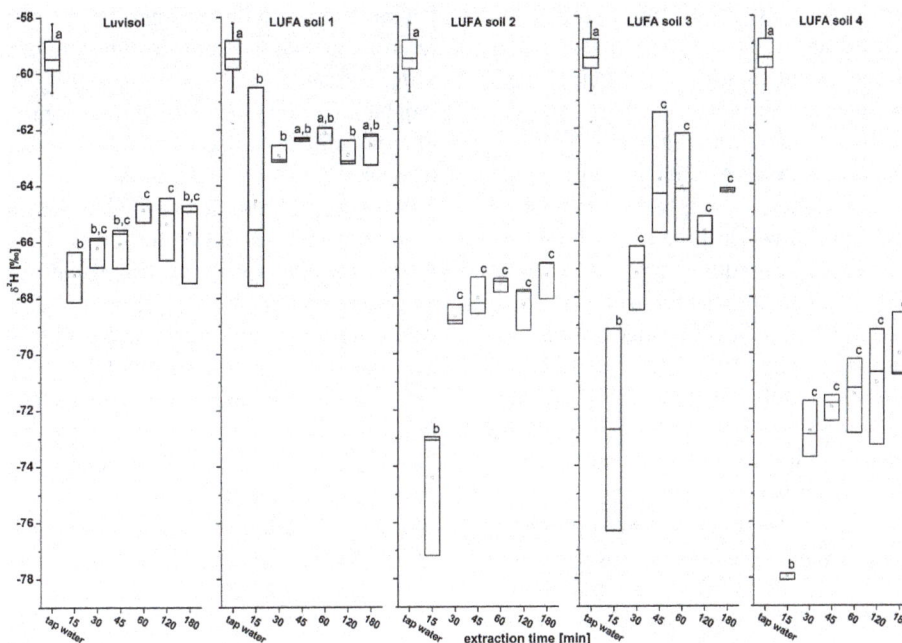

Figure 5. Comparison of $\delta^2 H$ values of untreated tap water with water recovered from five different dried soil substrates (Luvisol, LUFA soils 1 to 4) rehydrated with the same tap water and extracted for increasingly long time intervals (15 to 180 min). Different letters indicate significant differences ($p < 0.05$).

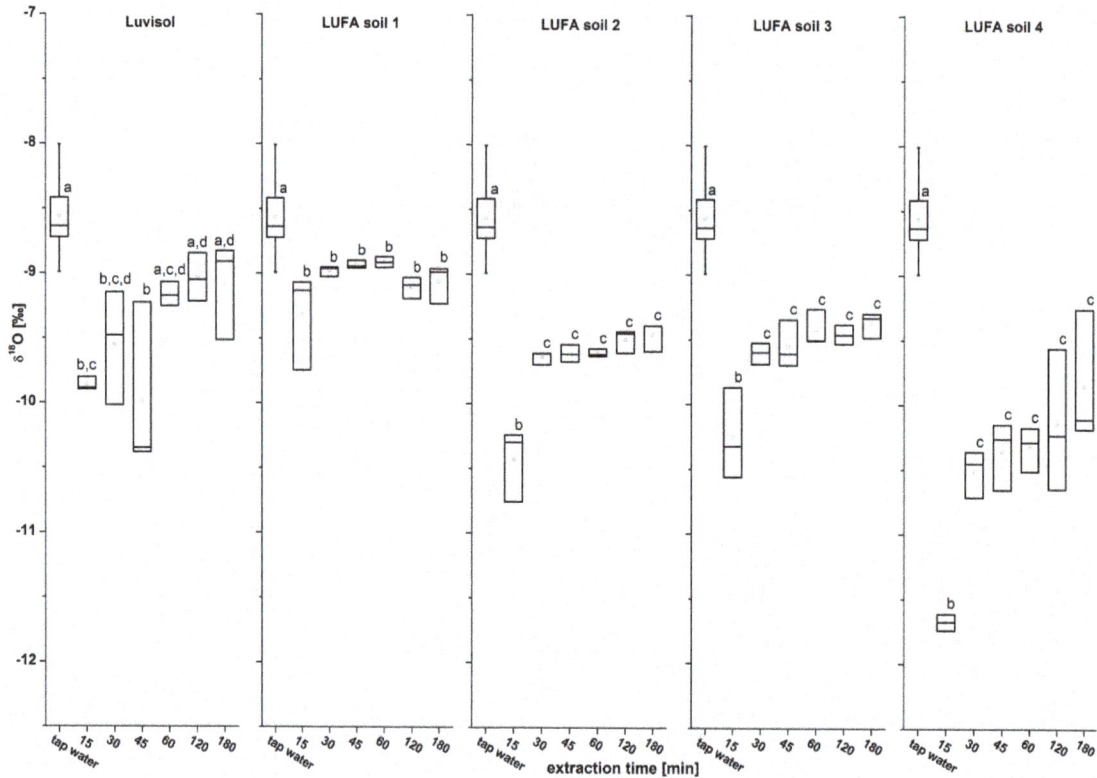

Figure 6. Comparison of $\delta^{18}O$ values of untreated tap water with water recovered from five different dried soil substrates (Luvisol, LUFA soils 1 to 4) rehydrated with the same tap water and extracted for increasingly long time periods (15 to 180 min). Different letters indicate significant differences ($p < 0.05$).

values between 15 to 180 min extraction time were found (Figs. 5 and 6). For 45, 60, and 180 min extraction time, δ^2H values of LUFA soil 1 were not significantly different from that of the introduced tap water. Moreover, the $\delta^{18}O$ values of LUFA soil 1 showed the smallest mean difference (0.49 ‰) from the tap water values. Extracted water isotopic signatures of LUFA soil 2 to 4 exhibited strongly depleted values for both isotopes at 15 min extraction time (Figs. 5 and 6). Furthermore, these soil types showed the largest differences in both δ^2H and $\delta^{18}O$ values in comparison to the isotopic signature of added water. No significant changes in the isotopic signatures of water extracted from LUFA soils 2 to 4 could be observed after 30 to 180 min extraction time, suggesting that 30 min extraction time was sufficient for those soils. Nevertheless, none of LUFA soil 2 to 4 water extracts reflected the original tap water isotopic values.

In contrast to the findings of Koeniger et al. (2011), whose extraction procedure was unable to fully recover water added to clayey and silty soils, tap water $\delta^{18}O$ values could be recovered from the Luvisol. The isotopic signatures for the silty and clayey LUFA soils 3 and 4 showed the largest range. In contrast, soils with a loamy texture (LUFA soil 2 and 4) showed the highest mean differences from the introduced isotopic signature for both isotopes.

Gravimetric soil water analysis revealed no significant differences in weight between soils before water addition, after water extraction, and after oven-drying of the extracted soils, indicating a complete extraction. Incomplete water extraction in terms of weight was only observed for one sample, with a recovery rate still as high as 99 %.

West et al. (2006) determined extraction times to obtain an unfractionated water sample for sandy soils of 30 min, and 40 min for clay soils, consistent with the results of Goebel and Lascano (2012) who recommended 30 min extraction duration for a sandy clay loam. Koeniger et al. (2011) applied even shorter extraction times (2.5 to 40 min), recovering the original water isotopic signature after 15 min. Our extraction duration of 180 min exceeded the extraction times of the abovementioned studies. The same was true for the applied pressure level (0.3 Pa), which was at the lower end of other vacuum extraction procedures ranging from 13 Pa (Goebel and Lascano, 2012), 8.0 Pa (West et al., 2006), 3.07 Pa (Koeniger et al., 2011), 1.3 Pa (Vendramini and Sternberg, 2007) to 0.13 Pa (Peters and Yakir, 2008). We conclude that either longer extraction times (> 180 min), or higher extraction temperatures (> 90 °C), or lower pressures (< 0.3 Pa), or a combination are required to achieve reliable results.

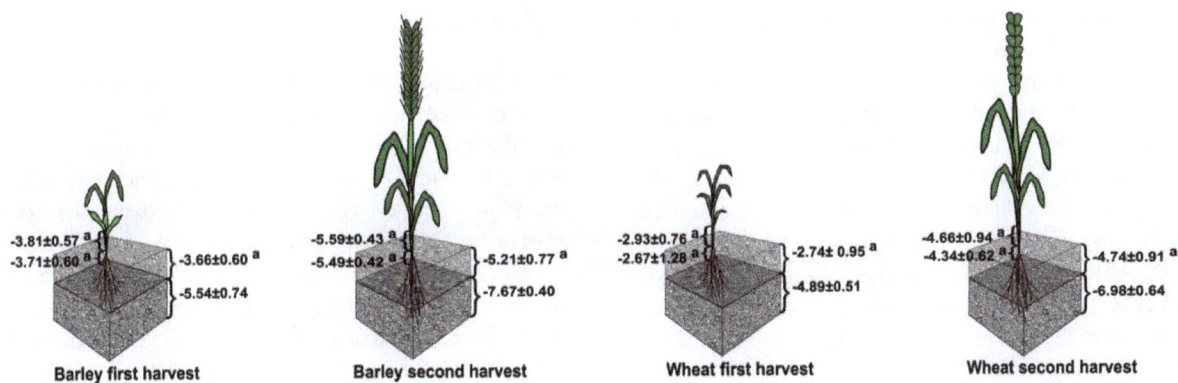

Figure 7. Means and standard deviations of oxygen isotopic composition in per mil in stems, root crowns, soil upper- and lower 8 cm for barley and wheat at two harvest times. Different letters indicate significant differences ($p < 0.05$).

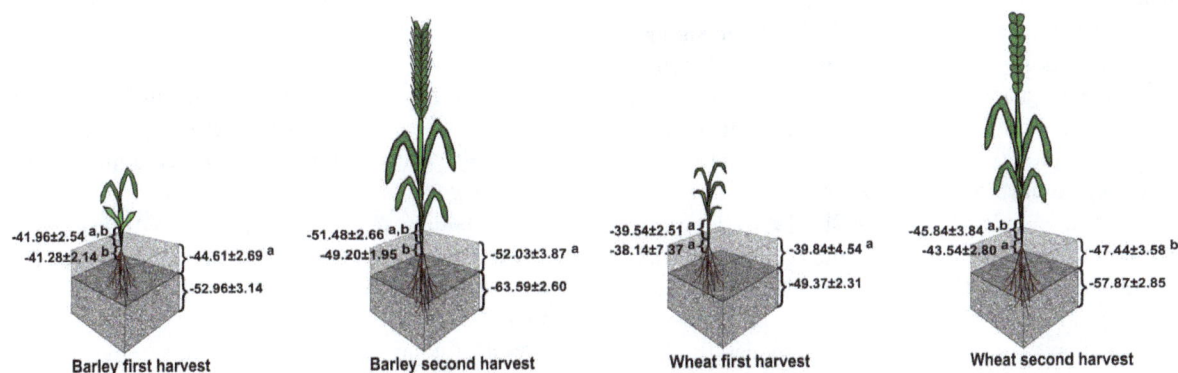

Figure 8. Means and standard deviations of hydrogen isotopic composition in per mil in stems, root crowns, soil upper- and lower 8 cm for barley and wheat at two harvest times. Letters indicate significant differences ($p < 0.05$).

To date, only a few studies have applied extraction temperatures $> 100\,°C$ (Walker et al., 1994; Araguás-Araguás et al., 1995). Despite using such a high temperature, Walker et al. (1994) could not recover the added water isotopic signature for dry and wet clays, sand, and gypseous sand. Nevertheless, high-temperature extractions yielded smaller deviations of isotopic signatures from the introduced water than low-temperature extractions (35 to 80 °C). In a later study, Araguás-Araguás et al. (1995) achieved recovery rates $> 98\,\%$ for pure sand by either increasing the temperature or the extraction time. Thus, we recommend analyzing the potential of high temperature extractions for a complete, fractionation-free retrieval of soil water, especially for clayey soils.

3.3 Applying cryogenic extraction to investigate crop water uptake

Quantitative analyses (MANOVAs) – performed for the identification of crop water uptake in various soil depths – revealed that barley and wheat were taking up their water from the upper 8 cm of the soil column at both harvest times. We found no significant differences in oxygen isotopic signatures between water extracted from stems, root crowns, and the upper 8 cm of the soil at first and second harvest for both crop species (Fig. 7).

For hydrogen isotopic data, no significant differences between stems and root crowns at both harvest times and for both species could be observed. However, only the hydrogen isotopic signatures of stem tissue water corresponded to that of water from the upper 8 cm of soil in the experiments with barley at both harvests, and for wheat at the second harvest (Fig. 8). For wheat both plant tissues reflected the upper soil isotopic signatures at first harvest.

Isotopic values were generally more enriched at first harvest for both species. Moreover, root crowns showed more positive δ values than stems for both isotopes, species, and harvest times (Figs. 7 and 8). Even though a larger rooting depth was observed at second harvest, the majority of roots was found in the upper evaporative soil layer. Therefore, the water from root crowns showed more positive δ^2H values than that of stems, which was in agreement with the findings of Barnard et al. (2006).

Since evaporative enrichment of soil water is a well-known process under field conditions (Barnard et al., 2006), we decided not to inhibit it and, therefore, did not cover the soil, contrary to Walker and Richardson (1991). Due to evaporation the δ^2H and δ^{18}O values of the soil water decreased with depth, while the intensity of this effect was mainly influenced by atmospheric conditions, such as temperature, relative humidity and the soil water content within the pots (Figs. 7 and 8). More bare soil and lower transpiration rates resulted in more enriched δ values in the upper part of the soil column at the first harvest for both isotopes and consequently also in the water-bearing plant tissues, confirming results by Barnard et al. (2006) for grass species. As a result, soil isotopic signatures did not reflect irrigation water signatures at the first nor at the second harvest.

In general, barley and wheat showed no species-specific or harvest-time effect on the water uptake depth. Stem material reflected more often the isotopic signature of the soil water than root crown tissue. Both observations are contrary to results presented by Barnard et al. (2006). To the author's knowledge there is no common explanation for that phenomenon, since water moves from soil to roots, and then to the stems without isotopic fractionation. Theoretically, the water can be sampled fractionation-free from plant tissues, and is assumed to have the same isotopic signature as the weighted average of soil water (Takahashi, 1998). However, taking the analytical accuracy for δ^2H into account, which is ± 0.6‰, the isotopic signatures of root crown and stem material are likely to be similar, resulting in no significant differences to the upper soil isotopic signature. Generally, δ^{18}O values yielded more consistent results for the identification of crop water uptake zones than δ^2H. Barnard et al. (2006) stated that the estimation of plant source water is typically based on the comparison of δ^{18}O values of plant tissues with that of soil or rain water. In previous studies on water uptake by grass and crop species, either oxygen (Wang and Yakir, 2000; Barnard et al., 2006; Durand et al., 2007; Nippert and Knapp, 2007; Asbjornsen et al., 2007; Rothfuss et al., 2010; Wang et al., 2012) or hydrogen isotopes (Dalton, 1988; Zegada-Lizarazu and Iijima, 2004; Walter and Morio, 2005; Yang et al., 2011) were analyzed. Only few studies analyzed both isotopes (Corbin et al., 2005; Wang et al., 2010; Rossatto et al., 2013). Furthermore, only a small number of out of these studies (Zegada-Lizarazu and Iijima, 2004; Barnard et al., 2006; Rossatto et al., 2013) performed statistical methods comparable to the here presented tests. In order to evaluate results with regard to plant water uptake depths, both isotopes should be analyzed. The calculations from δ^{18}O values should not differ significantly from the calculations obtained from δ^2H values, as verified by Rossatto et al. (2013).

4   Conclusions

The functionality tests on the extraction system demonstrated that the extraction system was vacuum-tight, and assured reproducibility, stable extraction conditions and reliable results. Alterations of isotopic signatures due to the extraction procedure itself, cross-contamination between the extraction lines or the high-purity nitrogen purging could be excluded.

Surprisingly, extractions of dried and rehydrated soils revealed significant differences in the isotopic composition of the added and the extracted water. While extraction time and pressure could be excluded, temperature seemed to be the crucial factor for the impossibility of recovering the isotopic signature of the added water. Therefore, the temperature effect on isotopic signatures of extracted water should be carefully scrutinized in future studies applying cryogenic extraction techniques, especially for soils with a high silt and clay content.

Applying cryogenic extraction to quantify water uptake of barley and wheat revealed that these crop species were taking up their water from the upper soil column even at a later growth stage. In order to verify results of plant water uptake calculations and, moreover, to avoid misinterpretation of plant water sources, a comparison of hydrogen and oxygen isotopic data should be performed, which has to generate the same outcome.

Present findings raise the question whether results from different extraction systems and conditions are comparable, for instance in the context of plant water studies. If isotopic data is applied to studies, whose focus is on the mobile water fraction available for plants, Araguás-Araguás et al. (1995) recommended adjusting extraction conditions to lower temperatures and shorter extraction times to keep the effect of weakly bound soil water as low as possible. Extraction errors due to the abovementioned extraction problems implicate an offset of measured isotope values due to fractionation as a consequence of incomplete extraction, or a mixture of different isotopic pools (Araguás-Araguás et al., 1995) when adding water to soil which then mixes with the remaining weakly bound water. Thus, soil water isotopic signatures obtained from vacuum extraction have to be critically compared with those of other water pools, such as precipitation (Kendall and McDonnell, 1998). The ultimate question therefore is: which water are we actually extracting under certain extraction conditions (temperature, pressure, and extraction time) from a specific soil type, and is this fraction of water the one that is utilized by plants or is it a mixture of water pools stored in the soil?

Table A1. Parts list of extraction system.

Item	Article description	Ordering number	Producer	Quantity	Measures
1	Laboratory-trolley	615911	Kaiser und Kraft Inc., Stuttgart, DE	1	W: 2000 mm, D: 800 mm, H: 940 mm
2	Vacuum pump, RV5	A65301903	Edwards Inc., München, DE	1	
3	KF flexible hose to vacuum pump	FX25K100-316	Vacom Inc., Jena, DE	1	DN 25, L: 1000 mm, Diameter: 40 mm
4	PIRANI® vacuum gauge, VAP 5-set	188-1130	Vacuubrand Inc., Wertheim, DE	1	
5	Vacuum gauge, DCP 3000 + VSK 3000	683170	Vacuubrand Inc., Wertheim, DE	2	
6	Water bath, JB aqua 18, standard	462-8136	Grant Instruments, Hillsborough, NJ, US	1	Vol: 18 L, W: 340 mm, D: 570 mm, H: 270 mm
7	Nitrogen cold trap	478-4302	Reichelt Chemietechnik Inc., Heidelberg, DE	4	Vol: 4 L, W: 280 mm, D: 190 mm, H: 110 mm
8	Y-connector, QSMY-3	153370	Festo Ltd., Esslingen, DE	5	Diameter: 3 mm
8	Teflon hose-connection to nitrogen gas source	741632	Riesbeck Inc., Biebergemünd, DE	1	Diameter: 3.2 mm
8	Reducing plug connection, QSM-6-4	153327	Festo Ltd., Esslingen, DE	1	Diameter: 6 mm × 4 mm
9	Vacuum manifold (stainless steel)	316TI-T10M-S-1.5M-6ME	Swagelok Company, Solon, OH, US	1	Diameter 10 mm, Wall: 1.5 mm, L: 1000 mm
10	Diaphragm valve	SS-DLS8MM	Swagelok Company, Solon, OH, US	12	Diameter 8 mm
11	KF flexible hoses to extraction-collection-units	FX16K100-304	Vacom Inc., Jena, DE	6	DN 16, L: 1000 mm, Diameter: 30 mm
12	KF clamping chain	710653-1	Rettberg Inc., Göttingen, DE	54	DN16
12	KF centering ring	1340150160	Rettberg Inc., Göttingen, DE	54	DN16
12	KF bored flange	KF16B19-316	Vacom Inc., Jena, DE	16	DN16
12	KF clamp ring	KF16C	Vacom Inc., Jena, DE	16	DN16
13	Extraction tube made of DN16 glass flange	1340130160; hand-made	Rettberg Inc., Göttingen, DE; Glass blowing, JLU Giessen, DE	18	Round-bottom, L: 120 mm
14	U-tube made of DN16 glass flanges	1340130160; hand-made	Rettberg Inc., Göttingen, DE; Glass blowing, JLU Giessen, DE	18	W: 180 to 200.5 mm, L: 180 to 200 mm, distance U-tube arms: 40.5 mm

Acknowledgements. We are thankful to Rolf Siegwolf from the Paul Scherrer Institute (Villigen, CH), who gave us valuable insight into cryogenic vacuum extraction techniques. Further appreciation goes to Günther Jennemann from the University of Applied Sciences Mittelhessen (Giessen, DE), who helped us preparing the technical drawings of the cryogenic vacuum extraction device. Thanks also to Paul Königer for the intensive discussions on results we observed. Natalie Orlowski acknowledges funding by the Friedrich-Ebert-Stiftung.

Edited by: G. S. Aluri
Reviewed by: two anonymous referees

References

Araguás-Araguás, L., Rozanski, K., Gonfiantini, R., and Louvat, D.: Isotope effects accompanying vacuum extraction of soil water for stable isotope analyses, J. Hydrol., 168, 159–171, 1995.

Asbjornsen, H., Mora, G., and Helmers, M. J.: Variation in water uptake dynamics among contrasting agricultural and native plant communities in the Midwestern U.S., Agr. Ecosyst. Environ., 121, 343–356, doi:10.1016/j.agee.2006.11.009, 2007.

Barnard, R. L., De Bello, F., Gilgen, A. K., and Buchmann, N.: The δ^{18}O of root crown water best reflects source water δ^{18}O in different types of herbaceous species, Rapid Commun. Mass. Sp., 20, 3799–3802, doi:10.1002/rcm.2778, 2006.

Barnes, C. J. and Allison, G. B.: Tracing of water movement in the unsaturated zone using stable isotopes of hydrogen and oxygen, J. Hydrol., 100, 143–176, doi:10.1016/0022-1694(88)90184-9, 1988.

Barrow, N. J. and Whelan, B. R.: A study of a method for displacing soil solution by centrifuging with an immiscible liquid, J. Environ. Qual., 9, 315–319, doi:10.2134/jeq1980.00472425000900020031x, 1980.

Batley, G. and Giles, M.: Solvent displacement of sediment interstitial waters before trace-metal analysis, Water Res., 13, 879–886, doi:10.1016/0043-1354(79)90223-9, 1979.

Brooks, J., R., Barnard, H., R., Coulombe, R., and McDonnell, J., J.: Ecohydrologic separation of water between trees and streams in a Mediterranean climate, Nat. Geosci., 3, 100–104, doi:10.1038/ngeo722, 2009.

Brunner, P., Li, H. T., Kinzelbach, W., Li, W. P., and Dong, X. G.: Extracting phreatic evaporation from remotely sensed maps of evapotranspiration, Water Resour. Res., 44, W08428, doi:10.1029/2007WR006063, 2008.

Butt, S., Ali, M., Fazil, M., and Latif, Z.: Seasonal variations in the isotopic composition of leaf and stem water from an arid region of Southeast Asia, Hydrolog. Sci. J., 55, 844–848, doi:10.1080/02626667.2010.487975, 2010.

Corbin, J. D., Thomsen, M. A., Dawson, T. E., and D'Antonio, C. M.: Summer water use by California coastal prairie grasses: Fog, drought, and community composition, Oecologia, 145, 511–521, doi:10.1007/s00442-005-0152-y, 2005.

Craig, H.: Standard for reporting concentrations of deuterium and oxygen-18 in natural waters, Science, 133, 1833–1834, doi:10.1126/science.133.3467.1833, 1961.

Dalton, F. N.: Plant root water extraction studies using stable isotopes, Plant Soil, 111, 217–221, 1988.

Dawson, T. E. and Ehleringer, J. R.: Streamside trees that do not use stream water, Nature, 350, 335–337, doi:10.1038/350335a0, 1991.

Dawson, T. E. and Ehleringer, J. R.: Isotopic enrichment of water in the "woody" tissues of plants: Implications for plant water source, water uptake, and other studies which use the stable isotopic composition of cellulose, Geochim. Cosmochim. Ac., 57, 3487–3492, doi:10.1016/0016-7037(93)90554-A, 1993.

Dawson, T. E. and Pate, J. S.: Seasonal water uptake and movement in root systems of Australian phraeatophytic plants of dimorphic root morphology: A stable isotope investigation, Oecologia, 107, 13–20, doi:10.1007/BF00582230, 1996.

Durand, J. L., Bariac, T., Ghesquière, M., Biron, P., Richard, P., Humphreys, M., and Zwierzykovski, Z.: Ranking of the depth of water extraction by individual grass plants, using natural ^{18}O isotope abundance, Environ. Exp. Bot., 60, 137–144, doi:10.1016/j.envexpbot.2006.09.004, 2007.

Eggemeyer, K. D., Awada, T., Harvey, F. E., Wedin, D. A., Zhou, X., and Zanner, C. W.: Seasonal changes in depth of water uptake for encroaching trees Juniperus virginiana and Pinus ponderosa and two dominant C_4 grasses in a semiarid grassland, Tree Physiol., 29, 157–169, doi:10.1093/treephys/tpn019, 2009.

Ehleringer, J. R. and Dawson, T. E.: Water uptake by plants: Perspectives from stable isotope composition, Plant Cell Environ., 15, 1073–1082, doi:10.1111/j.1365-3040.1992.tb01657.x, 1992.

Gat, J. R., Yakir, D., Goodfriend, G., Fritz, P., Trimborn, P., Lipp, J., Gev, I., Adar, E. and Waisel, Y.: Stable isotope composition of water in desert plants, Plant Soil, 298, 31–45, doi:10.1007/s11104-007-9321-6, 2007.

Gazis, C. and Feng, X.: A stable isotope study of soil water: Evidence for mixing and preferential flow paths, Geoderma, 119, 97–111, 2004.

Goebel, T. S. and Lascano, R. J.: System for high throughput water extraction from soil material for stable isotope analysis of water, Journal of Analytical Sciences, Methods and Instrumentation, 2, 203–207, doi:10.4236/jasmi.2012.24031, 2012.

Hsieh, J. C. C., Savin, S. M., Kelly, E. F., and Chadwick, O. A.: Measurement of soil-water δ^{18}O values by direct equilibration with CO_2, Geoderma, 82, 255–268, doi:10.1016/S0016-7061(97)00104-3, 1998.

Ingraham, N. L. and Shadel, C.: A comparison of the toluene distillation and vacuum/heat methods for extracting soil water for stable isotopic analysis, J. Hydrol., 140, 371–387, 1992.

Kendall, C. and McDonnell, J. J. (Eds.): Isotope tracers in catchment hydrology, First edition, Elsevier, Amsterdam, the Netherlands, 1998.

Koehler, G., Wassenaar, L. I., and Hendry, M. J.: An automated technique for measuring δD and δ^{18}O values of porewater by direct CO_2 and H_2 equilibration, Anal. Chem., 72, 5659–5664, doi:10.1021/ac000498n, 2000.

Koeniger, P., Marshall, J. D., Link, T., and Mulch, A.: An inexpensive, fast, and reliable method for vacuum extraction of soil and plant water for stable isotope analyses by mass spectrometry, Rapid Commun. Mass. Sp., 25, 3041–3048, doi:10.1002/rcm.5198, 2011.

Leen, J. B., Berman, E. S. F., Liebson, L., and Gupta, M.: Spectral contaminant identifier for off-axis integrated cavity output spectroscopy measurements of liquid water isotopes, Rev. Sci. Instrum., 83, 044305, doi:10.1063/1.4704843, 2012.

LGR: Los Gatos Research, http://www.lgrinc.com/, last access: 5 February 2013.

Lin, G. H. and Sternberg, L. da S. L.: Hydrogen isotopic fractionation by plant roots during water uptake in coastal wetland plants, in: Stable isotopes and plant carbon-water relations, edited by: Ehleringer, J. R., Hall, A. E., and Farquhar, G. D., Academic, San Diego, CA, USA, 497–510, 1993.

Lis, G., Wassenaar, L. I., and Hendry, M. J.: High-precision laser spectroscopy D/H and $^{18}O/^{16}O$ measurements of microliter natural water samples, Anal. Chem., 80, 287–293, doi:10.1021/ac701716q, 2008.

Liu, W., Liu, W., Li, P., Duan, W., and Li, H.: Dry season water uptake by two dominant canopy tree species in a tropical seasonal rainforest of Xishuangbanna, SW China, Agr. Forest Meteorol., 150, 380–388, doi:10.1016/j.agrformet.2009.12.006, 2010.

Liu, Y., Xu, Z., Duffy, R., Chen, W., An, S., Liu, S., and Liu, F.: Analyzing relationships among water uptake patterns, rootlet biomass distribution and soil water content profile in a subalpine shrubland using water isotopes, Eur. J. Soil Biol., 47, 380–386, doi:10.1016/j.ejsobi.2011.07.012, 2011.

McConville, C., Kalin, R. M., and Flood, D.: Direct equilibration of soil water for $\delta^{18}O$ analysis and its application to tracer studies, Rapid Commun. Mass Sp., 13, 1339–1345, doi:10.1002/(SICI)1097-0231(19990715)13:13<1339::AID-RCM559>3.0.CO;2-N, 1999.

Mubarak, A. and Olsen, R.: Immiscible displacement of soil solution by centrifugation, Soil Sci. Soc. Am. J., 40, 329–331, 1976.

Newman, B., Tanweer, A., and Kurttas, T.: IAEA Standard Operating Procedure for the Liquid-Water Stable Isotope Analyser: available at: http://www-naweb.iaea.org/napc/ih/documents/other/laser_procedure_rev12.PDF, last access: 10 April 2012.

Nippert, J. B. and Knapp, A. K.: Linking water uptake with rooting patterns in grassland species, Oecologia, 153, 261–272, doi:10.1007/s00442-007-0745-8, 2007.

Penna, D., Stenni, B., Šanda, M., Wrede, S., Bogaard, T. A., Gobbi, A., Borga, M., Fischer, B. M. C., Bonazza, M., and Chárová, Z.: On the reproducibility and repeatability of laser absorption spectroscopy measurements for $\delta 2H$ and $\delta 18O$ isotopic analysis, Hydrol. Earth Syst. Sci., 14, 1551–1566, doi:10.5194/hess-14-1551-2010, 2010.

Peters, L. I. and Yakir, D.: A direct and rapid leaf water extraction method for isotopic analysis, Rapid Commun. Mass Sp., 22, 2929–2936, doi:10.1002/rcm.3692, 2008.

Phillips, D. L. and Gregg, J. W.: Source partitioning using stable isotopes: coping with too many sources, Oecologia, 136, 261–269, doi:10.1007/s00442-003-1218-3, 2003.

Revesz, K. and Woods, P. H.: A method to extract soil water for stable isotope analysis, J. Hydrol., 115, 397–406, 1990.

Rossatto, D. R., Ramos Silva, L. de C., Villalobos-Vega, R., Sternberg, L. da S. L., and Franco, A. C.: Depth of water uptake in woody plants relates to groundwater level and vegetation structure along a topographic gradient in a neotropical savanna, Environ. Exp. Bot., 77, 259–266, doi:10.1016/j.envexpbot.2011.11.025, 2012.

Rossatto, D. R., Sternberg, L. da S. L., and Franco, A. C.: The partitioning of water uptake between growth forms in a Neotropical savanna: Do herbs exploit a third water source niche?, Plant Biol., 15, 84–92, doi:10.1111/j.1438-8677.2012.00618.x, 2013.

Rothfuss, Y., Biron, P., Braud, I., Canale, L., Durand, J.-L., Gaudet, J.-P., Richard, P., Vauclin, M., and Bariac, T.: Partitioning evapotranspiration fluxes into soil evaporation and plant transpiration using water stable isotopes under controlled conditions, Hydrol. Process., 24, 3177–3194, doi:10.1002/hyp.7743, 2010.

Rothfuss, Y., Braud, I., Le Moine, N., Biron, P., Durand, J.-L., Vauclin, M., and Bariac, T.: Factors controlling the isotopic partitioning between soil evaporation and plant transpiration: Assessment using a multi-objective calibration of SiSPAT-Isotope under controlled conditions, J. Hydrol., 442–443, 75–88, doi:10.1016/j.jhydrol.2012.03.041, 2012.

Sala, O. E., Jackson, R. B., Mooney, H. A., and Howarth, R. W. (Eds.): Methods in Ecosystem Science, Springer, New York, USA, 2000.

Schultz, N. M., Griffis, T. J., Lee, X., and Baker, J. M.: Identification and correction of spectral contamination in $^2H/^1H$ and $^{18}O/^{16}O$ measured in leaf, stem, and soil water, Rapid Commun. Mass Sp., 25, 3360–3368, doi:10.1002/rcm.5236, 2011.

Scrimgeour, C. M.: Measurement of plant and soil water isotope composition by direct equilibration methods, J. Hydrol., 172, 261–274, doi:10.1016/0022-1694(95)02716-3, 1995.

Song, X., Wang, S., Xiao, G., Wang, Z., Liu, X., and Wang, P.: A study of soil water movement combining soil water potential with stable isotopes at two sites of shallow groundwater areas in the North China Plain, Hydrol. Process., 23, 1376–1388, doi:10.1002/hyp.7267, 2009.

Stratton, L. C., Goldstein, G., and Meinzer, F. C.: Temporal and spatial partitioning of water resources among eight woody species in a Hawaiian dry forest, Oecologia, 124, 309–317, doi:10.1007/s004420000384, 2000.

Takahashi, K.: Oxygen isotope ratios between soil water and stem water of trees in pot experiments, Ecol. Res., 13, 1–5, doi:10.1046/j.1440-1703.1998.00240.x, 1998.

Thorburn, P. J. and Ehleringer, J. R.: Root water uptake of field-growing plants indicated by measurements of natural-abundance deuterium, Plant Soil, 177, 225–233, 1995.

Thorburn, P. J., Walker, G. R., and Brunel, J.-P: Extraction of water from Eucalyptus trees for analysis of deuterium and oxygen-18: laboratory and field techniques, Plant Cell Environ., 16, 269–277, doi:10.1111/j.1365-3040.1993.tb00869.x, 1993.

Unkovich, M., Pate, J., McNeill, A., and Gibbs, J.: Stable isotope techniques in the study of biological processes and functioning of ecosystems, Kluwer Academic Publishers, Dordrecht, the Netherlands, 2001.

Vendramini, P. F. and Sternberg, L. da S. L.: A faster plant stem-water extraction method, Rapid Commun. Mass Sp., 21, 164–168, doi:10.1002/rcm.2826, 2007.

Walker, C. D. and Richardson, S. B.: The use of stable isotopes of water in characterising the source of water in vegetation, Chem. Geol.: Isotope Geoscience section, 94, 145–158, 1991.

Walker, G. R., Woods, P. H., and Allison, G. B.: Interlaboratory comparison of methods to determine the stable isotope composition of soil water, Chem. Geol., 111, 297–306, 1994.

Walter, Z.-L. and Morio, I.: Deep root water uptake ability and water use efficiency of pearl millet in comparison to other millet species, Plant Prod. Sci., 8, 454–460, 2005.

Wang, P., Song, X., Han, D., Zhang, Y., and Liu, X.: A study of root water uptake of crops indicated by hydrogen and oxygen stable isotopes: A case in Shanxi Province, China, Agr. Water Manage., 97, 475–482, 2010.

Wang, P., Song, X., Han, D., Zhang, Y., and Zhang, B.: Determination of evaporation, transpiration and deep percolation of summer corn and winter wheat after irrigation, Agr. Water Manage., 105, 32–37, doi:10.1016/j.agwat.2011.12.024, 2012.

Wang, X. F. and Yakir, D.: Using stable isotopes of water in evapotranspiration studies, Hydrol. Process., 14, 1407–1421, 2000.

Wassenaar, L. I., Hendry, M. J., Chostner, V. L., and Lis, G. P.: High resolution pore water δ^2H and δ^{18}O measurements by H$_2$O (liquid)-H$_2$O (vapor) equilibration laser spectroscopy, Environ. Sci. Technol., 42, 9262–9267, 2008.

Wershaw, R. L., Friedman, I., Heller, S. J., and Frank, P. A.: Hydrogen isotopic fractionation of water passing through trees, in: Advances in organic geochemistry, International series of monographs on earth sciences, edited by: Hobson, G. D., Speers, G. C., and Inderson, D. E., 32, Pergamon Press, New York, USA, 55–67, 1966.

West, A. G., Patrickson, S. J., and Ehleringer, J. R.: Water extraction times for plant and soil materials used in stable isotope analysis, Rapid Commun. Mass Sp., 20, 1317–1321, doi:10.1002/rcm.2456, 2006.

West, A. G., Goldsmith, G. R., Brooks, P. D., and Dawson, T. E.: Discrepancies between isotope ratio infrared spectroscopy and isotope ratio mass spectrometry for the stable isotope analysis of plant and soil waters, Rapid Commun. Mass Sp., 24, 1948–1954, doi:10.1002/rcm.4597, 2010.

White, J. W. C., Cook, E. R., Lawrence, J. R., and Wallace S. B.: The D/H ratios of sap in trees: Implications for water sources and tree ring D/H ratios, Geochim. Cosmochim. Ac., 49, 237–246, doi:10.1016/0016-7037(85)90207-8, 1985.

Williams, D. G. and Ehleringer, J. R.: Intra- and interspecific variation for summer precipitation use in pinyon-juniper woodlands, Ecol. Monogr., 70, 517–537, doi:10.2307/2657185, 2000.

Yang, H., Auerswald, K., Bai, Y., and Han, X.: Complementarity in water sources among dominant species in typical steppe ecosystems of Inner Mongolia, China, Plant Soil, 340, 303–313, doi:10.1007/s11104-010-0307-4, 2011.

Zegada-Lizarazu, W. and Iijima, M.: Hydrogen stable isotope analysis of water acquisition ability of deep roots and hydraulic lift in sixteen food crop species, Plant. Prod. Sci., 7, 427–434, doi:10.1626/pps.7.427, 2004.

8

Early forest fire detection using low-energy hydrogen sensors

K. Nörthemann[1], J.-E. Bienge[2], J. Müller[2], and W. Moritz[1]

[1]Humboldt-Universität zu Berlin, Brook-Taylor-Str. 2, 12489 Berlin, Germany
[2]Johann Heinrich von Thünen-Institut, Alfred-Möller-Straße 1, 16225 Eberswalde, Germany

Correspondence to: K. Nörthemann (kai.noerthemann@hu-berlin.de)

Abstract. Most huge forest fires start in partial combustion. In the beginning of a smouldering fire, emission of hydrogen in low concentration occurs. Therefore, hydrogen can be used to detect forest fires before open flames are visible and high temperatures are generated. We have developed a hydrogen sensor comprising of a metal/solid electrolyte/insulator/semiconductor (MEIS) structure which allows an economical production. Due to the low energy consumption, an autarkic working unit in the forest was established. In this contribution, first experiments are shown demonstrating the possibility to detect forest fires at a very early stage using the hydrogen sensor.

1 Introduction

Forest fires in today's commercial and cultural landscapes cause large problems and damage. The risk of large forest fires has reduced in the last decades on the basis of technological progress and scientific knowledge. However, great fires appeared in Russia in 2010 and California, United States, in 2009 with massive financial loss and even death (Hirschberger, 2011). In such huge fires the forces of nature are recalcitrant. To avoid this damage a detection of the fire is required when the involved area is still small, and the fire process is still in the early stages. With changing climatic conditions, the potential risk for forest fires will increase in some regions. Since these currently unforeseen events can occur all over the earth, it is necessary to improve a system to warn of the early stages of forest fires.

Present forest fire detection systems are based on the detection of effects, which appear when an open fire already exists. Most of the systems determine heat, open flames or dust particles. Not so common is the detection of gases which occur by smouldering of organic matter before an open fire accrues. The occurrence of hydrogen during smouldering fire was published in the literature (Cofer III et al., 1988; Jackson and Robins, 1994; Grosshandler, 1997; Krause et al., 2006). Amamoto et al. (1990) pointed out that during the burning of

wooden building hydrogen levels are raised and that hydrogen is the first detectable event during such an experiment. The concentration of hydrogen created during smouldering is about 20 ppm, which was measured in a shielded environment. At the point when the smouldering forest fire should be detected, the expected concentration is even lower. To detect such an occurrence of hydrogen, the sensor must be reliable in measuring such small concentrations. Because of this we use a detector with a very low detection limit; doing so it is possible to measure hydrogen concentration below 5 ppm. Hydrogen occurs in the troposphere in a concentration of about 0.4–0.6 ppm. The yearly average variate for different locations on the planet and the fluctuation over the year is about 0.1 ppm (Yver et al., 2011). Subsequently, there is no constant hydrogen concentration; however, the change within the natural hydrogen amount is small and will not influence the fire detection using a hydrogen sensor. Carbon dioxide and methane, which are also present in the atmosphere, have no influence on the sensor signal. The changes of other gases like NO_2 (0.6–$31\,\mu g\,m^{-3}$), SO_2 (0.2–$15\,\mu g\,m^{-3}$), NH_3 (0.2–$28\,\mu g\,m^{-3}$) and O_3 (1.2–$166\,ppb$) are also negligible. These values were measured between 2009 and 2011 at different German forest sites (Fischer, 2013, oral note).

One of the latest comparisons of hydrogen sensors by Boon-Brett et al. (2010) reported the limits of commercial

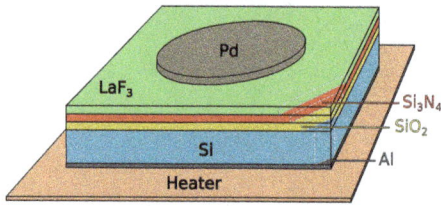

Figure 1. Sketch of the capacitive sensor structure mounted on a heater.

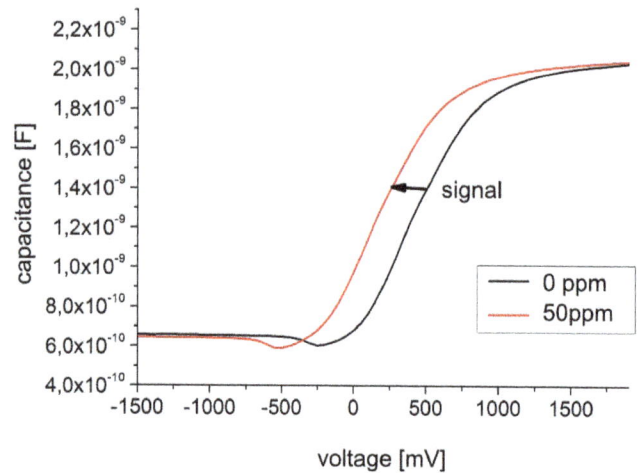

Figure 2. Changes in the capacitance voltage plot due to hydrogen uptake. When hydrogen occurs at the palladium surface, the graph is shifted to lower potential.

hydrogen sensor detection. None of the reported sensors was able to detect concentration below 10 ppm. Because of the low hydrogen concentration induced by smouldering fires, it was necessary to develop a much more sensitive sensor. To gain very sensitive measurements of hydrogen, creating an innovative concept and design of the hydrogen sensor was unavoidable. We developed a low-energy hydrogen sensor set-up for outdoor application. Here we present first results of field experiments in Scots pine stands. Using wood and grass, a smouldering fire was created in the forest. The sensors were placed at different distances from the fire location to monitor the signal which occurs due to a started fire.

2 Sensor

The sensor we developed is a metal/solid electrolyte/insulator/semiconductor (MEIS) structure. An explicit sketch is shown in Fig. 1. Silicon oxide and silicon nitride insulators are grown on a silicon wafer. The thickness of this insulator is 150 nm. Afterwards a 150 nm lanthanum trifluoride layer is grown by physical vapour deposition at high vacuum. The 20 nm palladium layer is produced by DC sputtering. This layer operates as gate metal. The structure is different to the Lundström sensor (Lundström et al., 1975) because of an additional superionic conductor between the insulator and the gate layer. Due to this additional layer, the dependence of the electrical signal on hydrogen concentration is changed. In contrast to a square root dependency of the Lundström sensor, our sensor shows a logarithmic concentration response. Details about the solid electrolyte layer can be found in Moritz and Krause (2004).

The final sensor structure used in our experiments was $Pd/LaF_3/Si_3N_4/SiO_2/Si$. This sensor chip is a capacitive element, and the capacitance depends on the voltage between gate (palladium) and bulk (silicon, ohmic contact). The capacitance was measured between the backside aluminium contact and the gate. This structure was bonded on top of a ceramic heater which is used for the activation of the sensor. The sensors were activated once a day to improve the response time and the signal. Details and effects of the activation process can be found in Linke et al. (2012). During operation the sensor does not need to be heated. The measurements were done at ambient temperature. Due to this

fact, the power consumption of the electronics is low, and operation with batteries is possible.

When hydrogen reacts with the sensor structure, the capacitance/voltage behaviour changes, which is shown in Fig. 2. Hydrogen molecules dissociate at the palladium surface. And hydrogen atoms are solved in the palladium and interact with the lanthanum trifluoride layer. Via this process, additional charges occur, and the chemical potential is modified. The result is a shift of the capacitance voltage plot to lower potential in comparison to absence of hydrogen. To determine the hydrogen concentration, only the voltage shift has to be measured. The voltage shift of this sensor shows a logarithmic dependency to the hydrogen concentration (Moritz et al., 2006). Due to this relation, it is possible to measure hydrogen in a very large concentration range. Concentrations as high as 10 % (100 000 ppm) and very low concentrations (< 5 ppm) can be measured. To detect tiny amounts of the tracing gas, this is necessary for an alarm system in forests where dilution is a major effect. Details of the response to hydrogen with these sensor systems were described by Moritz et al. (2006).

3 Experiments

For the field experiment an electronic design was developed which measures the capacitance of the sensor and adjusts the applied potential in feedback mode. These electronics measure all data autonomously and are able to store the collected data. With a commercial XBee RF (radio frequency) module, a wireless communication was implemented so that it was able to control all sensors and collect all data via a computer. Because of this wireless communication, it is practicable to arrange the sensors at different positions in a large area in the

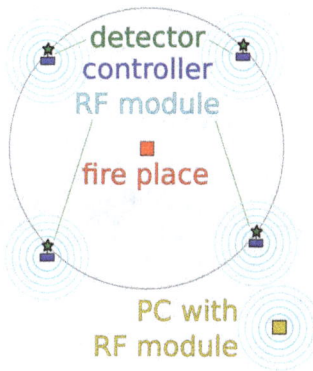

Figure 3. Experimental set-up in the forest. The sensors were placed around the fire place at different trees. The experiments were done with changing height and distance of the sensors. The communication with the sensors was done with an RF module.

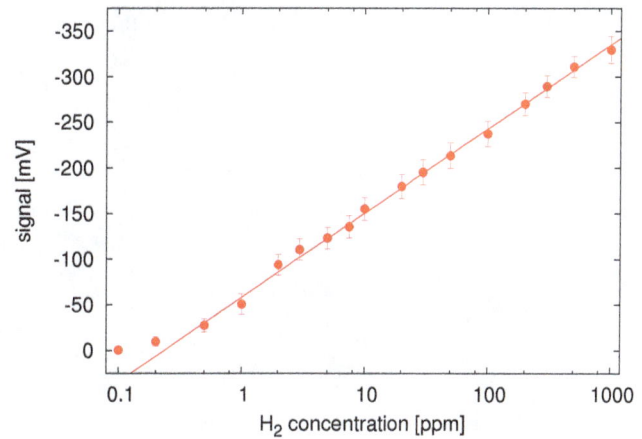

Figure 4. Sensor signal as function of the hydrogen concentration in the range from 0.1 to 1000 ppm. Shown is the signal average over 16 different sensors with standard deviation. The sensor has a logarithmic dependency between the hydrogen concentration and sensor signal. This relation allows the measurement of very low hydrogen concentrations.

forest. The experimental environment was established inside of a mixed forest stand. For safety reasons the smouldering fire was started in a metal bowl. For this first proof of concept, small fires including an area below 3 m^2 were burned. Around the fire place, the sensors were mounted at different trees. Details of the experimental set-up are illustrated in Fig. 3. For the different experiments, the heights of the sensors varied between 1 and 4 m. Also the distances between fire and sensors were changed in different measurements.

Figure 4 shows the sensor signal for different hydrogen concentrations. This graph illustrates the logarithmic dependency between the hydrogen concentration and the resulting sensor signal from below 1 to 1000 ppm. Due to this logarithmic relation, it is possible to measure very low hydrogen concentration with our sensor. With this signal–concentration relation, even tiny hydrogen occurrence results in a detectable signal response. So our sensor is able to trace tiny quantities of hydrogen with concentrations below 5 ppm, which is beneath the detection limits that Boon-Brett et al. (2010) reported in an overview of commercial hydrogen sensors. Thus a small lower detection limit is necessary for this application because hydrogen only occurs in slight concentrations during a smouldering fire and is additionally diluted in the forest. The detection of concentration up to 10 vol. % and the sensor stability are published elsewhere (Moritz et al., 2006; Lang et al., 2013).

4 Results

To investigate the thinning of the hydrogen in a smouldering fire, an experiment was carried out in a hall with a size of about 25 m × 12 m × 7 m. Nine small beech bricks were heated with a heating plate until fumes were formed. This set-up is related to the test fire TF2 described in DIN EN 54 (DIN, 2000). But all experimental details of the norm were not fulfilled. Only a set-up was used in which surely hydrogen is formed in a smouldering fire. At a distance of 3.5 m

from the fire place, sensors were mounted at different heights in this closed room. For details of the set-up, see Fig. 5a. The data of this test fire are shown in Fig. 5b. In all three sensors, signal differences due to the occurrence of the fire are detectable. The signal levels in this experiment of about 200 mV are in a range where no complicated signal processing is necessary. On the measured signal in Fig. 5b, some fluctuations are visible. Such fluctuations were also visible in the smoke of the smouldering fire. The first change is visible about 3 min after the heating plate was switched on. This time includes the heating of the heating plate, with a resistance heater and the following heating of the beech bricks. The response time strongly depends on the parameter of the heating plate and is not discussed here in detail. With this experiment we only prove that with the selected hydrogen sensor it is possible to detect a smouldering fire.

Forests with open canopies do not have limits or boundaries, resulting in hydrogen being diluted and widely spread in the air around. The treetop is only a tiny limit and will not prevent the rise of the hydrogen. Due to this, the size of the fire was increased to up to 2 m^2. With this the amount of hydrogen is increased, and the thinning starts at a higher concentration. Due to the fact that the fire was started with grass and moistened wood, also a lot of smoke was created. By watching this visible fire, an estimation can be predicted where the hydrogen should appear.

In a field experiment the wind direction has an influence, being inhomogeneous inside a forest. Major influence factors on the wind directions are stand structure, stand density, tree height and age. Trees produce major turbulence inside the forest. Figure 6 shows the different wind profiles of two positions inside the stand. Even if those measurement spots had only a distance of 200 m, both plots would still be

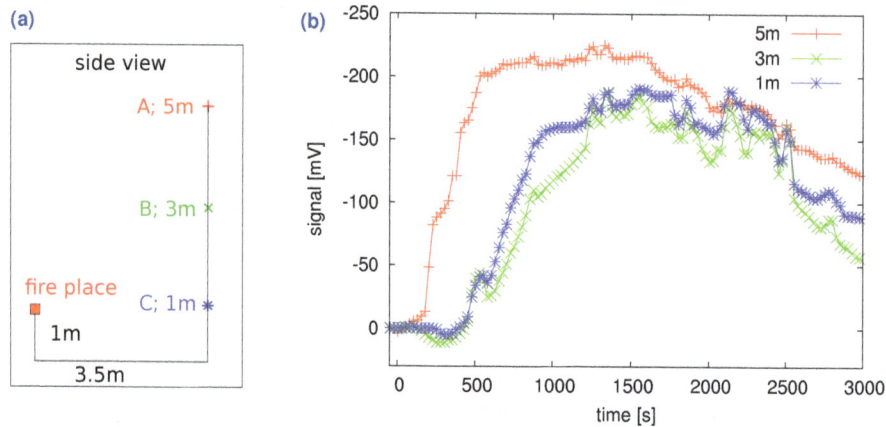

Figure 5. Indoor experiment in a hall with a size of $25 \times 12 \times 7\,\text{m}^3$. On a heating plate, wood was heated according to EN54; at three different positions hydrogen sensors were mounted, as illustrated in **(a)**. The measured signals are shown in **(b)**. All sensors detect a change due to the smouldering fire.

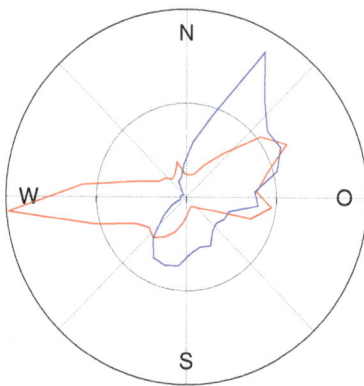

Figure 6. Frequency of the wind direction in the forest at two different positions. Both data sets were collected over one year. The measurement points had a lateral distance of 200 m. The difference is caused by the unequal structure of both standings (red: open structure; blue: closed structure).

completely different. Because of this it is not possible to get a uniform wind direction over the complete measurement area. The wind is also dependent on time. Due to this complexity, there are no indicators of the wind direction in the following plots.

The initial forest experiment was done on a small scale. A smouldering fire of approximately $0.5\,\text{m}^2$ was burned at a height of 50 cm above the ground. Eight sensors were positioned at a height of 1 m and 3 m in four different directions. The lateral distance between fire and sensors was 5 m. The arrangement is shown in Fig. 7a. In Fig. 7b the corresponding sensor signals are displayed. Six sensors show a signal due to the smouldering fire. The absence of a reaction at the sensors C and E is caused by the wind direction. It is evident that the height of the sensor has an influence on the signal. Due to the wind and crown at this experiment, the major effect was measured at a height of 1 m.

When the distance between sensor and fire place was enhanced, the sensors were mounted higher than 1 m, because the visible smoke of the fire was approximately at a height of 3 m and above. Mounting the sensor higher than 5 m was much more complicated, and inside of the treetop the RF communication would be inhibited.

An expanded experiment was done with the same fire size but a larger distance of the sensors to the fire. The sensors were placed in a circle with a radius of about 25 m, and the fire was located in the centre. The sensors were mounted to different trees at a height of about 3 to 4 m. The set-up is shown in Fig. 8. In this figure the sensor signals are also displayed. In the data of sensors B, C and H, a clear signal difference is visible. This change is caused by the hydrogen which occurs during the smouldering fire. This shows that the fire was detected by these three sensors. The difference of about 24 mV is slightly above the detection limit. This small signal is caused by the small fire of about $0.5\,\text{m}^2$. In this plot also a background fluctuation of about 4 mV is visible. This signal variation does not exist in the laboratory environment, when the sensors are measured in a known constant hydrogen concentration. In such a set-up the noise is below 0.2 mV. The reason could be variations of the hydrogen concentration in the forest; however, the shielding of the sensor can also cause this effect. In this first experiment the sensors were only covert by a metal plate to prevent direct light incidence. Reflections were only minimised by painting the housing and plate with black paint. The optimisation of the sensor shielding is now in progress to reduce the outside effects, and to be sure that only hydrogen can pass to the sensor and reduce the influence factors.

To prove the usability of the sensor in a realistic environment, enhanced distances were also measured. In Fig. 9a the set-up is described where the distance between fire and sensor was approximately 110 m. To measure hydrogen in this huge dilution due to the enormous volume, the size of the

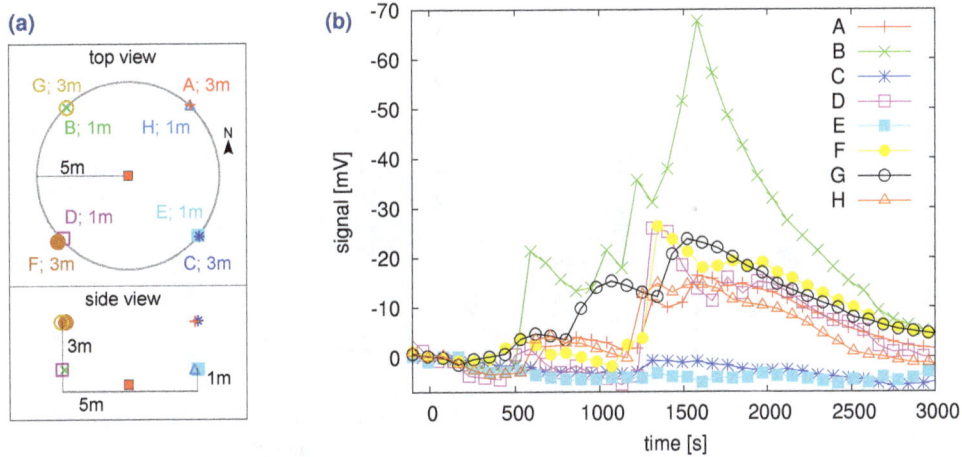

Figure 7. Experiment in the forest with distance between fire and sensors of 5 m. The sensors were placed in four directions at a height of 1 m and 3 m as illustrated in part **(a)**. In part **(b)** the corresponding sensor signals are plotted. The dependency of the wind direction is visible. Also the influence of the detector height can be observed. In this configuration the main signal was observed at a height of 1 m.

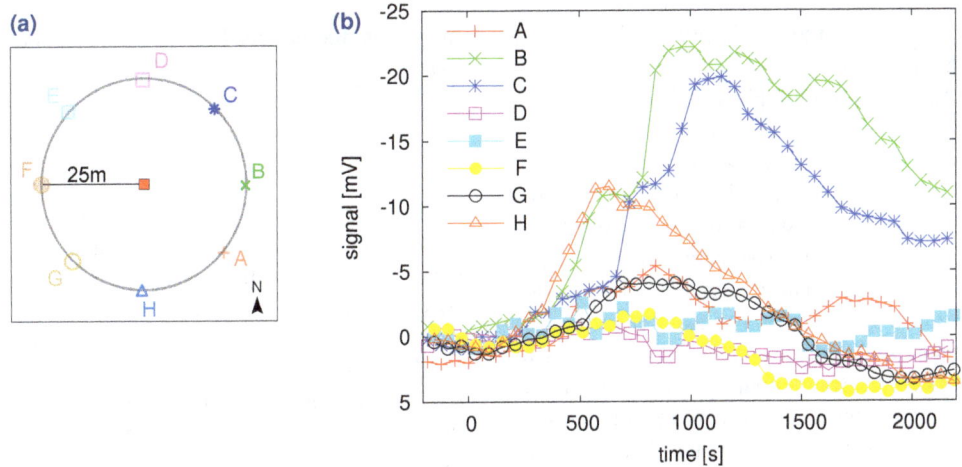

Figure 8. Experiment in the forest with a distance between fire and sensors of 25 m. The sensors were mounted at a height of 3 to 4 m. The position of the sensors is shown in part **(a)**. The measured sensor signal is displayed in part **(b)**. This graph illustrates very clearly a signal difference due to the fire at the sensors B, C and H.

fire was increased to about 2 m², which is still small at the scale of forest fires. The corresponding signal is displayed in Fig. 9b. In this experiment the sensors B, C and F show a significant change in the signal. For these sensors the distance to the fire was 25 and 105 m. Even at sensor H, with a distance of about 115 m, a small influence of the fire appeared.

5 Discussion and outlook

The measurements in the forest (see Figs. 7, 8 and 9) reveal that not all placed sensors show a signal difference after the fire was started. This is caused by the fact that the hydrogen is transported by the wind from the fire to the sensor. The wind direction therefore has an important influence where this propagation occurs. In Fig. 7 the sensors C and E show

no relevant change after the fire was started. This is caused due to the fact that the wind was blowing away from where the sensors were located to the fire place. All other sensors in this set-up detect the hydrogen from the small fire. It is because the distance of 5 m to the fire is still small. This short distance and open environment has also the effect that the most highly placed sensors do not show a major signal difference. The wind has a primary influence on the measured signal. The time between starting time of the smouldering fire and the first occurrence of a signal change was from 5 min to 30 min. When this response time is measured, the convection time in the forest has to be accounted for. With this experiment the occurrence of visible smoke at the sensor and measured hydrogen signal were the same. To equate the occurrence of hydrogen with the existence of smoke is only a

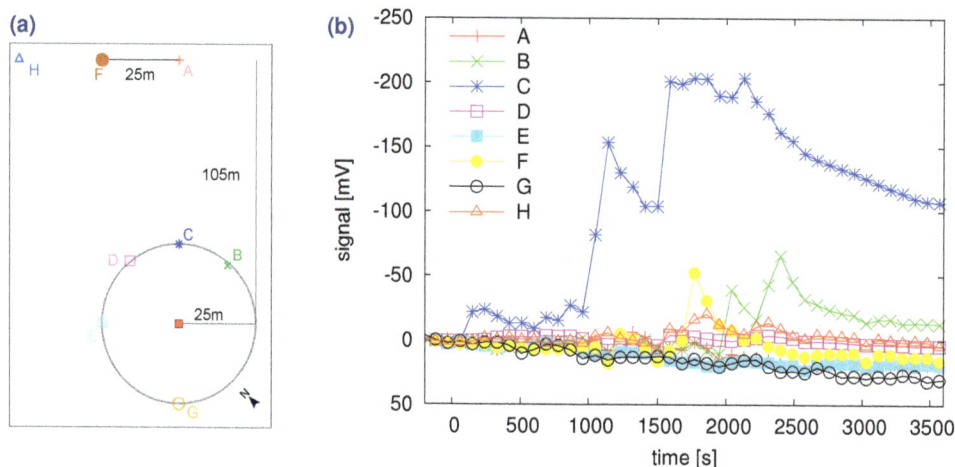

Figure 9. Experiment in the forest with a sensor distance between 25 and 115 m. In part **(a)** the experimental set-up is pictured. The sensor signals are displayed in part **(b)**. Three sensors show a significant difference due to the fire (C, B and F). In the signal of sensor H, a small change is visible.

rough guess. But it illustrates the possibility in which speed fire products can be detected.

A distance of 110 m between fire place and sensor (Fig. 9) is a large expanse; the signals of sensor F still show significant fluctuations which are also visible at other, more closely situated sensors. This shows that monitoring of forest fires with the presented sensors is possible. For the coverage of huge forest areas, the distance has to be increased where a fire is still detectable. A rough estimation with a sensor to sensor distance of 100 m gives a density of 1 sensor per ha. In this case 1 million sensors are needed to cover all the forests in the German federal state of Brandenburg. To reduce the number of necessary sensors, the distance between the sensors has to be increased. Further experiments are planned where distances between fire and sensors are increased. When the sensors are used outside of a controlled sand box, an advanced data analysis has to be developed to correlate the signals of different sensors. An autonomous process has to be found which decides independently to trigger a fire alarm. The measurements have shown that with increasing wind speed it becomes easier to detect hydrogen at distant positions. High wind speed is one of the factors favouring forest fires.

In this publication the sensors use a ceramic heater with a size of 7×7 mm^2. This geometry is not the optimal set-up. One of the topics which we are now working on is to deposit a heater directly around the palladium dot displayed in Fig. 1. Due to this the energy which is consumed during heating can be reduced by a factor more than 1000. Another part which can be optimised is the RF communication. For this first experiment we included an RF module where it is sure that the communication will work. Further investigation can reduce the additional energy consumption. By doing all these opti-

misations, the battery lifetime can be enhanced from weeks to several years.

6 Summary

In these first experiments we have shown that hydrogen is formed during smouldering fire, and this hydrogen is detectable outside in the forest with the newly developed sensors. The shielding of the detectors has to be optimised to avoid side effects which also change the signal. The response time was below 5 min from the time the fire was started. This time includes the convection of the hydrogen from the fire to the sensor over a distance up to 25 m. A significant change in the signal due to a fire was measured at a distance of about 110 m.

Acknowledgements. The authors would like to thank Michael Dallmer, Michael Milstrey and Michael Rothe for their support during the measurements. This work was financially supported by Zentrales Innovationsprogramm Mittelstand (ZIM) in the project InPriWa.

Edited by: M. Penza
Reviewed by: two anonymous referees

References

Amamoto, T., Tanaka, K., Takahata, K., Matsuura, S., and Seiyama, T.: A fire detection experiment in a wooden house by SnO$_2$ semiconductor gas sensors, Sensor. Actuat. B-Chem., 1, 226–230, doi:10.1016/0925-4005(90)80206-F, 1990.

Boon-Brett, L., Bousek, J., Black, G., Moretto, P., Castello, P., Hübert, T., and Banach, U.: Identifying performance gaps in hydrogen safety sensor technology for automotive and stationary applications, Int. J. Hydrogen Energ., 35, 373–384, doi:10.1016/j.ijhydene.2009.10.064, 2010.

Cofer III, W. R., Levine, J. S., Riggan, P. J., Sebacher, D. I., Winstead, E. L., Shaw Jr., E. F., Brass, J. A., and Ambrosia, V. G.: Trace Gas Emissions From a Mid-Latitude Prescribed Chaparral Fire, J. Geophys. Res., 93, 1653–1658, doi:10.1029/JD093iD02p01653, 1988.

DIN: EN 54-7, 2000.

Fischer, U.: Database of the Programme Co-ordinating Centre (PCC) of ICP Forests, 2013.

Grosshandler, W.: Towards the development of a universal fire emulator-detector evaluator, Fire Safety J., 29, 113–127, doi:10.1016/S0379-7112(96)00031-8, 1997.

Hirschberger, P.: Wälder in Flammen, http://www.wwf.de/fileadmin/fm-wwf/Publikationen-PDF/110727_WWF_Waldbrandstudie.pdf, 2011.

Jackson, M. and Robins, I.: Gas sensing for fire detection: Measurements of CO, CO_2, H_2, O_2, and smoke density in European standard fire tests, Fire Safety J., 22, 181–205, doi:10.1016/0379-7112(94)90072-8, 1994.

Krause, U., Schmidt, M., and Lohrer, C.: A numerical model to simulate smouldering fires in bulk materials and dust deposits, J. Loss Prevent. Proc., 19, 218–226, doi:10.1016/j.jlp.2005.03.005, 2006.

Lang, M., Banach, U., Nörthemann, K., Gerlitzke, A.-K., Milstrey, M., Kaufer, R., Woratz, M., Hübert, T., and Moritz, W.: Long-term stability of a MEIS low energy hydrogen sensor, Sensor. Actuat. B-Chem., 187, 395–400, doi:10.1016/j.snb.2012.12.081, 2013.

Linke, S., Dallmer, M., Werner, R., and Moritz, W.: Low energy hydrogen sensor, Int. J. Hydrogen Energ., 37, 17523–17528, doi:10.1016/j.ijhydene.2012.07.072, 2012.

Lundström, K. I., Shivaraman, M. S., and Svensson, C. M.: A hydrogen-sensitive Pd-gate MOS transistor, J. Appl. Phys., 49, 3876, doi:10.1063/1.322185, 1975.

Moritz, W. and Krause, S.: Solid state chemical sensors using LaF_3 thin layer structures, Recent Res. Devel. Solid State Ionics, 2, 243–279, 2004.

Moritz, W., Fillipov, V., Vasiliev, A., Cherkashinin, G., and Szeponik, J.: A Field Effect Based Hydrogen Sensor for Low and High Concentrations, ECS Transactions, 3, 223–230, doi:10.1149/1.2357262, 2006.

Yver, C. E., Pison, I. C., Fortems-Cheiney, A., Schmidt, M., Chevallier, F., Ramonet, M., Jordan, A., Søvde, O. A., Engel, A., Fisher, R. E., Lowry, D., Nisbet, E. G., Levin, I., Hammer, S., Necki, J., Bartyzel, J., Reimann, S., Vollmer, M. K., Steinbacher, M., Aalto, T., Maione, M., Arduini, J., O'Doherty, S., Grant, A., Sturges, W. T., Forster, G. L., Lunder, C. R., Privalov, V., Paramonova, N., Werner, A., and Bousquet, P.: A new estimation of the recent tropospheric molecular hydrogen budget using atmospheric observations and variational inversion, Atmos. Chem. Phys., 11, 3375–3392, doi:10.5194/acp-11-3375-2011, 2011.

9

Precise temperature calibration for laser heat treatment

M. Seifert[1], **K. Anhalt**[2], **C. Baltruschat**[2], **S. Bonss**[1], and **B. Brenner**[1]

[1]Fraunhofer IWS, Dresden, Germany
[2]Physikalisch-Technische Bundesanstalt, Berlin, Germany

Correspondence to: M. Seifert (marko.seifert@iws.fraunhofer.de)

Abstract. A new induction-heated fixed-point device was developed for calibration of temperature measurement devices typically used in laser heat treatment for the temperature range 1000–1500 °C. To define the requirements for the calibration method, selected measurement setups were compared as well as process data and results of industrial processes were analyzed. Computer simulation with finite element method (FEM) and finite difference method (FDM) was used to optimize the system components and processing parameters of the induction heating of fixed-point cells. The prototype of the fixed-point device was tested successfully, and the first measuring results are presented here. The new calibration method is expected to improve the quality and reproducibility of industrial heat treatment processes with temperature control.

1 Introduction

The technology of laser surface heat treatment for steel or cast iron parts with high-power diode lasers was developed in the late 1990s and has been established in industrial mass production for more than 10 yr (see, e.g., Bonss et al., 2003, 2005, 2009). Because of the high power density of the laser heat sources and the high processing temperatures often desired to be close to the melting temperature of the materials, precise temperature measurement and control is essential for keeping the process stable and ensuring reproducible quality of the parts. For the successful heat treatment of special high-alloyed steel grades or gray cast iron surfaces, the process temperature needs to be controlled within a band of only a few kelvin. Temperatures too high can cause heavy melting on the surface of the parts and temperatures too low result in lower surface hardness and lower heat penetration depths.

For the measurement of temperature, radiation thermometers and thermal imaging systems are used. One key problem is the nonlinear signal damping by the laser optics in the case of the co-axial view of the measurement device. Very critical is the pollution and wear of optical components during their lifetime, i.e., caused by processing fumes, and the connected drift of the measuring signal damping.

Therefore a fast and precise calibration method is required. Conventional blackbody or fixed-point devices for high tem-

peratures of up to 1500 °C are mostly too large and too slow for this task, not flexible enough and difficult to transport. On-site calibration is needed because the temperature measurement devices are mechanically and electrically integrated into very complex and large machine systems.

To close the gap between existing calibration methods (Hollandt et al., 2003; Anhalt, 2008) and the demands from the industry, a mobile induction-heated fixed-point device for fast calibration in the temperature range 1000–1500 °C was developed within the Euramet EMRP project HiTeMS (EMRP, 2011–2014). As a first step, different variants of temperature measurement configurations, typically used in industrial laser heat treatment, were investigated to define the requirements for the device. The geometry and arrangement of system components and the processing parameters of the induction-heated fixed-point device were optimized by a computer simulation with FEM and FDM.

Based on the results, a prototype of an induction-heated fixed-point device was developed and tested.

2 Temperature measurement in laser heat treatment

The temperature measurement device can be used lateral or co-axial to the laser optics (Fig. 1). When the pyrometer is mounted lateral to the laser optics, it is facing in the direction of the laser interaction zone at the part's surface at an angle to

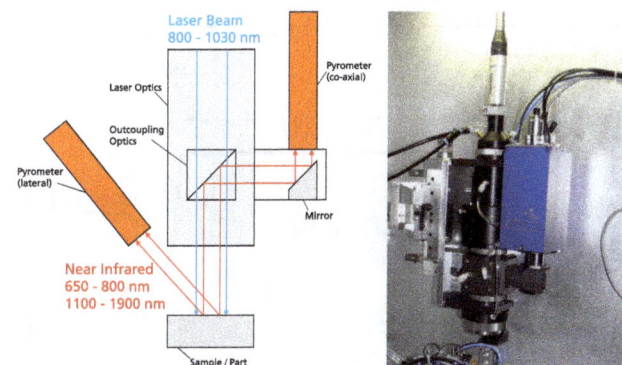

Figure 1. Left: schematic diagram – variants of assembling a temperature measurement device for laser heat treatment. Right: example of the thermal imaging system E-MAqS mounted on special ring optics for co-axial temperature measurement in industrial mass production of valves.

Table 1. Influence of contamination of pyrometer optics for temperature measurement with pyrometer MAURER KTR 1075 (manufacturer calibration) at PTB Berlin. The contaminated front lens has been used for several years for laser processing in a laterally assembled pyrometer viewing unprotected at the laser–metal interaction zone.

$T_{\mathrm{blackbody}}/°C$	$T_{\mathrm{clean}} - T_{\mathrm{contaminated}}/K$
1004	18
1113	18
1222	19
1300	26
1409	32

the laser beam axis. The measuring spot of the pyrometer has to be carefully adjusted to the position of maximum temperature. The measuring signal can get lost if edges of the part shadow the pyrometer measuring path, e.g., if heat treating the ground of a groove or a bore hole. In spite of the known disadvantages this variant is used for industrial mass production, the efforts to generate reproducible heat treatment results are high.

One main advantage of direct measurement is the fact that the measurement is not influenced by the laser optics (i.e., by signal damping) and that the calibration values of the manufacturer can be used. Nevertheless, contamination of the pyrometer optics can cause significant measuring errors (cf. Table 1) and calibration and adjustment is needed regularly.

Because of the known problems with a lateral pyrometer setup and measurement, it is preferable today to measure co-axially through the laser optics. For this task special optical components are used that let the laser radiation pass and reflect the near-infrared (NIR) measuring wavelength towards the pyrometer (Fig. 1). If the outcoupling plate is at the position of the collimated laser beam, the pyrometer is focused by the laser lenses and the shape and size of the measuring spot are changed. Using a pyrometer co-axially at the laser optics requires a correction of the characteristic curve of the pyrometer or even a completely new calibration in the mounted state. An example of co-axial temperature measurement in industrial mass production is shown in Fig. 1.

The investigation of typical measurement setups for laser heat treatment has shown that the best wavelength range for temperature measurement between 1000 and 1500 °C is 650 to 2100 nm. The wavelength band of the laser heat sources (800–1070 nm) is not available for a temperature measurement, because the laser radiation overlaps the NIR temperature radiation of the glowing object. To avoid measurement errors, special blocking filters with a high optical density at

the discrete laser wavelength or wavelength band have to be used.

Towards longer wavelengths the working wavelength of the pyrometer is limited by the optical properties of laser lenses, additional beam-shaping optics, shielding glass and outcoupling optics typically made of fused silica and optimized with special coatings.

Because the interaction time of the laser during a full heat treatment process is within the range of milliseconds to seconds, a short response time of the measuring sensor is required.

If all the preconditions for temperature measurement in laser heat treatment are taken into consideration, thermal radiation detectors based on the widely used silicon or InGaAs photodiodes are suited for laser heat treatment.

3 Requirements for accuracy of the temperature measurement

The requirements for the accuracy of the measurement devices are given by the dependency of the heat treatment result from the temperature. As a result of a typical industrial process, the hardness distribution across the hardening zone was measured; the observed surface hardness and hardening depth are reported below.

As an example, Fig. 2 illustrates the strong dependence of the local hardness on the maximum local temperature during the hardening process. The hardness–temperature curve shows a maximum at a temperature around 1200 °C. The decrease of the hardness towards higher temperatures is caused by stabilization of residual austenite that is not transformed into martensite after cooling down to room temperature. To generate a high hardness level of > 700 HV the absolute surface temperature has to be within a band of several tens of kelvin.

The dependency of the process temperature on the hardening result was analyzed for samples of carbon and tool steel and was investigated for selected industrial applications. Figure 2 shows one example. If the surface temperature exceeds the material-specific austenite start temperature

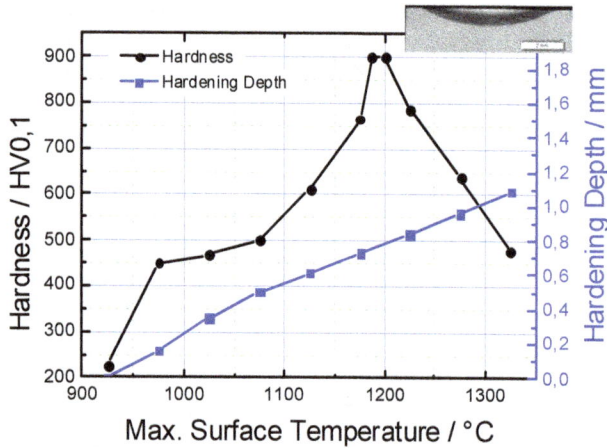

Figure 2. Surface hardness and depth of the hardening zone according to the surface temperature. Flat samples are made of tool steel X155CrMoV12.1, with the depth measured in the etched cross section (example at top of diagram).

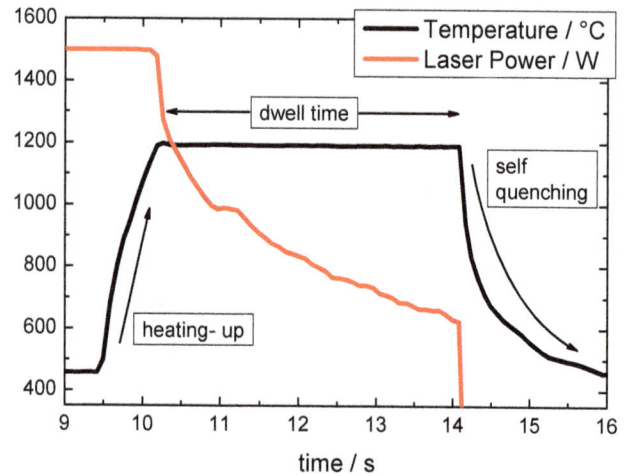

Figure 3. Process data of a temperature-controlled laser heat treatment process for the end face of a screw; temperature measured with pyrometer MAURER KTR 1075.

during the heat treatment process (~900 °C in Fig. 2), then the hardening process begins, and once cooled down, a hardening depth is generated. At surface temperatures above this threshold the resulting hardening depth shows a strong dependence on the maximum surface temperature. For typical laser heat treatment parameter sets with a processing speed of 100–500 mm min^{-1}, the hardening depth changes by several thousandths of a millimeter per kelvin.

The generation of a reproducible hardening result is not the only important criterion for reliable industrial production; good process stability is also important.

To ensure the optimum surface temperature during the process, precise temperature controllers are used. Depending on a part's material, geometry and surface quality, temperature fluctuations of ±1 to ±10 K are typical for industrial processes today (Fig. 3).

In particular cases a temperature deviation of +5 to +10 K can cause instable laser processes with oscillating temperature because of localized accelerated growth, delamination and the resulting overheating of oxide layers in combination with overreaction of the temperature controller.

In conclusion, accuracy and reproducibility of temperature measurement in laser heat treatment down to 5 K as well as a calibration method with an uncertainty significantly lower are required.

4   Calibration and inspection of temperature measurement devices

The calibration of radiation thermometers with blackbody or fixed-point radiators is an established method at national metrology institutes for the primary realization of the currently valid international temperature scale ITS-90 (Hollandt et al., 2003). Selected temperature measurement devices that

are typically used in laser heat treatment were tested and analyzed at the PTB Berlin using the variable high-temperature blackbody HTBB3200pg (Friedrich and Fischer, 2000). The uncertainty ($k = 2$) of the blackbody device is 2 K. Long-term stability of the temperature by about ±0.3 K can be ensured by temperature control of the graphite heating element.

Different influencing factors on the indicated temperature value were investigated, e.g., contamination of the optical components, variation of the working distance, defocusing of the optics and the influence of different aperture size on the blackbody emitter. From this investigation it follows that the temperature measurement devices typically used in laser heat treatment have the potential to measure temperatures with accuracy down to a few kelvin. However a variety of influencing factors can cause total measurement errors on the order of several tens of kelvin.

Table 1 illustrates the effect of contaminated pyrometer optics, and Fig. 4 depicts the effect of varying the furnace aperture between 6 and 30 mm, the so-called size-of-source effect.

To achieve the aimed accuracy level with temperature measurement devices that are electrically, optically and mechanically fully integrated into complex machines at industrial sites, on-site calibration and inspection with a mobile calibration device is needed. Only the in situ calibration in the final machine setup guarantees a high precision of the temperature measurement. For this reason a portable fixed-point calibration device, based on copper and metal-carbon eutectic high-temperature fixed points, was developed for the use in the temperature range between 1084 and 1492 °C. In order to achieve a compact design and fast operability, the fixed point is inductively heated. The design and operational characteristics of this calibration device were investigated using FEM and FDM simulation.

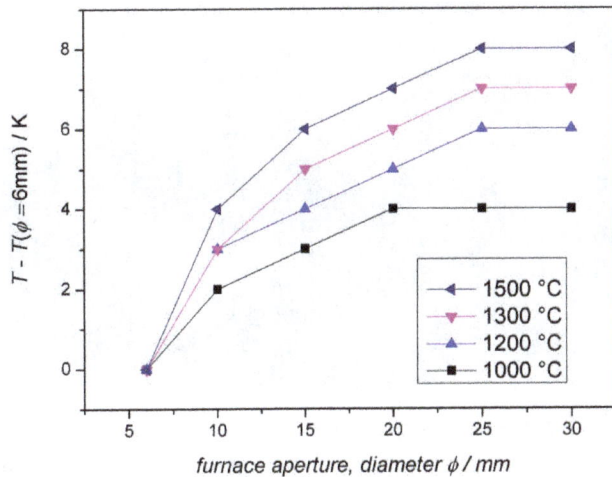

Figure 4. Influence of the size-of-source effect (SSE) on the temperature characteristics of a thermal imaging system based on a CCD camera.

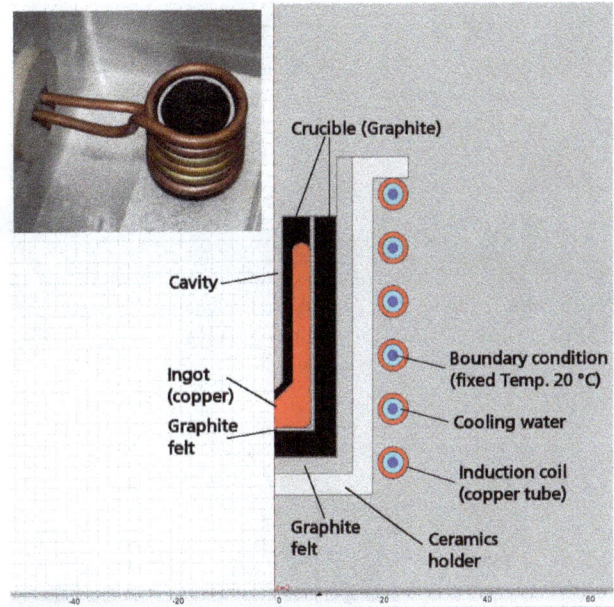

Figure 5. Components of radial symmetrical computer model for induction heating of fixed-point cells. Dimension of the cell: diameter 25 mm × length 45 mm. Photograph shows the technical implementation.

5   Results of FEM and FDM simulation

Induction heating is an established technology for heating up electrically conductive materials to high temperatures, and it is expected to be suited for fixed points.

The fixed-point cells for the high temperature range above 1000 °C are typically heated by a Joule-heated graphite furnace (Friedrich and Fischer, 2000). Following this design, in principle, a theoretical optimization of the induction coil design and the composition and dimension of components are aimed at achieving a good homogeneity of the temperature field across the fixed-point cell. This is essential to achieve a homogeneous melting of the metal alloy and a sharp and precise inflection point in the melt or freeze temperature curves (Figs. 11, 12).

Figure 5 shows the main components of the COMSOL FEM computer model and a photograph of the technical implementation. Graphite felt is used for thermal insulation of the inner components that are heated up to 1500 °C. A fixed-point cell and graphite felt are installed in a special ceramic holder which electrically insulates the graphite components from the induction coil. This setup is expected to provide optimum efficiency and good thermal long-term stability. In the radial, symmetrical FEM model, the physical effects of electromagnetic heating and heat transfer by heat conduction and radiation were taken into consideration. The induction heating was simulated with a constant current flowing through the copper coil. To achieve a steady state, the typical heating time was about 1000 s.

In the FEM simulation a strong influence of the induction frequency on the temperature field was found. The high penetration depth of the electromagnetic fields at low frequencies causes significant direct heating of the metal alloy and high temperature gradients across the fixed-point cell.

A radial symmetrical FDM model was developed to investigate the influence of the distribution of the induction heat sources and the resulting temperature fields on the melting process of the metal alloy. Up to about 8000 single finite elements were used to achieve a good spatial resolution and to minimize errors caused by the size of the finite elements itself. The induction heating was not simulated directly. The absorbed heat portion of each element was defined manually taking the FEM results into account. The physical effects of heat transfer by conduction, heat losses by radiation at the free crucible surfaces and melting of the metal alloy were implemented; all other physical effects neglected. To simulate the effect of melting, the following special variant was chosen: if the temperature of the finite element exceeds the melting temperature, the incoming heat portions during each time step are summed up without increasing the temperature of the element until the melting enthalpy of the material is reached. Then the element is marked as fully molten, the material parameters are changed and the further increase of the temperature is unblocked.

The simulation results show that the cavity temperature is stabilized by the melting metal alloy even in the case of inhomogeneous induction heating. However, the temperature gradient across the cavity wall is strongly influenced by the distribution of the inner heat sources. The inhomogeneous melting of the alloy causes a distortion of the temperature curve around the melting temperature and a shift of the inflection point, depending on the field of view of the pyrometer or thermal imaging system.

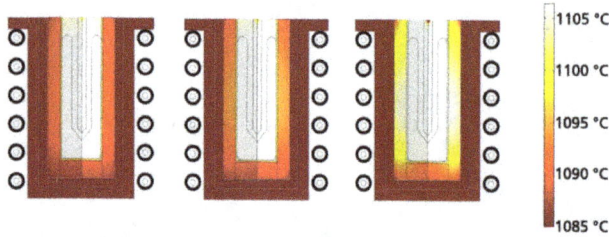

Figure 6. FEM simulation of temperature field across a fixed-point cell after induction heating (variation of induction frequency: left, 125 kHz; middle, 250 kHz; and right, 1000 kHz).

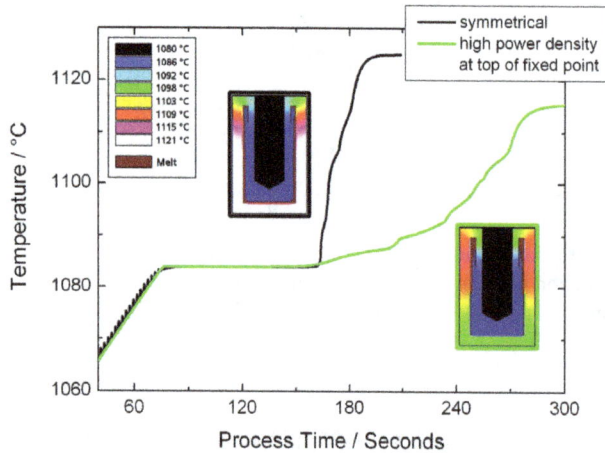

Figure 7. Dependence of cavity temperature of copper fixed-point cell on distribution of the induction heat sources, and the result of FDM simulation. Thermal images show the temperature field during the melting plateau.

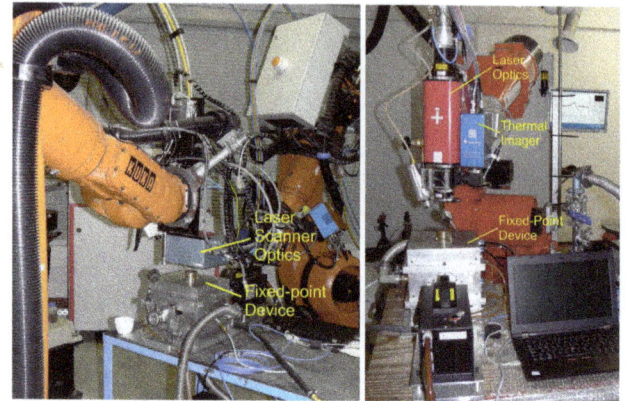

Figure 8. Prototype of the induction-heated fixed-point device, and setups for on-site calibration of laser processing heads mounted on industrial robot systems. Left: calibration of scanner optics with an integrated thermal imaging system at a machine for industrial laser hardening of steam turbine blades (Fraunhofer IWS Dresden, Germany). Right: calibration of zoom optics with variable laser beam width at a machine for job-shop laser hardening (ALOtec Dresden GmbH, Germany).

If the location of the inner induction heat sources is varied, the run-out of the temperature curve changes drastically after the melting plateau (Fig. 7). The following general relationship was found: if the density of the induction heat sources is higher at the bottom of the cell, then the melting plateau is short and the temperature curve has a steep run-out. In contrast, a high induction power density at the top of the cell causes a long plateau and long and soft run-out of the temperature curve. The knowledge of this behavior can be used for adjustment of the position of the fixed-point cell relative to the induction coil.

6   Design and realization of the fixed-point device

The heat source for fixed-point heating is an induction coil with six loops that works at a frequency up to 420 kHz. The fixed-point cell itself is insulated with graphite felt and placed in a ceramic crucible in order to avoid direct electrical contact with the induction coil (Fig. 5).

The device has a water-cooled and vacuum-sealed housing with ports for connection to a vacuum pump, shielding gas

supply and a special lead-through of the copper tubes of the induction coil. All equipment can be installed in a 19 in. rack for good system mobility. For safety reasons, all important system parameters such as shielding gas flow, water flow and temperature, and residual oxygen content are measured and observed. Figure 8 shows typical test setups used for on-site measurements at industrial laser-hardening machines.

One important result of the computer simulation with FEM and FDM was the high sensitivity of the local temperature field across the fixed-point cell on the induction frequency and the position of the cell relative to the induction coil. It is possible to further improve the homogeneity of the temperature field by installation of an additional metallic or graphite tube between the fixed-point cell and induction coil. Figure 9 schematically shows the cross section of this configuration. To integrate the additional tube, which is used as heating element, the diameter of the induction coil was increased to 54 mm. The lab setup was tested successfully by using Cu fixed-point cells. Because a large percentage of the electromagnetic fields are blocked by the tube element, the fixed-point cell is heated indirectly by the glowing tube and thus the influence of inductor geometry and induction frequency on the temperature gradient across the fixed-point cell is reduced. If a metallic tube is used as the heating element, the lifetime is limited because the chemical reaction with carbon at high temperatures influences the mechanical and thermal properties. An additional aperture is recommended for calibration of temperature measurement devices which are sensitive to the size-of-source effect. In this case it is important to block temperature radiation from the top face of the cell and let only radiation from the cavity pass through. Otherwise, changes in the surrounding furnace temperature,

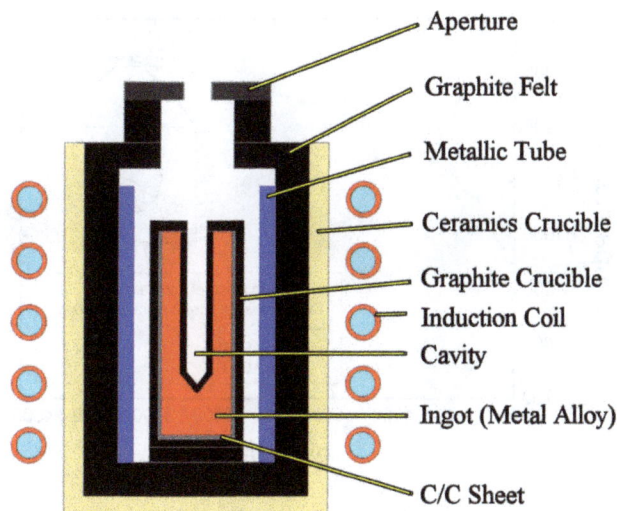

Figure 9. Schematic drawing of the setup for indirect induction heating of a fixed-point cell with a metallic tube as the heating element (outer diameter of ceramic crucible: 50 mm).

Aperture
Graphite Felt
Metallic Tube
Ceramics Crucible
Graphite Crucible
Induction Coil
Cavity
Ingot (Metal Alloy)
C/C Sheet

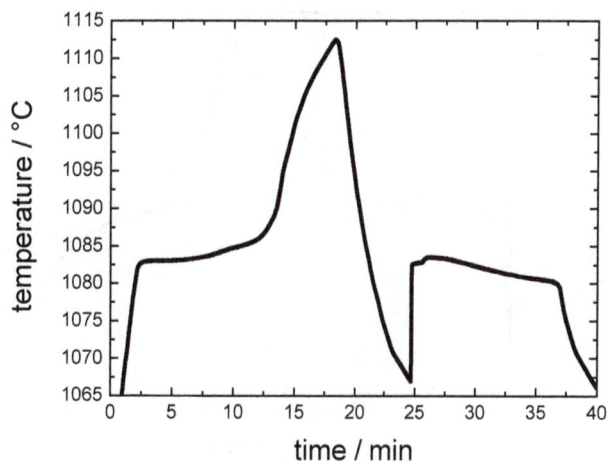

Figure 10. Melt and freeze cycle of a Cu fixed-point cell heated with an induction furnace, wrong inductor position and insufficient thermal insulation at bottom of crucible. Melting temperature: 1084.62 °C. Measured with the thermal imaging system E-MAqS (Fraunhofer IWS Dresden) with a measuring wavelength of 740 nm (compare with typical diagram in Fig. 11).

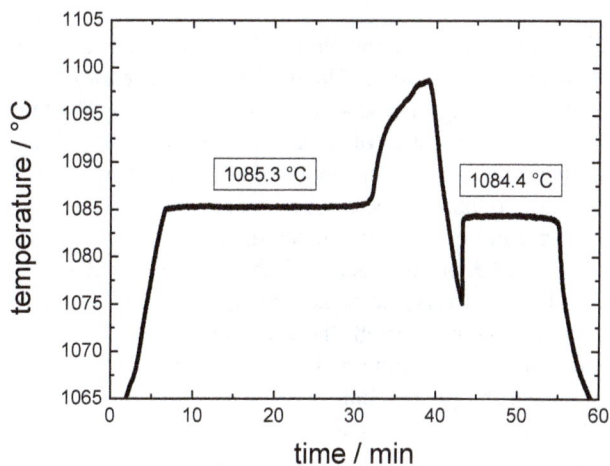

Figure 11. Melt and freeze cycle of a Cu fixed-point cell heated with an induction furnace containing a stainless steel tube as the heating element. Melting temperature: 1084.62 °C. Measured with the thermal imaging system E-MAqS (Fraunhofer IWS Dresden) with a measuring wavelength of 740 nm.

which occur especially when overheating or undercooling the fixed-point cell, will be sensed by the radiation thermometer during the melt or freeze plateau and a temperature measurement error cannot be excluded; consequently the uncertainty of the temperature measurement after calibration is increased.

7 Experimental results

The induction-heated fixed-point device was tested with Cu, Fe–C and Co–C fixed-point cells. The fixed-point cells cover the temperature range from 1084 to 1323 °C. Different factors influencing the accuracy and reproducibility of the fixed-point plateau temperatures were investigated, i.e., the position of the fixed-point cell relative to the induction coil, the arrangement of the thermal insulation (Fig. 10) and the parameters for overheating and undercooling of the cell in order initiate the melt and freeze of the metal alloy. One melt and freeze cycle in the case of direct induction heating takes about 30 min (±10 min) depending on the thermal properties of the metal alloy and the volume of the filling of the cell.

Figure 10 shows the influence of high temperature gradients across a Cu fixed-point cell on the characteristics of the melt and freeze plateau. The fixed-point cell was positioned in the lower half of the induction coil and the thermal insulation at the bottom of the ceramic crucible was insufficient. The resulting temperature gradient along the axis of the fixed-point cell caused strong distortion of the melt and freeze temperature curves (compare to theoretical curve in Fig. 7).

An inductively heated tube as an additional heating element to improve the homogeneity of the heating was tested with a Cu fixed-point cell in a lab setup within a special vacuum chamber. The heating element chosen was a tube made of stainless steel. Even if the tube reacts with carbon atoms of the surrounding graphite felt, the heat resistance up to the melting temperature of the material (eutectic temperature of the Fe–C alloy after saturation with carbon atoms) was expected theoretically and was proved in long-term tests. The induction frequency used, about 100 kHz, is normally too low for direct fixed-point heating. Nevertheless, very flat melt and freeze plateaus were measured with a thermal imaging system typically used in industrial laser heat treatment (Fig. 11).

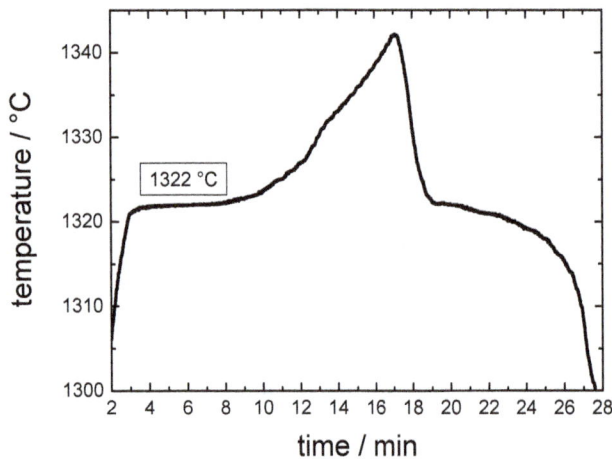

Figure 12. Melt and freeze cycle of an inductively heated Co–C fixed-point cell (cavity diameter 5 mm) after on-site calibration of the laser processing head at an industrial machine for laser hardening of steam turbine blades (Fraunhofer IWS Dresden, Germany). The inflection point of the melt plateau agrees with reference temperature (linear pyrometer LP3 at PTB Berlin) within ±0.5 K.

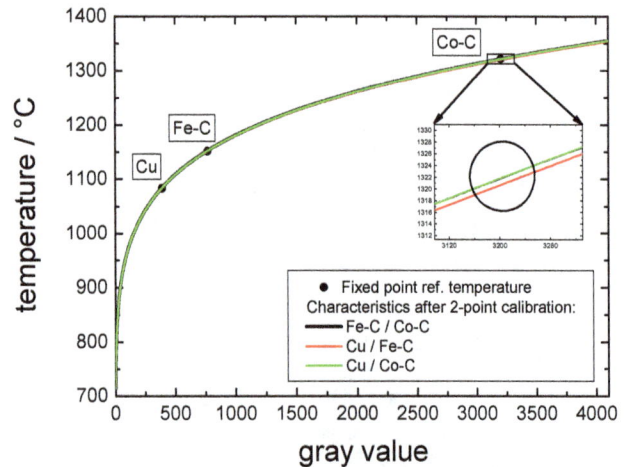

Figure 13. Temperature characteristics of the thermal imaging system after calibration with three inductively heated fixed-point cells. Co-axial temperature measurement was performed through zoom laser optics at an industrial machine for job-shop laser hardening (ALOtec Dresden GmbH, Germany). Maximum temperature deviation over full temperature range: 1.3 K.

The prototype of the mobile fixed-point device was tested on-site for the calibration of industrial laser processing heads currently working in production (compare Fig. 8). The temperature characteristics of three different laser optical systems (scanner optics, zoom optics and 90° mirror optics) with integrated co-axial temperature measurement were inspected and corrected. As a result, temperature deviations up to several tens of kelvin were detected; these were mainly caused by calibration errors, the measurement uncertainty of the radiation thermometer itself, the exchange of optical components after their lifetime and the contamination of optical components by process fumes. As an example, the exchange of the shielding glass of a laser scanning head caused a jump in the temperature measuring value by 10 K. To ensure the best possible accuracy, the glass was exchanged between two melt and freeze cycles of a Co–C fixed point cell (1322 °C) without changing the adjustment and processing parameters. For this case, the reproducibility of the cell temperatures is typically better than 1 K, and the influence of the shielding glass on the temperature signal was measured directly.

After calibration of the industrial laser processing heads, the temperature characteristics, which were generated with inductively heated fixed points, were tested again. All fixed-point temperatures could be reproduced with an accuracy of 1 to 2 K (compare Fig. 12), representing the typical measuring error of the devices under ideal conditions.

Figure 13 shows the typical temperature characteristics of a thermal imaging system that is used industrially for job-shop laser hardening. It was proved that a two-point calibration with two different fixed-point cells is sufficient to achieve an agreement of the temperature characteristics with a third fixed-point temperature within ±1 K.

8 Conclusions

As a first step, the requirements for temperature measurement in laser heat treatment were investigated. Based on the results, a new induction-heated fixed-point device for calibration was developed. The prototype was constructed and manufactured by the PTB Berlin and tested with Cu, Fe–C and Co–C fixed-point cells for the calibration of temperature measurement devices typically used in laser heat treatment. The mobility of the calibration equipment was tested at two industrial sites. Typical temperature deviations in real production and the uncertainty of the calibration procedure were investigated. The pollution and exchange of optical components of laser processing heads and temperature measuring devices were found to be the most critical sources for temperature measuring errors.

By using inductively heated fixed-point cells, industrial laser processing heads were calibrated on-site with a precision that was previously unachievable. The inflection points of the melt and freeze temperature curves agree with the theoretical values within a band of ±1–2 K and are within the measurement uncertainty of the radiation thermometers used.

Further experimental investigations are planned to minimize the uncertainty of the new calibration method. For this task the potential of metallic tubes for indirect induction heating of fixed-point cells could be demonstrated. It is expected that the temperature-controlled processing of this kind of induction furnace will improve the homogeneity of the temperature field during the heating of difficult cell geometries and fillings as well as the reproducibility of the inflection points of the melt and freeze plateaus.

The new possibilities for precise and reproducible temperature measurement in laser heat treatment processes are expected to improve the quality of industrial processes and open the door to new applications for difficult materials whose properties strongly depend on the surface temperature in laser processing.

Acknowledgements. The work was undertaken in the content of the EURAMET joint research project "HiTeMS" (High temperature metrology for industrial applications).

The EMRP is jointly funded by the EMRP participating countries within EURAMET and the European Union.

Thanks go to PTB Berlin (Germany) and NPL Teddington (UK) for training in high-temperature fixed points and supply of calibration and measuring devices, pictures and measuring data.

Industrial laser-hardening machines for on-site tests of the fixed-point device were kindly provided by Fraunhofer IWS Dresden (Germany) and ALOtec Dresden GmbH (Germany).

Edited by: N.-T. Nguyen
Reviewed by: two anonymous referees

References

Anhalt, K.: Radiometric measurement of thermodynamic temperatures during the phase transformation of metal-carbon eutectic alloys for a new high-temperature scale above 1000 °C, dissertation, urn:nbn:de:kobv:83-opus-19712, http://opus.kobv.de/tuberlin/volltexte/2008/1971/, 2008.

Bonss, S., Seifert, M., Brenner, B., and Beyer, E.: Precision Hardening with High Power Diode Lasers using Beam Shaping Mirror Optics, Proc. SPIE, San Diego, USA, 4973, 86–93, 2003.

Bonss, S., Seifert, M., Hannweber, J., Karsunke, U., and Beyer, E.: Low Cost Camera Based Sensor System for Advanced Laser Heat Treatment Processes, ICALEO 2005 Paper #1804, 851–855, 2005.

Bonss, S., Hannweber, J., Karsunke, U., Seifert, M., Brenner, B., and Beyer, E.: Precise Laser automotive die hardening, Heat treating progress 9 (2009), March/April, 44–47 ISSN: 1536-2558, 2009.

EMRP A169 (REG): RESEARCHER GRANT CONTRACT NO. IND01-REG1, JRP: IND01 HITEMS, http://projects.npl.co.uk/hitems/, 2011 - 2013.

Friedrich, R. and Fischer, J.: New spectral radiance scale from 220 nm to 2500 nm, Metrologia, 37, 539–542, 2000.

Hollandt, J., Friedrich, R., Gutschwager, B., Taubert, D., and Hartmann, J.: High-accuracy radiation thermometry at the National Metrology Institute of Germany, the PTB High Temperatures – High Pressures, 35/36, 379–415, 2003.

10

Simulative verification of a novel semi-active broadband energy harvester

K. Retan[1], **A. Graf**[1], **and L. Reindl**[2]

[1]ZF Friedrichshafen AG Graf-von-Soden-Platz 1 88046 Friedrichshafen, Germany
[2]IMTEK Lehrstuhl für Elektrische Mess- und Prüfverfahren Georges-Köhler-Allee 106, 79110 Freiburg, Germany

Correspondence to: A. Graf (alexander.graf@zf.com)

Abstract. This paper presents a semi-active broadband vibrational-energy harvesting system. Based on a non-resonant rotational generator, electronic circuitry was used to overcome the physical start-up restrictions. Due to the functional design it remains an energy harvester suitable for battery-less devices. For the first time a vibrational energy harvester is presented that allows standardization and thus higher volume production. A system layout, simulation, and measurement data will be shown.

1 Introduction

Wireless sensor systems have become an essential component in the field of structural health monitoring (SHM). These devices offer continuous information regarding the mechanical stability of structures ranging from automobiles to large buildings. Using a wireless transceiver, data can be gathered and used to evaluate structural integrity, which allows for accurate predictions of the structure's life expectancy and facilitates the timely repair of damaged components (Yang et al., 2007; mon, 2013).

Due to the limited lifespan of battery powered systems, energy harvesting systems have been developed, which transform environmental energy into electrical power. The classic implementation of this idea is the solar cell, which generates an electric potential from solar radiation. This energy can then be used to power system components directly or recharge a battery, which can dramatically improve the system's life expectancy. In addition to solar power, thermal gradients and vibrational forces can be exploited. The optimal energy solution depends on the application environment.

Where sunlight or thermal gradients are sporadic or nonexistent, vibrational energy can be used to generate power needed for wireless systems. Ambient vibrations are present in a variety of environments, ranging from structures such as bridges and buildings to industrial machines and household appliances. This form of energy harvesting generally uses the mechanical energy from a vibrating body to excite a spring-mass system, which then oscillates with the frequency of the vibration source. The systems exhibit high power output at vibration frequencies very close to the harvester's resonant frequency but are very inefficient outside this small frequency range. Examples of commercial resonant energy harvesters include the PMG Free Standing Harvester from Perpetuum (Per, 2012), as well as the Volture from MIDE (Stephen, 2006).

The application environment, however, can exhibit a wide range of vibration frequencies, necessitating a factory tuning of the harvester to match the device's resonant frequency with that of the environment. Additionally, the dominant vibration frequency is not necessarily static, which disqualifies these devices from use in environments in which the vibration frequency does not remain within a very narrow bandwidth.

To increase the operating bandwidth of kinetic energy harvesters, recent efforts have focused on incorporating nonlinear effects into the harvester system. Xing et al. (2011) recently published results from a cantilever harvester constructed of a high-permeability magnetic material (Xing et al., 2011). The cantilever was mounted horizontally inside a solenoid. Two permanent magnets were placed near the tip such that tip oscillations would cause a periodic

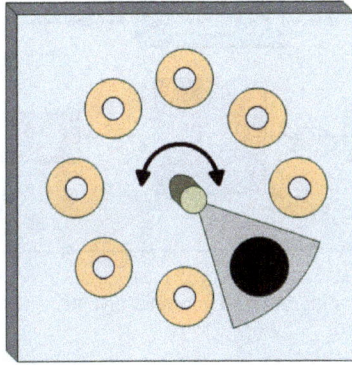

Figure 1. Pendulum-based non-resonant energy harvester developed by Spreemann et al. (2009).

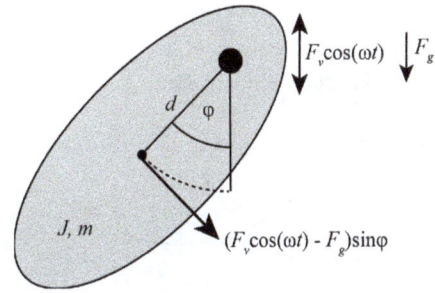

Figure 2. Physical pendulum described by its mass (m), moment of inertia (J) and center of mass (d).

magnetization reversal in the cantilever. The combination of magnetostatic and elastic effects increased the bandwidth of the system to 10 Hz while maintaining a peak power density of 1.07 mW cm^{-3}.

Additionally, fully non-resonant harvesting concepts have been investigated. Bowers and Arnold (2009) constructed a rolling-magnet generator, which consisted of a spherical magnet enclosed in a spherical cavity, which was then wrapped in copper wiring. The device was designed to generate energy from human motion, which is characteristically aperiodic and varies greatly over time. A voltage of up to 700 mV was generated, which equated to a power density of up to 0.5 mW cm^{-3}.

Finally, Spreemann et al. (2009) developed a kind of pendulum harvester (cp. Fig. 1) which operated using complete rotations rather than small-amplitude oscillations. In terms of broadband characteristics, this harvester was very successful, exhibiting an operating frequency range from 30 to 80 Hz at output powers between 0.4 and 3 mW. A significant design flaw, namely the inability to establish this rotational state automatically, prevented the generator from becoming a viable energy harvesting solution.

Therefore, this paper focuses on the start-up functionality and describes the corresponding system layouts.

2 Basic pendulum theory

To understand the broadband characteristics of pendulum-based energy harvesters a basic knowledge of their dynamic behavior is required. The equation of motion for a damped mathematical pendulum excited by vibrational forces is given by (Gitterman, 2010):

$$\ddot{\varphi} + c\dot{\varphi} + \frac{g}{l}\sin\varphi = F\cos(\omega t), \qquad (1)$$

where φ, $\dot{\varphi}$ and $\ddot{\varphi}$ denote the angular position, velocity and acceleration, respectively. The constants c and l represent the damping coefficient and length of the pendulum, respectively, while g stands for the gravitational constant. The term

$F\cos(\omega t)$ represents the force driving the pendulum, which is assumed to be sinusoidal with amplitude F and angular frequency ω. This equation describes a pendulum which consists of a point mass at the end of a massless pendulum arm and is driven by a force tangential to pendulum's path.

In order to describe a physical pendulum driven by a vibrational force (see Fig. 2), several modifications were made to this equation. First, the pendulum was described as an object with nonzero dimensions using its moment of inertia. Second, the vibrational force is restricted to a purely vertical motion, which means it is no longer tangential to the path of rotation. The coupling of the driving force into the pendulum system is a function of the pendulum's current position and is described by the prefactor $\sin(\varphi)$. Finally, the sum of all driving or damping moments originating in the transducer is represented by M_T. These moments will be investigated in detail in Sect. 2.2. After making these modifications, the resulting equation of motion can be written as

$$\ddot{\varphi}J + M_T + mgd\sin\varphi = -A\omega^2 md\cos(\omega t)\sin(\varphi), \qquad (2)$$

with the constants m, J, and d representing the pendulum's mass, moment of inertia and the distance of the center of mass from the pivot, respectively. A and ω characterize the amplitude and the frequency of the vibration.

Given a pendulum which is initially motionless and modest vibration intensity, the pendulum oscillates around the resting position. For small oscillations, the angular position is close to zero, and thus prefactor $\sin(\varphi)$, which determines the coupling of the vibration forces into the pendulum system, is small. For energy harvesting, this is an inherently inefficient generation state, as very little mechanical energy is transferred into the system. As the amplitude of the vibrations increases, so does the amplitude of the pendulum oscillations, and thus also the coupling factor as it approaches the maximum value of one.

In order to maximize the efficiency of the pendulum as an energy harvester, the coupling of the mechanical energy of a given vibration into the pendulum system must be maximized. This is realized by matching the phase of the pendulum's oscillations with that of the vibrations, which implies

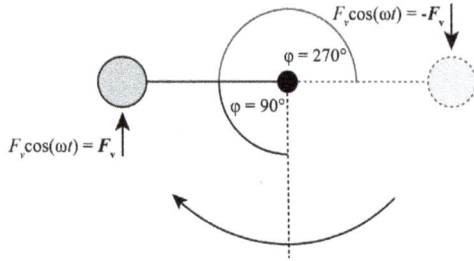

Figure 3. Values of φ for which the transfer of vibrational energy is maximized.

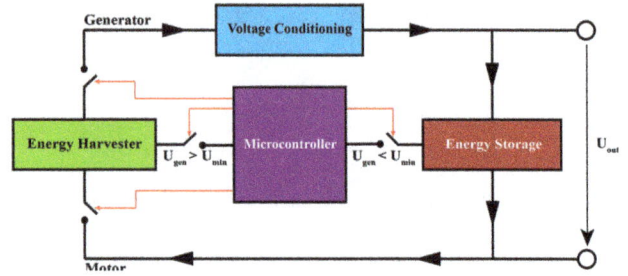

Figure 4. Block diagram of the system layout.

$$\varphi = \omega t + \frac{\pi}{2}. \tag{3}$$

Physically, the phases of the vibrations and the pendulum are matched when the pendulum performs complete rotations with a frequency equal to that of the vibrations and a phase shift of $\pi/2$. When the vertical vibration force reaches its maximum amplitude at $\omega t = 0$, the pendulum is completely horizontal, enabling maximum transfer of vibrational moment into the system as depicted in Fig. 3.

At this point, the transfer of energy is efficient enough that the vibrations can sustain the pendulum's rotation for certain levels of damping. As Spreemann et al. (2009) discovered, energy transfer in this state is so efficient that rotation can be maintained while simultaneously extracting meaningful amounts of energy from the system in the form of electricity. This rotational state is therefore the preferred state for pendulum-based energy harvesting, as it maximizes energy transfer into the system to the point that surplus energy can be extracted with an appropriate transduction mechanism.

The pendulum's equation of motion, like most nonlinear differential equations, does not possess a closed solution (Gitterman, 2010), and the classical linearization assuming small deflection angles is not valid in this case. Therefore, numerical simulations of the pendulum's behavior were conducted to investigate its suitability as an energy harvesting solution. Simulations of the pendulum's rotational behavior will be covered in Sect. 2.1.

2.1 System concept

The proposed energy harvester concept requires a level of system complexity which exceeds that of traditional resonant energy harvesters. The components required for the proposed semi-active harvester are shown in a block diagram in Fig. 4. As stated previously, in order for the harvester to efficiently generate energy its angular velocity must be matched to the dominant vibration frequency. The system must therefore possess the ability to both accelerate the pendulum and generate power. The simplest and most elegant solution available, which was also selected for the proposed energy harvesting system, consists of a DC-motor with an attached pen-

dulum. By controlling the direction of energy flow to and from the motor, it can be used initially to accelerate the pendulum and subsequently as the transduction mechanism for energy harvesting.

Additionally, control electronics and a suitable algorithm are necessary to regulate the harvester's rotational speed, to ensure its rotational stability and to control energy flow to and from the energy storage. Generated voltages must be conditioned for energy storage, necessitating the use of a voltage converter. Control electronics can be powered either directly from the harvester or from the energy storage, depending on the harvester's output voltage, which is a function of its operating frequency.

2.2 Modeling of components

2.2.1 DC motor

A simplified model was used for the DC motor. Inductive effects were intentionally omitted due to the motor's small electrical time constant, which is given by

$$\tau_e = \frac{L_{\mathrm{motor}}}{R_{\mathrm{motor}}}, \tag{4}$$

with L_{motor} and R_{motor} representing the motor's inductance and internal resistance, respectively. The DC motors considered for the energy harvester typically have a resistance in the order of 100 Ω and induction of several hundred μH, yielding time constants of less than 5 μs. This is very small compared to the vibration frequencies expected in the target application, which should not exceed 100 Hz. Inductive effects would therefore have a negligible influence on the motor's behavior and are not necessary for modeling. Motor rotation induces a voltage V_{EMF}, which is proportional to the rotational velocity according to

$$V_{\mathrm{EMF}} = k_{\mathrm{E}} \dot{\varphi} \frac{60}{2\pi}, \tag{5}$$

where V_{motor}, $\dot{\varphi}$ and k_{E} denote the motor voltage, motor velocity in radians per second and the back EMF constant in V rpm^{-1}, respectively. The constant k_{E} is an experimentally determined, intrinsic property of every DC motor and is listed in its datasheet. In motor operation this voltage opposes the applied voltage, effectively increasing the internal resistance of the motor. In generator operation this is the measured output voltage at a given rotational velocity. The motor current is then given by

$$I_{motor} = \frac{V_{op} - V_{EMF}}{R_{motor} + R_{load}} = \frac{V_{op} - k_E\dot\varphi}{R_{motor} + R_{load}} \cdot \frac{60}{2\pi},$$ (6)

where I_{motor}, V_{op}, R_{motor} and R_{load} represent the motor's current, operating voltage, internal resistance and load resistance, respectively. During motor operation the load resistance can be ignored, which simplifies Eq. (6) to

$$I_{motor} = \frac{V_{op} - V_{EMF}}{R_{motor}}.$$ (7)

During generator operation the operating voltage is zero, which reduces Eq. (6) to

$$I_{motor} = \frac{V_{EMF}}{R_{motor} + R_{load}}.$$ (8)

A second constant k_I, which is termed the current constant and is given in units $(A\,mN\,m^{-1})$, relates this current flow to motor torque. As with the back-EMF, k_I is specific to each motor and is listed in its datasheet. Using Eq. (6) and constant k_I, the current-induced motor moment can be expressed as

$$M_E = I_{motor} \cdot \frac{1}{k_I} = \frac{V_{op} - k_E\dot\varphi}{R_{motor} + R_{load}} \cdot \frac{60}{2\pi} \cdot \frac{1}{k_I},$$ (9)

where M_E denotes the electrical contribution to the motor moment. When the DC motor is operated as such during the energy harvester's acceleration phase, this moment causes rotational acceleration. When the motor is operated at a generator, this moment opposes the vibration-induced rotation and thus can be seen as a viscous damping force, as it is proportional to the motor's rotational speed. In addition to the electrical moment (M_E), a mechanical damping (M_M) was included to model the friction in the motor bearings. This friction moment is largely independent of rotational speed and thus is included as a constant moment (c_m) which always opposes the direction of rotation.

$$M_M = c_m \cdot sgn(-\dot\varphi)$$ (10)

The sum of the electrical and mechanical moments in the DC motor,

$$M_T = M_E + M_M,$$ (11)

can then be inserted into Eq. (2), yielding the complete pendulum energy harvester equation of motion which was used in the simulation model.

2.2.2 Voltage conditioning

A boost converter was necessary to increase the generator voltage to the point that it could be stored in a lithium ion

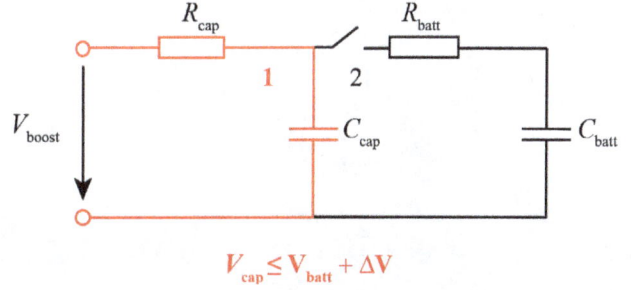

Figure 5. Initial state of the battery charging circuit.

battery. Boost converters are circuits which utilize switching transistors and inductors to increase voltage. These circuits typically operate at MHz frequencies, and thus their dynamic behavior has little impact on the relatively low-frequency energy harvester system. For the sake of simplicity, a boost converter was modeled which generated a fixed output voltage (V_{out}) and exhibited a certain efficiency (η), which is defined as

$$\eta = \frac{P_{out}}{P_{in}}.$$ (12)

In reality, the efficiency on the input voltage, decreasing as it approaches the minimum input voltage, below which the converter ceases to function. For modeling purposes, an average efficiency was estimated for the harvester's output voltage range using efficiency values from component datasheets.

2.2.3 Energy storage and charging circuit

The energy storage and charging circuit simulation blocks were implemented using the differential equation for capacitor charging, which is given by

$$V_c(t) = i(t)R + \frac{1}{C}\int_0^t i(\tau)d\tau,$$ (13)

where V_c, $i(t)$, R and C denote the capacitor voltage, capacitor current series resistance and capacitance, respectively. Capacitance and resistance can be approximated using an exponential fit of the charging characteristic found in the battery's datasheet.

The charging circuit was designed to have a minimum number of active components while still effectively charging the battery. In the proposed energy harvesting system, the output voltage of the boost converter is used to charge the battery. For the simulation model, an output voltage of 5 V was selected. To charge the thin-film battery, the charging voltage must exceed the current battery voltage (V_{batt}) while not exceeding the maximum voltage rating (V_{max}) of the battery in order to avoid damaging the battery.

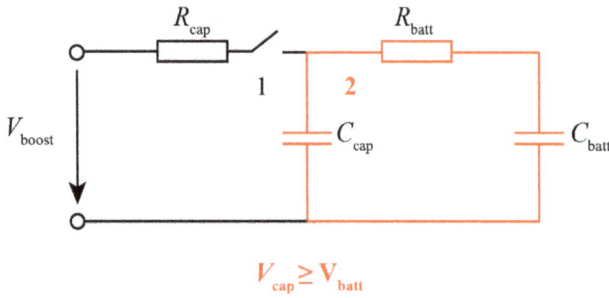

Figure 6. Transition state of the battery charging circuit.

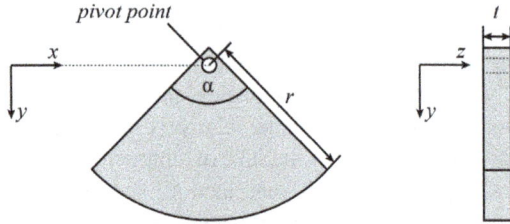

Figure 7. Geometry and orientation of the rotational mass.

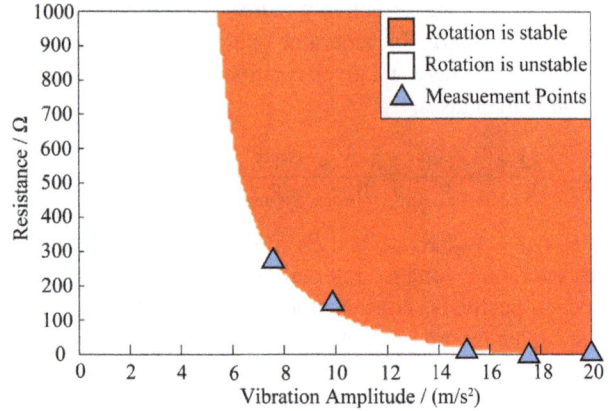

Figure 8. Verification of the simulation model.

To this end, the charging circuit depicted in Figs. 5 and 6 was used. It consists of the energy storage component with charging voltage $V_{batt} = 4.1$ V, a capacitor and two switches. Initially, switch 1 is closed and switch 2 is open. The capacitor is thus charged to a value $V_{batt} + \Delta V$ above the charging voltage and below the maximum rating. The switch states are then reversed, which results in the capacitor transferring its charge to the battery until its voltage is equal to or below that of the battery voltage V_{batt}.

2.3   Model verification

The primary goal of preliminary simulations was the verification of the simulation model's accuracy. To this end, a simple pendulum generator was constructed using a cylindrical segment pendulum (cp. Fig. 7) and a Faulhaber DC motor.

By varying the opening angle (α), the radius (r) and the thickness (t) of the cylindrical segment, pendulums with various moments of inertia and centers of mass could be designed. The pendulum was designed to maximize vibration torque according to

$$M_{vib} = F_{vib}d\sin\varphi, \tag{14}$$

where F_{vib}, d, and φ represent the instantaneous vibration force, the pendulum's center of mass and its angular position respectively. The center of mass (d) and the total mass (m) were thus chosen as large as possible without exceeding the motor bearing's tolerance of radial forces as described by:

$$F_{r,max} = md\dot{\varphi}^2, \tag{15}$$

where $\dot{\varphi}$ and m denotes the mass and angular velocity of the pendulum. For simulation verification a 1016 012 G DC

motor from Faulhaber was selected due to its high radial-load tolerance of 5 N (fau, 2012). The pendulum was designed such that its radial force would not exceed 5 N at the chosen maximum operating frequency of 100 Hz. Based on this design criterion a steel pendulum with the dimensions $r = 10$ mm, $\alpha = 120°$ and $t = 3$ mm was constructed, yielding a maximum radial force of 4.93 N at 100 Hz. Using an electromagnetic shaker as a vibration source, the maximum current output of the generator was investigated at an operating frequency of 20 Hz and for amplitudes between 0.75 and 2 g.

The pendulum was accelerated into a stable rotation using a finger flick. During the acceleration phase the generator load was 1 MΩ, resulting in minimal electrical damping which facilitated phase matching. Once rotations were stable, the load was reduced from 1 kΩ in 10Ω steps until the pendulum fell out of rotation due to excessive electrical damping. This load was deemed the minimum tolerable electrical load for a given vibration frequency and amplitude (cp. Fig. 8).

Subsequently, the dimensions of the constructed pendulum as well as the damping characteristics of the selected DC motor, which are given in the corresponding datasheet (fau, 2012), were incorporated into the simulation model. Simulations were then conducted for which the pendulum was given an initial angular velocity and phase which were matched to the vibration frequency. As with the generator measurements, the vibration frequency was set to 20 Hz, while the amplitude was varied between 0 and 2 g. The electrical load was also varied between 0 and 1 kΩ. A simulation time length of 10 s was chosen after it had been observed that this interval was adequate to determine whether or not the pendulum's rotation was truly stable at the given electrical load. Upon cessation of the simulation, the pendulum's angular velocity was investigated to determine whether or not the pendulum was still rotating.

Simulation results as well as the measurement results are shown in Fig. 8. The plot regions which are marked in red indicate stable rotation, whereas the white regions indicate a cessation of rotation. The red-white transition indicated the

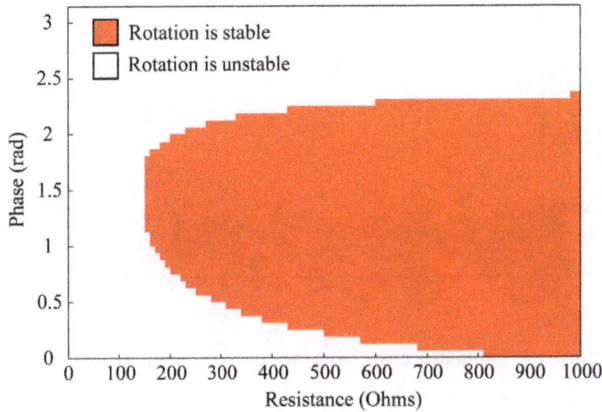

Figure 9. Investigation of the phase matching requirement as a function of the load resistance.

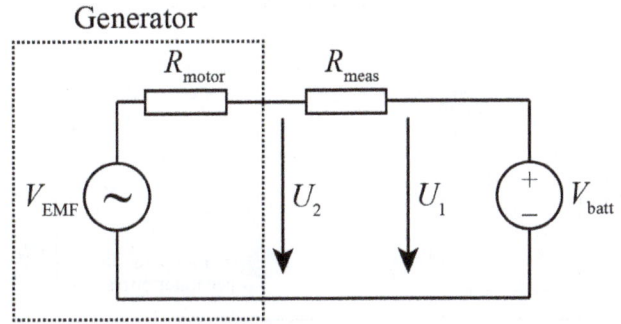

Figure 10. Equivalent circuit for the determination of V_{EMF}

minimum electrical load for which rotations could be maintained for a given vibration amplitude. The good agreement between the simulation and measurement results was interpreted as a verification of the accuracy of the simulation model, which could therefore be used to further investigate the dynamics of the pendulum generator.

3 Simulation and measurement results

In this section the simulation model will be used to investigate different operational phases of the generator system. Based on these data, hardware was developed and tested.

3.1 Simulation of the start-up phase

After verifying the accuracy of the simulation model, additional system components necessary for generator start-up and energy storage can be included. As stated previously, the proposed energy harvester concept requires a substantial energy investment in order to accelerate the pendulum to its operating frequency. Thus an energy storage component is necessary to provide the initial energy investment and subsequently store generated energy. For this simulation model, the energy storage component was modeled after an MEC225 Thinergy thin-film lithium-ion battery from Infinite Power Solutions (Inf, 2012). Thin-film batteries generally exhibit much lower leakage currents than capacitors with comparable storage capacities, which increases the maximum tolerable system down-time between periods of vibration activity.

Additionally, a control strategy was necessary to match the motor speed to the vibration frequency. Initial simulations indicated that phase matching tolerances between the motor and vibrations become increasingly stringent as electrical loads decrease, which affect input energy and dissipated power, respectively (cp. Fig. 9). A reduction in vibration amplitude was also observed to decrease the tolerances for phase

matching. Therefore, a very accurate control strategy could potentially allow for motor operation at lower vibration amplitudes. However, the benefits obtained from an accurate but complex control strategy must be weighed against the potential increases in the operation current. For a preliminary analysis, a two-point control strategy was used, which could be implemented on a very low current budget.

Therefore, the back EMF was calculated from voltage measurements first:

$$V_{EMF} = U_1 - \frac{U_1 - U_2}{R_{meas}} \cdot (R_{motor} + R_{meas}), \qquad (16)$$

where R_{meas}, R_{motor}, U_1 and U_2 represent the current measurement resistor, internal motor resistance and two potential measurements shown in Fig. 10, respectively.

From the back EMF and the back EMF constant found in the motor's datasheet, the motor speed can be calculated according to

$$f_{motor} = \frac{V_{EMF}}{60 \cdot 2\pi}. \qquad (17)$$

The control strategy itself functions as shown in Fig. 11. The vibration frequency (f_{target}) represents the target average motor velocity, with the instantaneous motor velocity oscillating periodically around this value. Upper and lower velocity tolerances were then defined, which allow for the natural oscillatory variation around f_{target}. Starting from a standstill, the motor is first accelerated to the upper motor speed tolerance, at which point it is switched off. The velocity then decelerates through the tolerance range. If phase matching occurs during this deceleration, the motor remains within the velocity tolerances. Should the motor velocity decrease beyond the lower velocity tolerance, it is accelerated again and allowed to decelerate through the tolerance range again. This process is repeated until the motor achieves phase matching with the vibrations.

Once the control strategy was incorporated into the simulation model, simulations of the motor start-up phase were conducted to confirm the control strategy's effectiveness. Simulations were conducted at vibration frequencies of 20, 40 and 60 Hz for a vibration amplitude of 1 g. Results are

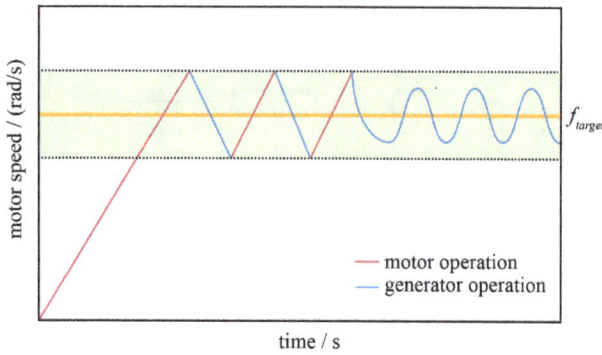

Figure 11. Desired transient response of the accelerated pendulum.

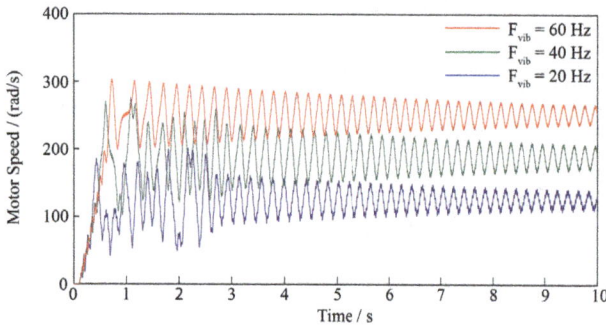

Figure 12. Simulated motor speed with respect to frequency and time.

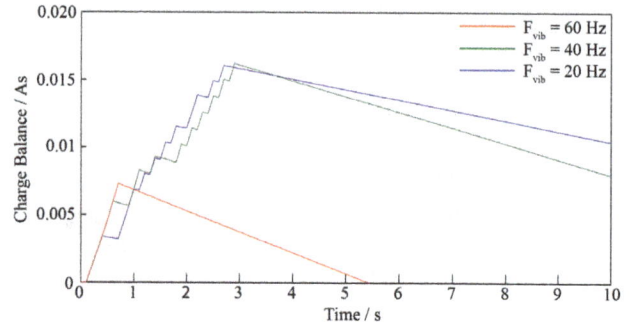

Figure 13. Simulated charge balance of the start-up phase.

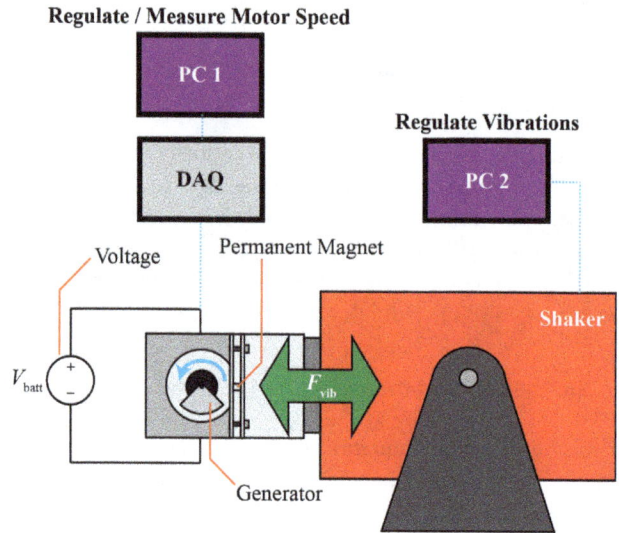

Figure 14. Setup of the measurement equipment.

shown in Figs. 12 and 13. The diagram in Fig. 12 shows the velocity of the motor over time, which increases initially to the target frequency and subsequently enters a stable rotation, as indicated by the stable average velocity. The diagram in Fig. 13 shows the balance of charge following through the motor.

A positive slope indicates current flowing to the motor, whereas a negative slope indicates that the motor is generating current that is then flowing to the energy storage. This plot can be used to determine how much energy is invested to accelerate the pendulum as well as how long the generator must run to recuperate the energy investment. These charge balance plots show that for a vibration amplitude of 1 g, phase matching occurs relatively quickly. The current flowing to the energy storage, which is indicated by the slope of the plot, is 0.8, 1.15 and 1.3 mA for $f_{vib} = 20$, 40 and 60 Hz, respectively. The varying slopes of the charge balance curves arise from the fact that the generator's output voltage is proportional to its rotation frequency. Extrapolating the charge balance curves, the charge invested to start the generator can be recuperated in 20, 10.7 and 5.4 seconds for the respective rotation frequencies.

3.2 Measurements

After verifying the control strategy's effectiveness using simulation results, the strategy was then implemented using a NI USB-6259 BNC data acquisition device (DAQ) from National Instrument, which was connected to a computer PC1. The control software was programmed in C. A voltage source set to 4.1 V was used to emulate the MEC225 battery. A 1k resistor was then used as an electrical load, which approximated the battery's resistive load. As a vibration source, an electromagnetic shaker was again used, which was controlled by a second computer PC2. The complete measurement setup can be seen in Fig. 14.

To facilitate the comparison with the simulation results, measurements were conducted a 20, 40 and 60 Hz for a vibration amplitude of 1 g. Measurement results are shown in Figs. 15 and 16.

As evident from the results, phase matching was achieved for all three frequencies. Compared to the simulations, the energy expended for acceleration was recuperated much more quickly. This is due to the fact that a simple load

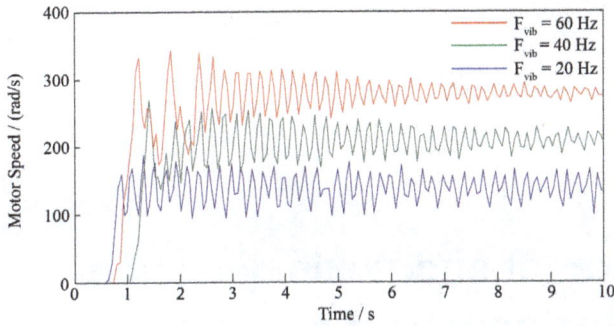

Figure 15. Measured motor speed with respect to frequency and time.

Table 1. Generator output power over Frequency.

Frequency (Hz)	20	30	40	50
Power (mW)	16.9	10.6	5.8	6.8

resistance with no capacitive behavior was used, allowing for much larger motor currents. Additionally, no boost converter was necessary for voltage preparation, which tends to reduce the output current in order to raise the output voltage. Despite these differences, these measurements experimentally demonstrate the effectiveness of the simple two-point control strategy for motor speed regulation.

Finally, the motor's maximum power output was investigated at various frequencies (cp. Table 1). The maximum power was defined as the minimum resistive load for which the generator could remain in rotation at a given vibration frequency and amplitude. The maximum output power of 16.9 mW was measured for 20 Hz. The reduced power output at frequencies above 20 Hz is likely due to the measured frequency dependence of the friction in the motor bearings.

4   Conclusions

The simulation and measurement results presented in Sect. 3 confirm the functionality and broadband characteristics of pendulum-based energy harvesters. In addition, the proposed mechanism for accelerating the pendulum to its operating frequency was also demonstrated to be viable. Using the proposed control strategy, phase matching between the pendulum rotation and vibration frequency was achieved over a frequency range from 20–50 Hz, allowing subsequent energy generation at power outputs greater than 5 mW. This output power level is more than adequate for state-of-the-art wireless sensor systems. In conclusion, the semi-active pendulum-based energy harvester was proved to be a viable broadband vibration energy harvesting solution. Future research will be focused on the development of an autonomous pendulum harvester, which will possess the

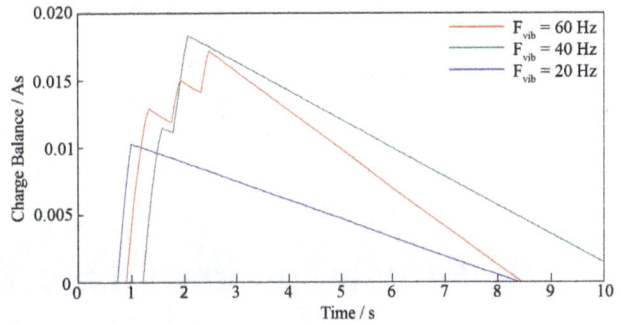

Figure 16. Measured charge balance of the start-up phase.

ability to autonomously detect ambient vibrations on an extremely low current budget and subsequently begin energy harvesting at the dominant vibration frequency.

Edited by: U. Schmid
Reviewed by: two anonymous referees

References

Bowers, B. J. and Arnold, D. P.: Spherical, rolling magnet generators for passive energy harvesting from human motion, J. Micromech. Microeng., 19, 1–7, 2009.

Serie 1016 DC-Kleinstmotoren mit Edelmetallkommutierung, Faulhaber, www.faulhaber.com, 2012.

Gitterman, M.: The Chaotic Pendulum, World Scientific Publishing Co. Pte. Ltd., 5 Toh Tuck Link, Singapore 596224, 2010.

MEC225 Solid-State, Flexible, Rechargeable Thin-Film Micro-Energy Cell, Infinite Power Solutions, www.infinitepowersolutions.com, 2012.

Wireless Sensor Power Management, http://www.monnit.com/blog, 2013.

PMG FSH Free Standing Harvester, www.perpetuum.com, 2012.

Spreemann, D., Manoli, Y., Folkmer, B., and Mintenbeck, D.: Non-resonant Vibration Conversion, Non-resonant Vibration Conversion, 16, 169–173, 2009.

Stephen, N.: On energy harvesting from ambient vibration, Journal of Sound and Vibration, 293, 409–425, 2006.

Xing, X., Yang, G. M., Liu, M., Lou, J., Obi, O., and Sun, N. X.: High power density vibration energy harvester with high permeability magnetic material, J. Appl. Phys., 109, 109–111, 2011.

Yang, W., Lynch, J., and Law, K.: A Wireless Structural Health Monitoring System with Multithreaded Sensing Devices: Design and Validation, Structure and Infrastructure Engineering: Maintenance, Management, Life-Cycle Design and Performance, 3, 103–120, 2007.

11

Novel fully-integrated biosensor for endotoxin detection via polymyxin B immobilization onto gold electrodes

A. Zuzuarregui[1], **S. Arana**[1,2], **E. Pérez-Lorenzo**[1,2], **S. Sánchez-Gómez**[3,*], **G. Martínez de Tejada**[3], **and M. Mujika**[1,2]

[1]CEIT and Tecnun (University of Navarra), Paseo Manuel Lardizábal No. 15, 20018 San Sebastián, Spain
[2]CIC microGUNE, Goiru Kalea 9, 20500 Arrasate-Mondragon, Spain
[3]Department of Microbiology and Parasitology, University of Navarra, Irunlarrea 1, 31008, Spain
[*]current address: Immunoadjuvant unit, Bionanoplus, Noain, Spain

Correspondence to: A. Zuzuarregui (anazuzuarregui@gmail.com)

Abstract. In this paper an electrochemical endotoxin biosensor consisting of an immobilized lipopolysaccharide (LPS) ligand, polymyxin B (PmB), is presented. Several parameters involved both in the device fabrication and in the detection process were analyzed to optimize the ligand immobilization and the interaction between PmB and LPS, aiming at increasing the sensitivity of the sensor. Different electrochemical pre-treatment procedures as well as the functionalization methods were studied and evaluated. The use of a SAM (self-assembled monolayer) to immobilize PmB and the quantification of the interactions via cyclic voltammetry allowed the development of a robust and simple device for in situ detection of LPS. Thus, the biosensor proposed in this work intends an approach to the demanding needs of the market for an integrated, portable and simple instrument for endotoxin detection.

1 Introduction

Sepsis is one of the most dreaded medical conditions, since it claims 750 000 lives per year in the United States (Wang et al., 2010). This pathology may lead to extensive injury of vascular endothelium, which often results in a more severe syndrome known as septic shock (Annane et al., 2005). About 50–60 % of septic shock episodes are related to infections by Gram-negative bacteria, and more concretely to the lipopolysaccharide (LPS or endotoxin), a major component of the bacterial outer membrane (Chaby, 1999). The human body is extremely sensitive to endotoxin and minimum amounts of this substance injected into it can cause serious effects to the system (e.g., fever, intravascular blood clotting and multiorgan failure) (Cohen, 2002).

Endotoxins are ubiquitous in the environment and they may be present in medical implants as well as in drugs for parenteral administration; so it is necessary to ensure that the content of endotoxin in these items does not exceed certain limits. As set by the American and European Pharmacopoeias, these limits range from 0.2 to 5 endo-

toxin units (EU)/kg body/hour for intravenous injections and 2.15–20 EU for medical devices (European Pharmacopeia, 2005; USPC, 2005), being 1 EU approximately equivalent to 100 pg of *E. coli* LPS. Therefore, the implementation and validation of efficient endotoxin detection techniques is of extreme importance.

There are two main validated methods for endotoxin detection nowadays: the test *Limulus* amebocyte lysate (LAL) and the in vitro pyrogen test (IPT). The LAL test, the most recommended by pharmacopoeias, is based on the coagulation cascade that the LPS triggers upon interaction with the horseshoe crab (*Limulus polyphemus*) amebocytes. The IPT measures the interleukin-1β secreted by human blood cells in the presence of LPS (Daneshian et al., 2008). Above this, there are several commercialized kits for endotoxin detection such as Pyrogent™, Endosafe® and EndoLISA®. The former two systems are based on LAL reagents, whereas in the third kit the amount of endotoxin is quantified monitoring the recombinant Factor C (rFC). All these methods, despite being sensitive, require long incubation times to render

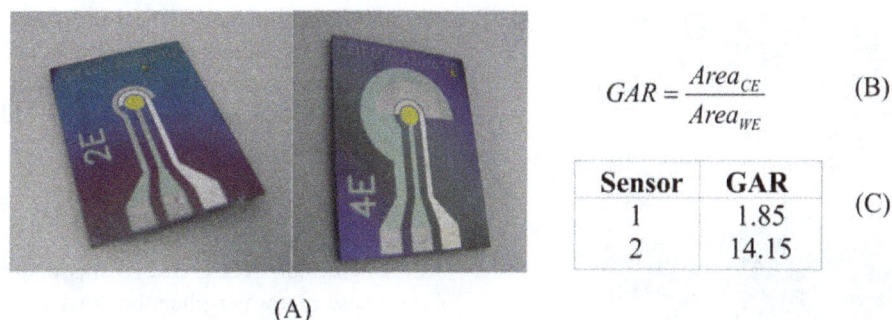

$$GAR = \frac{Area_{CE}}{Area_{WE}} \quad \text{(B)}$$

Sensor	GAR
1	1.85
2	14.15

(C)

(A)

Figure 1. (A) Fabricated chips: sensor 1 and sensor 2, **(B)** geometry area ratio, **(C)** GAR values.

results and are costly since they use colorimetric or fluorometric techniques.

Groups all over the world are focusing their research on alternative sensing methods for endotoxin detection to outweigh those disadvantages (Hreniak et al., 2004; Limbut et al., 2007; Priano et al., 2007). The electrochemical approach based on sensors or biosensors stands out among all of them. Most of these studies rely on electrochemical impedance spectroscopy (EIS) for the LPS detection (Cho et al., 2012; Ding et al., 2007; Heras et al., 2010; Rahman et al., 2013). For instance, the sensors developed by Ding and his co-workers and Rahman and its group use EIS to quantify the amount of LPS through immobilized PmB (polymyxin B). Other electrochemical techniques, such as differential pulse voltammetry (DPV) or cyclic voltammetry (CV), have also been used for the implementation of endotoxin detectors (Kato et al., 2007; Yeo et al., 2011).

As far as we know, all these methods use macroelectrodes or setups, making the portability almost impossible. The alternative we propose is based on the application of microbiosensors to endotoxin detection. These devices allow real-time analysis of biological reactions and can be applied to nearly any field. This kind of sensors, fabricated by means of microsystem techniques, are analytical devices that consist of two main components: a bioreceptor and a transducer that turns the detection event into an electrical output which is instantly measured (Berganza et al., 2007). The integration of the three electrodes and the use of thin-film technologies provide the developed system with the miniaturization and portability needed to carry out field experiments while decreases both the fabrication and operation costs.

It is known that some polycationic molecules such as the antibiotic PmB bind with high affinity to LPS, which is negatively charged at neutral pH (Brandenburg, 2002). In addition to being cationic, PmB has an amphiphilic nature due to the simultaneous presence in the molecule of the polycationic ring and the hydrophobic chain. This enables the stoichiometric PmB–LPS interaction and makes their binding reversible, under appropriate conditions (Morrison and Jacobs, 1976). All these features make PmB a promising candidate ligand for LPS in a bioreceptor aimed at detecting endotoxin (Chang, 1997). To ensure the correct immobilization of the ligand and to avoid nonspecific adsorption of endotoxin onto the surface of the sensors, PmB was covalently linked to self-assembled monolayers (SAMs) previously formed on the gold electrodes (Ignat et al., 2010). SAMs are organized structures of organic molecules that allow the efficient and simple immobilization of different compounds used for biological detection (Pillay et al., 2009). Once SAMs are formed, they remain strongly attached to the surface through their terminal thiol group and provide a well-defined and stable interface for ligand immobilization (Ansorena et al., 2011).

The aim of this study is the fabrication and characterization of an alternative method for endotoxin quantification. The microsensor proposed here is based on the detection of LPS via CV using PmB immobilized through a SAM as ligand. The integration of the whole device in a compact cell and the use of electrochemical techniques provide the basis for the development of a novel and in situ endotoxin detection method that the market is demanding.

2 Material and fabrication methods

2.1 Sensor fabrication

As an electrochemical biosensor, the device developed in this work consists of three integrated microelectrodes: a working electrode (WE), a counter electrode (CE) and a pseudo-reference electrode (RE). To analyze the influence of the geometry of the CE in the performance of the WE, two sensors differing in CE geometry were fabricated and tested (Fig. 1a). To quantify the influence of the area in the measurements, the geometry area ratio between CE and WE (GAR) was calculated according to Fig. 1b. The GAR values for the two sensors are shown in Fig. 1c.

Biosensors were fabricated employing standard microsystem processes on 4 in oxidized silicon wafers. The CE was made of platinum by DC sputtering (Edwards ESM-100) and patterned as shown in Fig. 1a with a thickness of 200 nm. The WE, made of gold, was deposited by RF sputtering (Edwards ESM-100) and shaped as a disk of 1.8 mm of diameter

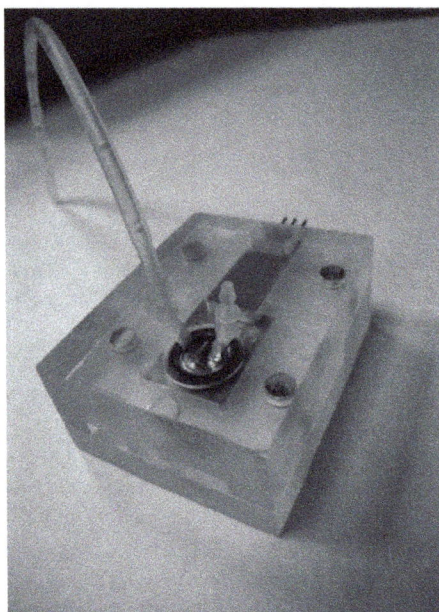

Figure 2. Biosensor mounted into the measurement cell. The inlet and outlet tubes can be appreciated

with a thickness of 150 nm. With the aim of integrating the reference electrode inside the structure of the microdevice, a concentric semi-circle of silver was deposited by DC sputtering (Edwards ESM-100). This layer of 1000 nm of thickness works as a pseudo-reference electrode (Añorga and Arana, 2011). To protect all the inactive parts of the biosensors, a 600 nm coating of SiO_2 was deposited by plasma enhanced vapor deposition (Oxford Plasmalab 80 plus). In order to fit the required sizes of the passivation layer, sensors were subjected to a wet SiO_2 etching.

2.2 Reaction cell fabrication

To minimize the influence of environmental conditions in the experiments and to assure the cleaning and sterilization needed, a measurement cell was designed and fabricated (Fig. 2). This device is made of methacrylate and has a 40 µL chamber where the assays take place. The base and cover of the cell are kept together with magnets and the microchamber is sealed by a toric joint. This microchamber can be filled and emptied using two fluidic connectors attached to the top of the cell.

2.3 Reagents

Sulfuric acid (H_2SO_4), pyrogen free water, polymyxin B sulfate salt (PmB) and phosphate buffer solutions (PBS) were purchased from Sigma-Aldrich. Perchloric acid ($HClO_4$), acetone 99.5 % pure and ethanol 99.5 % pure were supplied by Panreac. Mercaptopropionic acid (MPA), 1-ethyl-3-(3-dimethylaminopropyl) carbodiimide (EDC) and N-

hydroxysuccinimide (NHS) were purchased from Thermo Scientific. Trichlorethylene was obtained from Alden. Hellmanex II was supplied by Hellma. Ultra pure water of resistivity 18.2 MΩ was obtained from a Milli-Q Water System (Millipore Corp.).

2.4 Lipopolysaccharide preparation

LPS of *E. coli* ATCC 35218 was obtained from the aqueous phase of a water-phenol extract according to a published procedure (Leong et al., 1970). To remove traces of nucleic acids or proteins that could interfere with endotoxin detection, LPS extracts were dialyzed, lyophilized and purified following published protocols (Hirschfeld et al., 2000). The LPS from *E. coli* is the endotoxin used as reference by the regulatory agencies.

3 Experimental process

The whole electrochemical experiments were monitored with the Autolab Potentiostat PGSTAT 128N using the Nova 1.6 software version (Eco Chemie).

3.1 Sample preparation

Prior to conducting any procedure and to avoid the interference from potential contaminants, the sensors were subjected to a standard cleaning process consisting of three 5 min sonication steps, first in trichlorethylene, then in acetone and finally in ethanol.

To clean and prepare the sensors for the ligand immobilization, the gold electrodes (1.8 mm diameter) were electrochemically activated via cyclic voltammetry (CV). For this purpose, the potential was scanned from 0 to 1.2 V with a 0.1 V s^{-1} scan rate. Two electropolishing solutions were tested: sulfuric acid 0.05 M and perchloric acid 1 M. After each voltammetry assay, electrodes were rinsed with deionized water and dried with N_2.

SAMs were formed on the gold surface via a thiol group. The electrodes were incubated in MPA 1 mM for 2 h and then rinsed with ethanol to remove nonspecific bindings. To ensure ligand immobilization it is necessary to activate the SAM. For this purpose, the electrodes were incubated first with EDC 46 mM for 1 h and then with NHS 46 mM for 1 h. After each step, the microchamber was washed with water to remove unbound molecules.

Once the SAM was activated, the ligand was immobilized onto the working electrode surface. For this aim, the chamber was incubated with 100 µL of a 100 µg mL solution of PmB for 2 h. Before the subsequent processes, the chamber was rinsed to minimize nonspecific binding of the ligand.

LPS molecules form biologically active aggregates and the rate of aggregation can differ between experiments. To reduce this source of variability endotoxin containing solutions were subjected to three consecutive cycles of heating (10 min at 56 °C) and cooling (2 min at −20 °C) employing a vortex mixing between steps. To quantify the LPS detection by the functionalized sensor, 100 μL of the solution were placed in the measurement chamber for 30 min. Three different concentrations of analyte were tested: 100, 10 and 1 μg mL^{-1}.

As control assays, experiments without SAM were carried out to analyze the need for ligand immobilization. In these assays, PmB was immobilized directly onto the gold surface and then the LPS was added.

After performing the assays, the sensors, the cell chamber and the connectors were washed with Hellmanex II and then incubated with the same compound for 30 min. Finally, both sensor and cell were rinsed thoroughly with deionized water and dried with N$_2$.

3.2 Cyclic voltammetry measurements

The influence of the electrode's geometry, the gold pretreatment and the validation of the detection system were carried out by means of cyclic voltammetry (CV). All the assays were performed in the same environmental conditions: at room temperature (23 °C) and in PBS buffer (pH 7.4). The CV measurements were carried out in the range −0.2 to 0.7 V, with a scanning rate of 0.1 V s^{-1}.

Each assay consisted of three measurements. First of all, the behavior of the working electrode prior to any immobilization step was characterized (Bare Gold). Then another CV reading after the PmB immobilization was performed to monitor any potential change. Finally, the response of the sensor after the LPS incubation was measured, to analyze the interaction of the two molecules and quantify the amount of endotoxin bound.

4 Results and discussion

As previously mentioned, the purpose of this work was to design, fabricate and test an endotoxin biosensor. The results obtained are summarized in three sections focused on the following issues: the analysis of the influence of the CE geometry, the study of different solutions for gold pre-treatment and, finally, the validation of the performance of the biosensor.

4.1 Effect of the geometry of the counter electrode

The fabrication and patterning of the CE has a significant influence on the WE performance and on the degradation of the CE itself (due to the current densities passing through the platinum layer) as some authors have shown (Kim et al., 2004; Radev et al., 2010). This fact led us to design and fabricate two types of sensors varying the CE area. To compare the performance of the two developed sensors (Fig. 1a), experiments involving the formation of the SAM followed by the addition of PmB (100 μg mL^{-1}) and LPS (100 μg mL^{-1})

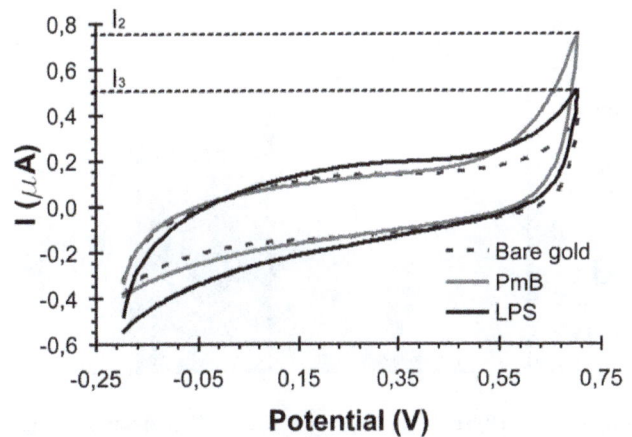

Figure 3. Cyclic voltammogram (−0.2–0.7 V) of the bare gold, the immobilized PmB (100 μg mL^{-1}) and the detected LPS (100 μg mL^{-1}).

were carried out with the two types of biosensors. Figure 3 shows the results obtained in each assay.

As depicted in Fig. 3, the immobilization of PmB induced an increase in the signal measured, more than likely due to the conductive properties of this peptide. However, after the addition of LPS, the current dropped, reflecting both neutralization of the positive charge of the PmB by the LPS and the presence of electrically insulating groups (lipid A) in the lipopolysaccharides.

Considering that no redox couple is added, and therefore there is no possibility for electron transfer between the solution and the electrode, the amount of endotoxin detected is quantified in terms of the changes in the measured current after the immobilization of the PmB (measurement I_2) and once the LPS is detected (measurement I_3). The current changes are calculated according to Eq. (1) and shown in Fig. 4.

$$\Delta I = \frac{I_2 - I_3}{Area_{WE}} \tag{1}$$

Change in the measured current (ΔI) where $Area_{WE}$ is the active area of the working electrode. These results imply that sensor 2 has a more sensitive response to LPS than sensor 1, since the change in current is significantly higher in the former. This conclusion is consistent with the fact that sensor 2 has a higher GAR value and that an increase of the GAR value improves the efficiency of the WE and prevents the deterioration of the CE. Therefore, sensor 2 was selected for subsequent assays.

4.2 Influence of the electropolishing solution

To investigate the effect of the medium used for the pre-treatment of the working electrode, cyclic voltammetry experiments (0–1.2 V) were performed using two solutions:

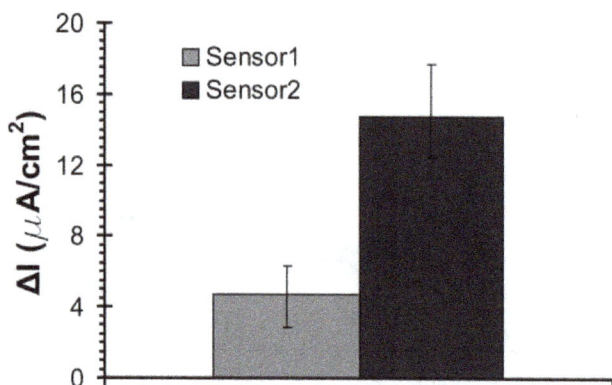

Figure 4. Analysis of the influence of the CE geometry. Change of the measured current in the two types of sensors due to the LPS ($100\,\mu g\,mL^{-1}$) interaction with the PmB ($100\,\mu g\,mL^{-1}$) immobilized via SAM.

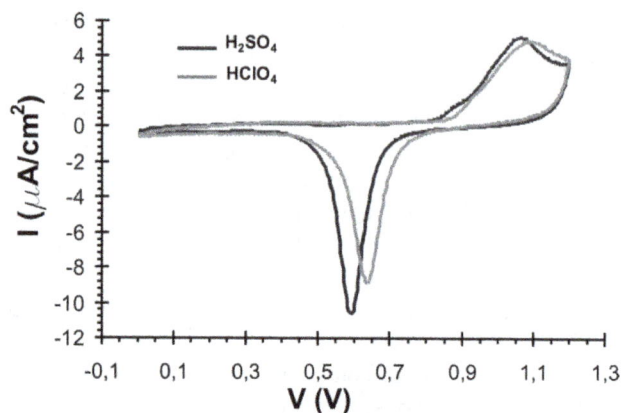

Figure 5. Cyclic voltammogram of the gold electrodes after H_2SO_4 and $HClO_4$ electropolishing.

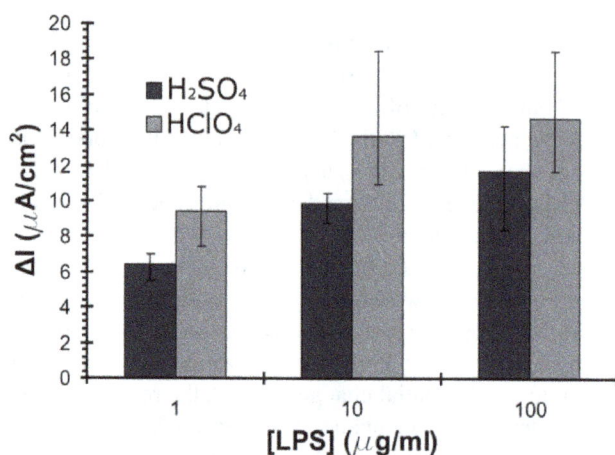

Figure 6. Influence of the pre-treatment solutions in the LPS detection. Change of the measured current in the two types of electropolished sensors due to the LPS interaction with the PmB ($100\,\mu g\,mL^{-1}$) immobilized via SAM.

sulfuric acid (0.05 M) and perchloric acid (0.05 M). To compare both methods, once the pre-treatment was finished, all sensors were cycled once more in sulfuric acid solution, obtaining the usual voltammograms (Fig. 5). The assays for each solution were carried out with at least five different sensors to confirm the results.

To study the surface roughness of the gold layers the electrochemical surface area (ESA) of the electrodes after the different pre-treatment methods has been determined. The measurement of the oxygen adsorption has been chosen as indicator of the microscopic surface area of gold (Hoogvliet et al., 2000). This determination was done integrating the gold oxide reduction peak from the voltammetry curves referred to the electrode area (Q_{exp}). The standard reference charge of gold electrodes is $390\pm10\,\mu C\,cm^{-2}$ (Q_{std}) (Trasatti and Petrii, 1991). The ESA is the ratio between the experimental charge of the gold electrodes and the theoretical one; whereas the roughness factor is that value expressed per unit of geometric surface area (Carvalhal et al., 2005).

$$ESA = Q_{exp}/Q_{std} \qquad (2)$$

The electrochemical surface area determined for the electrodes pre-treated with sulfuric acid is $3.52 \times 10^{-2} \pm 1.6 \times 10^{-3}\,cm^2$ (mean value and standard deviation of five electrodes) and the roughness factor (R_f) takes the value of 1.75 ± 0.08. On the other hand, the electrodes pre-treated with perchloric acid had an ESA of $2.83 \times 10^{-2} \pm 0.9 \times 10^{-3}\,cm^2$ (mean value and standard deviation of five electrodes) and a roughness factor (R_f) of 1.41 ± 0.05. These results improve the values obtained by other research groups (Bonroy et al., 2004; Carvalhal et al., 2005) with high reproducibility and confirm the increase in the active area that provides the gold electrodes with better biosensing features.

Despite these observations, we decided to study whether the pre-treatment of the sensor with either H_2SO_4 or $HClO_4$ could influence its subsequent ability to interact first with

PmB and then with LPS. For this purpose, sensors were first cycled with either H_2SO_4 or $HClO_4$, and once the SAM formation and the PmB immobilization were performed, the amount of LPS bound to the sensor was quantitatively assessed. In these assays, PmB was added to the chamber at a concentration of $100\,\mu g\,mL^{-1}$ and then different concentrations of LPS ($100\,\mu g\,mL^{-1}$, $10\,\mu g\,mL^{-1}$ and $1\,\mu g\,mL^{-1}$) were tested. The results of this group of assays are shown in Fig. 6.

The data depicted in Fig. 6 confirm previous results with the voltammograms. Despite the fact that the absolute values obtained with the $HClO_4$ pre-treatment are higher, this method results in an increase of the experimental variability. This makes almost impossible to discriminate between the sensor responses to $10\,\mu g\,mL^{-1}$ of LPS and to $100\,\mu g\,mL^{-1}$. In contrast, the use of H_2SO_4 results in lower variability and in clear dose-response behavior that allows differentiating the three concentrations of LPS tested. Ding and his co-workers stated that the pre-treatment with perchloric acid leads to

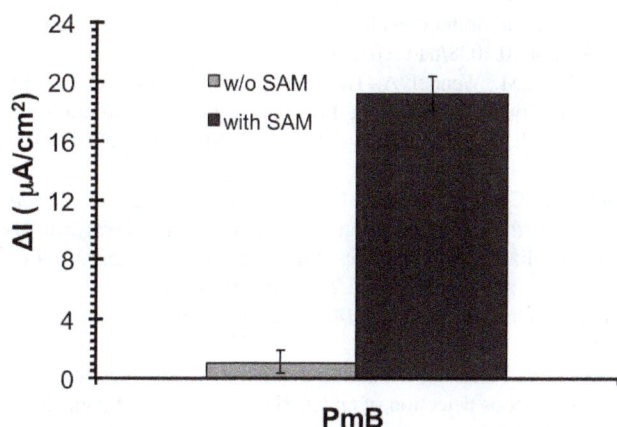

Figure 7. Effect of the SAM in the ligand immobilization. Change of the measured current due to the PmB immobilization with and without SAM.

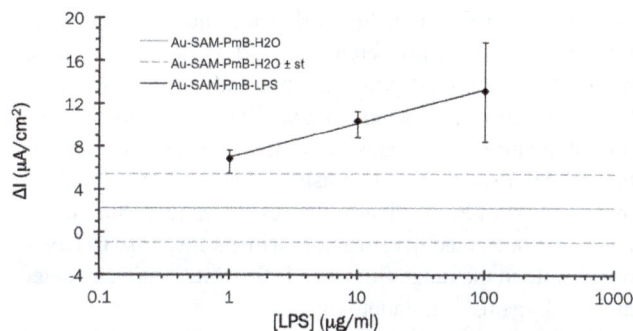

Figure 8. Quantitative detection of LPS by the biosensor.

better results than the sulfuric acid in their paper (Ding et al., 2007). They analyze the performance of the sensor through CV in $Fe(CN)_6^{-3/-4}$ measuring the obtained current, showing the voltammogram of a single sensor. Those results are similar to the ones obtained in this work as they get a larger response of the sensor; however, an increase in the current does not imply a better efficiency as shown in Fig. 6. These results together with the ESA and R_f values calculated from the cyclic voltammograms led us to select the sulfuric acid pre-treatment for subsequent assays.

4.3 Test and validation of the device

4.3.1 Effect of the SAM in the PmB immobilization and LPS detection processes

In order to test if the formation of SAM was necessary for an efficient PmB immobilization, the amount of PmB bound to sensors with and without SAM was quantified. In these assays, PmB was added at $100 \,\mu g\,mL^{-1}$ (Fig. 7).

Sensors with SAM bound approximately 10 times more PmB than their counterparts without SAM and the current increase due to PmB in sensors without SAM was almost imperceptible, demonstrating the importance of SAM formation for the proper PmB immobilization and presumably for an appropriate LPS detection. Without SAM, the PmB does not attach well to the gold surface and it might get washed off during the measurement process.

This behavior can be extrapolated to the LPS molecules as they do not include any sulfide group and have the mentioned amphiphilic character similar to the PmB. So, it can be presumed that the endotoxins do not get attached to the gold surface that could be available and do not interfere in the analysis. Regarding the possible binding to the SAM, LPS molecules do not have available amine groups so, as the immobilization takes places through a peptide bond, the reaction between the lipopolysaccharide and the self-assembled

monolayer is unlikely to happen. Besides this, the negative charge of the endotoxins and the SAM results in a repelling force that prevents the LPS binding.

4.3.2 Quantitative analysis of LPS detection

To demonstrate if the device allows detecting LPS quantitatively, the response of the optimized sensor (with SAM and PmB) to decreasing concentrations of LPS was measured. Besides this, control assays with pyrogen-free water were made to establish the detection limit (grey continuous and dashed lines). The results of these assays are presented in Fig. 8.

Since the variability of the measurements decreases as the concentration of the analyte diminishes (from 35 to 12 %), these results show that the developed biosensor is suitable for LPS detection, particularly at low LPS concentrations. Interestingly, as shown in Fig. 8 the change in current is in a linear relationship with the LPS concentration in the range of concentrations tested. The LOD (limit of detection), calculated as the concentration that gives a signal three times the standard deviation of the background signal of the system, is $32.89 \,\mu g\,mL^{-1}$. Although lower detection limits have been reported for endotoxin detection; to the best of our knowledge, all the methods require macroelectrodes that prevent the miniaturization of the device and hamper their portability. Therefore, the microbiosensor developed in this work is a proof of concept to the future implementation of integrated electrochemical biosensors for endotoxin detection.

5 Conclusions

Biosensors are rapid, sensitive and portable devices that can be used for the detection of biomedically relevant analytes, as well as for other biological or environmental applications. In this paper, a new LPS biosensor was designed, fabricated and tested for endotoxin detection. The microdevice was developed using standard microelectronic processes and patterned with the optimum size for each of the three electrodes.

Electrochemical polishing is an essential pre-treatment step for the cleaning and activation of gold electrodes. For the application shown in this work, we demonstrated that sulfuric acid outperforms perchloric acid as pre-treatment solution. We confirmed previous observations indicating that SAM formation is a method that provides stable and efficient immobilizations of ligands on surfaces. In fact, we showed that SAM formation was indispensable for the correct immobilization of PmB. Finally, our results indicate that the microdevice presented here provides a linear response to LPS in the sample in the range from 1 to $100\,\mu g\,mL^{-1}$ and can detect at least $1\,\mu g\,mL^{-1}$ of endotoxin.

Therefore, although further research is needed to improve the sensitivity of detection and to study its specificity, the fully integrated and miniaturized biosensor designed lays the foundation for the implementation of a useful tool for clinical and industrial applications.

Edited by: N.-T. Nguyen
Reviewed by: two anonymous referees

References

Annane, D., Bellissant, E., and Cavaillon, J.-M.: Septic shock, Lancet, 365, 63–78, doi:10.1016/S0140-6736(04)17667-8, 2005.

Añorga, L. and Arana, S.: G01N27/407D2. Patent: Pseudo-electrodo de referencia de película delgada y procedimiento para su fabricación, WO/2011/054982, 1, 2011.

Ansorena, P., Zuzuarregui, A., Pérez-Lorenzo, E., Mujika, M., and Arana, S.: Comparative analysis of QCM and SPR techniques for the optimization of immobilization sequences, Sensor. Actuat. B-Chem., 155, 667–672, 2011.

Berganza, J., Olabarria, G., García, R., Verdoy, D., Rebollo, A., and Arana, S.: DNA microdevice for electrochemical detection of Escherichia coli 0157:H7 molecular markers, Biosens. Bioelectron., 22, 2132–2137, doi:10.1016/j.bios.2006.09.028, 2007.

Bonroy, K., Friedt, J.-M., Frederix, F., Laureyn, W., Langerock, S., Campitelli, A., Sára, M., Borghs, G., Goddeeris, B., and Declerck, P.: Realization and characterization of porous gold for increased protein coverage on acoustic sensors, Anal. Chem., 76, 4299–306, doi:10.1021/ac049893u, 2004.

Brandenburg, K.: Biophysical investigations into the interaction of lipopolysaccharide with polymyxins, Thermochim.a Acta, 382, 189–198, doi:10.1016/S0040-6031(01)00731-6, 2002.

Carvalhal, R. F., Sanches Freire, R., and Kubota, L. T.: Polycrystalline Gold Electrodes: A Comparative Study of Pretreatment Procedures Used for Cleaning and Thiol Self-Assembly Monolayer Formation, Electroanalysis, 17, 1251–1259, doi:10.1002/elan.200403224, 2005.

Chaby, R.: Strategies for the control of LPS-mediated pathophysiological disorders, Drug Discov. Today, 4, 209–221, available at: http://www.ncbi.nlm.nih.gov/pubmed/10322288, 1999.

Chang, H.: Detection of lipopolysaccharide binding peptides by the use of a lipopolysaccharide-coated piezoelectric crystal biosensor, Anal. Chim. Acta, 340, 49–54, doi:10.1016/S0003-2670(96)00520-X, 1997.

Cho, M., Chun, L., Lin, M., Choe, W., Nam, J., and Lee, Y.: Sensitive electrochemical sensor for detection of lipopolysaccharide on metal complex immobilized gold electrode, Sensor. Actuat. B-Chem., 174, 490–494, doi:10.1016/j.snb.2012.09.017, 2012.

Cohen, J.: The immunopathogenesis of sepsis, Nature, 420, 885–891, doi:10.1038/nature01326, 2002.

Daneshian, M., Wendel, A., Hartung, T., and von Aulock, S.: High sensitivity pyrogen testing in water and dialysis solutions, J. Immunol. Methods, 336, 64–70, doi:10.1016/j.jim.2008.03.013, 2008.

Ding, S., Chang, B., Wu, C., Chen, C., and Chang, H.: A new method for detection of endotoxin on polymyxin B-immobilized gold electrodes, Electrochem. Commun., 9, 1206–1211, doi:10.1016/j.elecom.2006.12.029, 2007.

European Pharmacopeia: 2.6.14. Bacterial endotoxins, Test, 1, 0–7, 2005.

Heras, J. Y., Pallarola, D., and Battaglini, F.: Electronic tongue for simultaneous detection of endotoxins and other contaminants of microbiological origin., Biosens. Bioelectron., 25, 2470–2476, doi:10.1016/j.bios.2010.04.004, 2010.

Hirschfeld, M., Ma, Y., Weis, J. H., Vogel, S. N., and Weis, J. J.: Cutting edge: repurification of lipopolysaccharide eliminates signaling through both human and murine toll-like receptor 2., J. Immunol. (Baltimore, Md.: 1950), 165, 618–622, 2000.

Hoogvliet, J., Dijksma, M., Kamp, B., and van Bennekom, W. P.: Electrochemical pretreatment of polycrystalline gold electrodes to produce a reproducible surface roughness for self-assembly: a study in phosphate buffer pH 7.4, Anal. Chem., 72, 2016–2021, 2000.

Hreniak, A., Maruszewski, K., Rybka, J., Gamian, A., and Czywewski, J.: A luminescence endotoxin biosensor prepared by the sol–gel method, Opt. Mater., 26, 141–144, doi:10.1016/j.optmat.2003.11.013, 2004.

Ignat, T., Miu, M., Kleps, I., Bragaru, A., Simion, M., and Danila, M.: Electrochemical characterization of BSA/11-mercaptoundecanoic acid on Au electrode, Mater. Sci. Eng. B, 169, 55–61, doi:10.1016/j.mseb.2009.11.021, 2010.

Kato, D., Iijima, S., Kurita, R., Sato, Y., Jia, J., Yabuki, S., Mizutani, F., and Niwa, O.: Electrochemically amplified detection for lipopolysaccharide using ferrocenylboronic acid., Biosens. Bioelectron., 22, 1527–1531, doi:10.1016/j.bios.2006.05.020, 2007.

Kim, S. K., Hesketh, P. J., Li, C., Thomas, J. H., Halsall, H. B., and Heineman, W. R.: Fabrication of comb interdigitated electrodes array (IDA) for a microbead-based electrochemical assay system., Biosens. Bioelectron., 20, 887–894, doi:10.1016/j.bios.2004.04.004, 2004.

Leong, D., Diaz, R., Milner, K., Rudbach, J., and Wilson, J. B.: Some structural and biological properties of Brucella endotoxin., Infect. Immun., 1, 174–82, 1970.

Limbut, W., Hedström, M., Thavarungkul, P., Kanatharana, P., and Mattiasson, B.: Capacitive biosensor for detection of endotoxin, Anal. Bioanal. Chem., 389, 517–525, doi:10.1007/s00216-007-1443-4, 2007.

Morrison, D. C. and Jacobs, D. M.: Binding of polymyxin B to the lipid A portion of bacterial lipopolysaccharides, Immunochemistry, 13, 813–818, 1976.

Pillay, J., Agboola, B. O., and Ozoemena, K. I.: Electrochemistry of 2-dimethylaminoethanethiol SAM on gold electrode: Interaction with SWCNT-poly(m-aminobenzene sulphonic acid), electric field-induced protonation–deprotonation, and surface pKa, Electrochem. Commun., 11, 1292–1296, doi:10.1016/j.elecom.2009.04.028, 2009.

Priano, G., Pallarola, D., and Battaglini, F.: Endotoxin detection in a competitive electrochemical assay: synthesis of a suitable endotoxin conjugate., Anal. Biochem., 362, 108–116, doi:10.1016/j.ab.2006.12.034, 2007.

Radev, I., Cho, Y.-H., Koutzarov, K., Sung, Y.-E., and Tsotridis, G.: The effect of the geometric area ratio between working and counter PEM electrodes for electrochemical hydrogen reactions, Int. J. Hydrogen Energ., 35, 12449–12453, doi:10.1016/j.ijhydene.2010.08.079, 2010.

Rahman, M. S. A., Mukhopadhyay, S. C., Yu, P. L., Goicoechea, J., Matias, I. R., Gooneratne, C. P., and Kosel, J.: Detection of Bacterial Endotoxin in Food: New Planar Interdigital Sensors based Approach, J. Food Eng., 114, 346–360, doi:10.1016/j.jfoodeng.2012.08.026, 2013.

Trasatti, S. and Petrii, O. A.: Real surface Area Measurements in Electrochemistry, Pure Appl. Chem., 63, 711–734, doi:10.1351/pac199163050711, 1991.

USPC: Chapter 85: Bacterial Endotoxins Test, 2005.

Wang, H. E., Devereaux, R. S., Yealy, D. M., Safford, M. M., and Howard, G.: National variation in United States sepsis mortality: a descriptive study, Int. J. Health Geogr., 9, 1–9, doi:10.1186/1476-072X-9-9, 2010.

Yeo, T. Y., Choi, J. S., Lee, B. K., Kim, B. S., Yoon, H. I., Lee, H. Y., and Cho, Y. W.: Electrochemical endotoxin sensors based on TLR4/MD-2 complexes immobilized on gold electrodes, Biosens. Bioelectron., 28, 139–145, doi:10.1016/j.bios.2011.07.010, 2011.

12

Capacitive strain gauges on flexible polymer substrates for wireless, intelligent systems

R. Zeiser[1], T. Fellner[1,*], and J. Wilde[1]

[1]Department for Microsystems Engineering, University of Freiburg, Freiburg, Germany
[*]now at: Robert Bosch GmbH, Reutlingen, Germany

Correspondence to: R. Zeiser (roderich.zeiser@imtek.uni-freiburg.de)

Abstract. This paper presents a novel capacitive strain gauge with interdigital electrodes, which was processed on polyimide and LCP (liquid crystal polymer) foil substrates. The metallization is deposited and patterned using thin-film technology with structure sizes down to 15 µm. We determined linear strain sensitivities for our sensor configuration and identified the most influencing parameters on the output signal by means of an analytical approach. Finite-element method (FEM) simulations of the strain gauge indicated the complex interaction of mechanical strains within the sensitive structure and their effect on the capacitance. The influence of geometry and material parameters on the strain sensitivity was investigated and optimized. We implemented thin films on 50 µm thick standard polymer foils by means of a temporary bonding process of the foils on carrier wafers. The characterization of the strain sensors after fabrication revealed the gauge factor as well as the cross sensitivities on temperatures up to 100 °C and relative humidity up to 100 %. The gauge factor of a sensor with an electrode width of 45 µm and a clearance of 15 µm was -1.38 at a capacitance of 48 pF. Furthermore, we achieved a substantial reduction of the cross sensitivity against humidity from 1435 to 55 ppm %$^{-1}$ RH when LCP was used for the sensor substrate and the encapsulation instead of polyimide. The gauge factor of a sensor half-bridge consisting of two orthogonal capacitors was 2.3 and the cross sensitivity on temperature was reduced to 240 ppm K^{-1}. Finally, a sensor system was presented that utilizes a special instrumentation Integrated Circuit (IC). For this system, performance data comprising cross sensitivities and power consumption are given.

1 Introduction

Wireless sensor networks are presently under investigation for the condition monitoring of moving machine parts in industrial environments or in mobile systems. The sensor nodes must be suited to measure, store and transmit information about the state of the structure they are mounted on. Typical data of interest are temperature, pressure, acceleration and strain in the surface of the part. Sensor elements applied in such networks should measure the required data with high resolution, often with high (> 1 kHz) frequency and low power consumption (< 5 mW). A wired interconnection for power supply and data acquisition is not suitable for sensors on moving parts such as rotating shafts. Additionally, the supply of the sensor systems has to be independent from

the main power net of a drive system to guarantee that data will be collected even during a power "black out" for a limited period of time. The strain in the surface of a machine element is an important parameter for condition monitoring. On the basis of strains it is possible to calculate stresses and to detect possible overload of the machine. In stationary applications, resistive strain gauges are usually used for strain measurement. These transducers are supplied and read out with wires. Conventional resistive strain gauges have disadvantages in wireless applications due to their low impedance and therefore high power consumption.

Capacitive sensors are a promising alternative for resistive gauges in the field of wireless strain measurement. Available modern instrumentation ICs can be used to measure

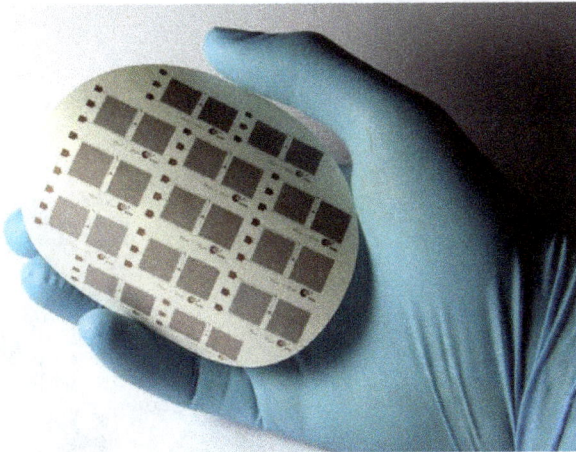

Figure 1. Capacitive strain gauges on LCP foil substrate, directly after processing, and detachment from a carrier wafer.

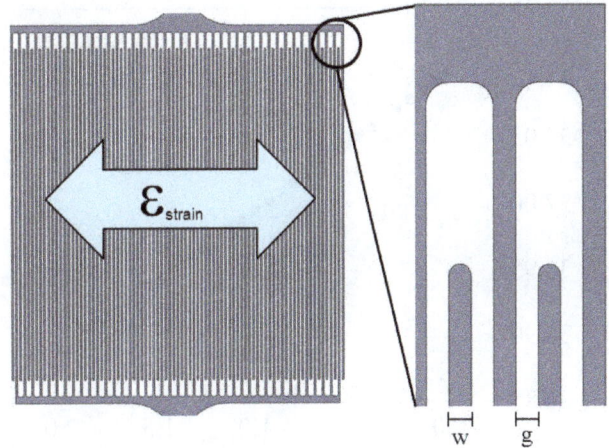

Figure 2. Schematic view of the sensor configuration with interdigital electrodes. Left: working principle of strain detection. Right: geometric electrode parameters – width w and distance g.

capacitances with high resolution, at high frequencies with a minimum of power consumption.

In the literature, several concepts for capacitive strain sensors were presented recently (Matsuzaki et al., 2007; Aebersold et al., 2006; Arshak et al., 2000). Our approach combines sensor structures processed with thin-film technology and flexible foil substrates with integrated instrumentation ICs. The capacitive sensor of this study is based on interdigital electrodes with structure heights of about 300 nm, fabricated on flexible polymer foils serving as sensor substrates. Figure 1 depicts capacitive sensor elements on a liquid crystal polymer (LCP) foil after thin-film processing. In this paper we will analyze the substrate material aspects, present the fabrication process and reveal results for both strain sensitivity and cross sensitivities of our strain gauges.

2 Theory

In this chapter an analytical model for the sensitivity of the presented strain gauges is given. The sensor principle is an interdigital capacitor with electrodes in a single plane, as exhibited in Fig. 2. The transfer of strain into the sensor structure leads to a change in the clearances of the interdigital electrodes. For an arrangement of the electrodes perpendicular to the strain in the test specimen, the capacitance decreases compared to the status without tensile strain. Igreja et al. (2004) gave an approximation for the computation of the capacitance for interdigital electrodes with Eq. (1) (Igreja et al., 2004). We computed capacitances and strain sensitivities for different interdigital configurations with this analytical formula. Mamishev et al. (2004) presented another method of capacitance computation based on complex transformation (Mamishev et al., 2004.).

The capacitance C of two interdigital electrodes can be obtained via following equation:

$$C = \varepsilon_0 \cdot \varepsilon_r \cdot L \cdot \frac{K(k)}{K(k')} . \qquad (1)$$

$K(k)$ is the elliptic integral of the first type with the modulus k and the complementary modulus k'. h is the electrode height, g is the distance between two electrodes, w is the electrode width and L is the active electrode length. ε_r represents the material-dependent relative permittivity and ε_0 the dielectric constant of free space.

The modulus of $K(k)$ is

$$k = \sin\left(\frac{\pi}{2}\eta\right) \qquad (2)$$

and the complementary modulus

$$k' = \sqrt{1 - k^2} . \qquad (3)$$

Here η stands for the geometric relation of electrode width to the electrode pitch,

$$\eta = \frac{w}{w + g} . \qquad (4)$$

Different geometrical configurations of a sensor element have been calculated. The obtained result for a sensor arrangement with electrode width $w = 45\,\mu m$, height $h = 300\,nm$ and distance $g = 15\,\mu m$ is presented in Fig. 3. In the following, these geometrical arrangements will be denoted as PI-45/15 and LCP-45/15.

The influence of strain on the electrode distance g is given as follows:

$$\varepsilon_{mech} = \frac{\Delta g}{g} . \qquad (5)$$

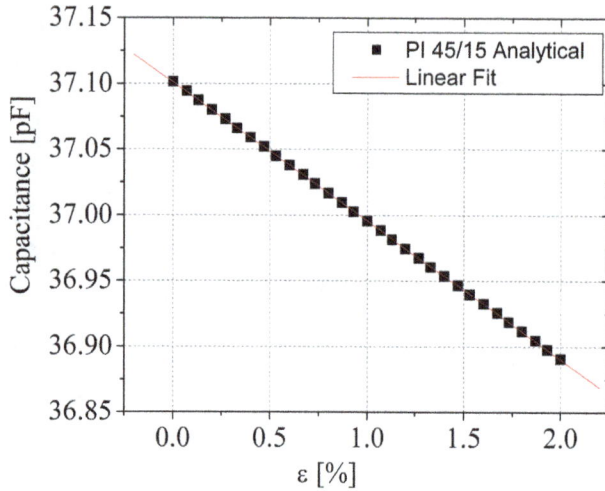

Figure 3. Results for the analytical evaluation of the sensor capacitance of a PI 45/15 as a function of strain.

Equation (6) computes the total capacitance of the interdigital configuration,

$$C_{\text{total}} = (N - 3)\frac{C_I}{2} + 2 \cdot \frac{C_I C_E}{C_I + C_E} \text{, for } N > 3 \text{,} \qquad (6)$$

where N is the number of electrode fingers; C_I is the capacitance of two inner electrodes, calculated the same as C in Eq. (1); and C_E stands for exterior electrodes and is computed with the modulus k_E of the elliptic integral $K(k_E)$ as follows:

$$k_E = \frac{2\sqrt{\eta}}{1 + \eta} \text{.} \qquad (7)$$

Figure 3 depicts the results for the capacitance of the interdigital configuration as a function of strain, calculated with the analytical approach. For low strain values, the change in the capacitance is linearly dependent on the change in the electrode distance.

The capacitance of this electrode configuration is 37.1 pF.

The gauge factor G_F is one of the principal specification values for strain gauges. For capacitive sensors the gauge factor can be determined as follows:

$$G_F = \frac{\Delta C}{C_0 \cdot \varepsilon_{\text{mech}}} \text{.} \qquad (8)$$

For the presented configuration PI 45/15, the computed gauge factor is −0.3.

2.1   FEM simulation of capacitance and strain sensitivity

In the previous section we verified the linear dependency of strain for the interdigital strain gauge. For investigation of the complex interaction of mechanical strains in a three-dimensional material compound and their effect on the capacitance, the analytical approach is unsatisfactory.

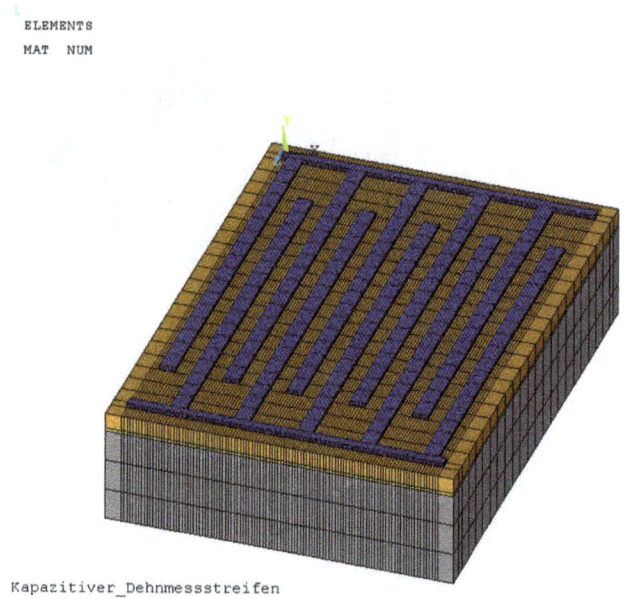

Kapazitiver_Dehnmessstreifen

Figure 4. Meshed 3-D FEM model of the sensor structure without encapsulation.

The finite-elements method (FEM) was chosen for the numerical calculation of the sensor capacitance and strain sensitivity. We performed coupled mechanical and electrostatic simulations with the finite-element-simulation software AN-SYS. Figure 4 exhibits the meshed 3-D FEM model of the sensor electrodes on a foil substrate, mounted on a test specimen with an adhesive layer. The picture illustrates the geometric and material-dependent arrangement of the brick elements. In the mechanical part of the simulation, different strains in the specimen were defined as the mechanical loads of the system.

The FEM model consists of three-dimensional, eight-node elements of the type SOLID185. After the mechanical simulation, the resulting deformed elements were transformed into electrostatic elements of type SOLID69. The materials parameters, utilzied for the simulations, are presented in Table 1.

The FEM software computes the capacitance of the strain gauge on the basis of the distribution of the electrostatic field in the dielectric material for a given voltage. Figure 5 shows the contour plot of results for an electrostatic simulation of the finger structures. The concentration of the electrostatic field between the electrodes is illustrated in this figure. One can deduce from this illustration that the effect of capacitance change due to strain is maximized in areas where the maximum electric field is located.

The strain dependency of the sensor output signal turned out to be linear, as predicted by the analytical approach for our sensor configuration. Figure 6 exhibits the results of a simulation for a strain gauge with an electrode width of 45 μm and a distance of 15 μm.

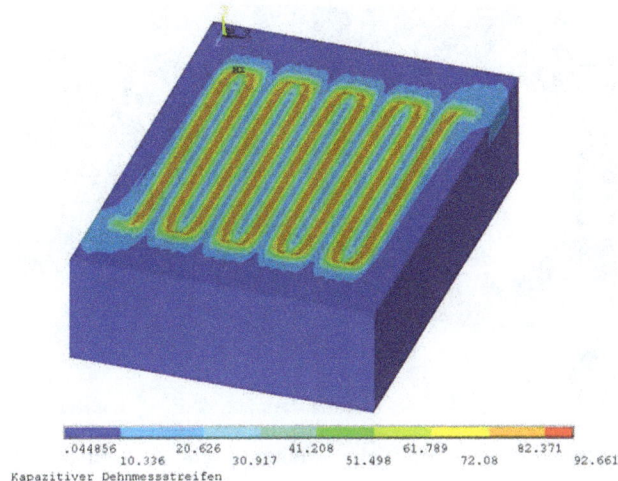

Figure 5. Distribution of the electrostatic field [in $V\,m^{-1}$] in the dielectric material for a voltage difference of 5 V at the interdigital electrodes.

When we varied different design parameters, we found a maximum strain sensitivity of $-0.566\,pF\,\%^{-1}$ within the fabrication limits for our sensor. A capacitance of 46 pF was obtained for a sensor area of $1\,cm^2$. The corresponding gauge factor G_F of this strain gauge is -1.23.

Due to a certain strain concentration between the electrodes and a simultaneous decrease in the substrate and encapsulation layer thickness in the loaded case, the gauge factor of the simulation was 4 times higher than when predicted with the analytical approach in the previous section.

We analyzed the influence of geometry and material-dependent parameters of the strain gauge and optimized the system with regard to high strain sensitivity. The optimum structure for a sensor area of $1\,cm^2$ within our process limitations was an electrode of width $w = 45\,\mu m$, height $h = 300\,nm$ and distance $g = 15\,\mu m$.

Several basic design rules for the sensor layout could be established on the basis of the FEM simulations:

- The sensor substrate should not be thicker than the pitch of the electrodes, because the sensitivity correlates inversely directly with the substrate thickness.

- The gauge factor increases with smaller interdigital distances of the electrodes. For a sensor area of $1\,cm^2$, the sensor structures should be in the range of 10 to $50\,\mu m$.

- As the the thickness of the electrode metallization increases, the characteristic curve of capacitance against strain changes from linear to a hyperbolic behavior.

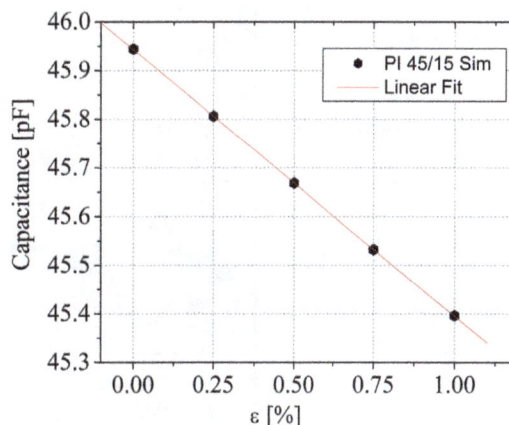

Figure 6. Results for the mechanical–electrostatic FEM simulation of a PI 45/15 sensor capacitance as a function of strain.

3 Fabrication

3.1 Flexible polymer substrate materials

The choice of the substrate material is crucial for the fabrication of a capacitive sensor. The mechanical and dielectric properties will affect the sensor output characteristics significantly. The principal influences on the capacitance are due to the dielectric properties. Polymer materials are well-suited substrates because a maximum strain transfer can be accomplished due to their low Young modulus and their low back stress on the structure. Hence, the strain is transmitted into the sensor structure with low gauge losses.

The principal foil-based substrate material used in our evaluation is polyimide, which is a standard substrate material used for both resistive strain gauges and flexible PCBs (printed circuit boards). Thin foils of polyimide down to $10\,\mu m$ are available with good thickness conformity and high mechanical strength. Other outstanding material properties are its high thermal and chemical stability. Due to the polymerization route, a high absorption of moisture of approximately 3 % is one disadvantage of polyimide as a sensor material (Melcher et al., 1989).

The second foil material utilized in this study is LCP. This material is mostly known as a material for premolded packages for MEMS sensors. LCP is also applied in the field of high-frequency electronics, mainly because of its low dielectric loss (Zhang, 2006). The value for moisture absorption of LCP is about 0.04 %, and approximately 75 times lower than for polyimide (Rogers Corporation, 2008).

3.2 Process flow

To obtain the structures with the necessary low electrode pitches, it is necessary to use photolithography for the processing. In this work a lift-off process was chosen for patterning because of its outstanding properties such as accuracy and reproducibility. Structures in the micrometer range

a) Temporary bonding of polymer substrate	
b) Lithography with negative photo resist	
c) O$_2$-plasma surface activation and PVD of Cr + Cu/Au	
d) Lift-off process	
e) Coverage of sensor structures with protective resist	
f) Electroplating of Ni + Au for bond pads	
g) Encapsulation with acrylic resin/LCP	

Substrate		Photo resist		Ni + Au	
Adhesive		Cr + Cu/Au		Acrylic-varnish	
Carrier Wafer		Photo resist 2			

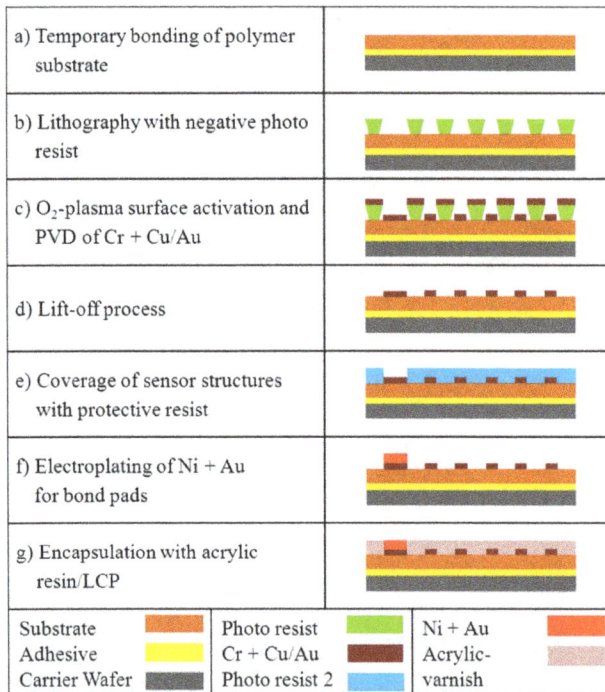

Figure 7. Process flow for the fabrication of capacitive strain gauges.

Figure 8. LCP- (left) and polyimide- (right) foils temporarily bonded onto silicon carrier wafers.

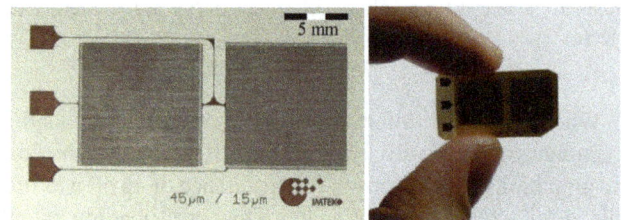

Figure 9. Sensor half bridge (design 1) on LCP substrate (left) and with a protective encapsulation layer of LCP (right).

are possible with this process. Unfortunately, a basic requirement for photolithography is a very flat and rigid substrate. Therefore, a method for processing polymer foils with the approved thin-film technologies in IMTEK's clean room had to be developed in this work. To this end, the foils are temporarily bonded onto carrier substrates with an adhesive that can be removed residue-free after processing is complete.

Figure 7 exhibits a schematic sequence of the sensor structures at every fabrication step, with a brief description.

For the sensor substrates a polyimide foil of type Kapton HN from DuPont and a LCP foil of type Ultralam 3000 from Rogers were used. Step (a) is the temporary bonding of a 50 μm polymer foil on a standard silicon wafer, used as a carrier. In this step the carrier wafer is spin-coated with a two-component adhesive and soft-baked at 130 °C. In Fig. 8 temporarily bonded foils of LCP and polyimide are exhibited.

Step (b) comprises spin coating and structuring by photolithography of a negative photoresist from Micro Resist Technology, type ma-N 1420 on the polymer foil. In step (c), an O$_2$ plasma is applied for surface activation and cleaning and subsequently 30 nm Cr and 350 nm Cu or Au are deposited by the PVD process sputtering. Subsequently the electrodes are patterned by lift-off in step (d). In (e) the sensor electrodes are protected with a photoresist, the interconnection pads are uncovered for an electroplating process in step (f). Thicker films of Ni plus Au were electrodeposited locally to achieve a stable and solderable metallization fin-

ish. The sensor structure on polyimide was encapsulated with an acrylic varnish. This varnish was spin-coated on the foil substrate for protection against humidity and particles in step (g).

Due to a low amount of curing agent in the glue, after processing, the foil substrate can be detached from the carrier by heating the assembly up to 20 °C above the soft-bake temperature without damage. Figure 1 depicts a removed LCP foil carrying sensor structures. Finally, individual sensors are separated from the sheet.

Different variations of the capacitive strain gauges were designed, fabricated and tested in this work. Three design levels were realized: designs 1, 2 and 3. In the first phase, single capacitors and half-bridges of design 1 were investigated for their sensitivity against strain, temperature and humidity. Geometry parameters such as electrode width, length and distance were varied and the measured capacitances were compared to the values calculated analytically and numerically.

Most of the sensors on polyimide and LCP substrate were coated with an acrylic resin of type Plastik 70 from CRC. Also, a transducer type that is fully encapsulated in LCP was manufactured by laminating an additional LCP sheet to the sensor. Both sensor types on LCP are depicted in Fig. 9.

In the second evaluation phase, half-bridges of the sensors design 1 and 2 were bonded directly on a test substrate. As

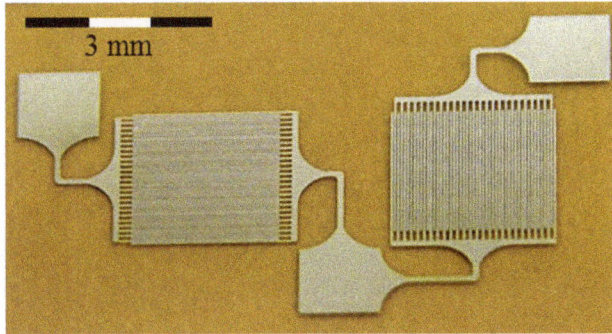

Figure 10. Sensor half-bridge (design 2) on polyimide substrate without encapsulation.

Table 1. Materials parameters applied in the FEM simulation (DuPont, 2008; Rogers Corporation, 2009).

Material	E modulus [GPa]	Poission's ratio	Dielectric constant
Substrate: polyimide	2.5	0.34	3.4
Substrate: LCP	2.25	0.32	2.9
Specimen: steel	220	0.3	–

the two sensor structures are arranged perpendicularly, under the effect of strain, tension will be induced in one structure and compression by transverse contraction in the other. The differential measurement of the capacitances can increase the strain sensitivity and compensate for cross sensitivities. If both capacitances change equally due to temperature, then no change in the difference is detected. Figure 10 depicts an example of a half-bridge configuration for design 2.

In the third phase of this study, we designed a half-bridge sensor layout referred to as design 3, for the integration of an instrumentation IC directly on the sensor substrate, as shown in Fig. 11.

4 Experimental

4.1 Set-up for strain- and cross-sensitivity measurement

The experimental setup for the characterization of the strain gauges is presented in this section. A ribbon made of high-strength spring steel with an ultimate elongation of 1 % was designed and fabricated to serve as a test specimen. For the characterization, the sensor prototypes were mounted by means of an acrylic adhesive. Subsequently the ribbon was clamped into a Zwick Z010 universal testing machine with a maximum force capacity of 10 kN in order to generate different levels of strain in the microrange in the steel ribbon.

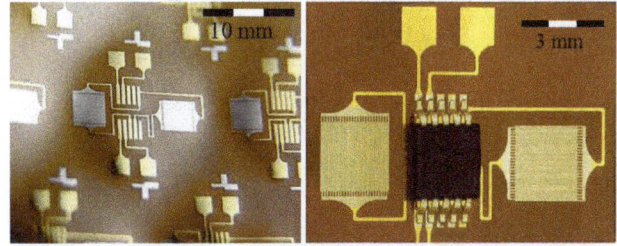

Figure 11. Design 3: capacitive strain sensors (design 3) on polyimide with solderable Ni/Au pads (left), strain gauge with a reflow-soldered 10 I/O instrumentation IC (right).

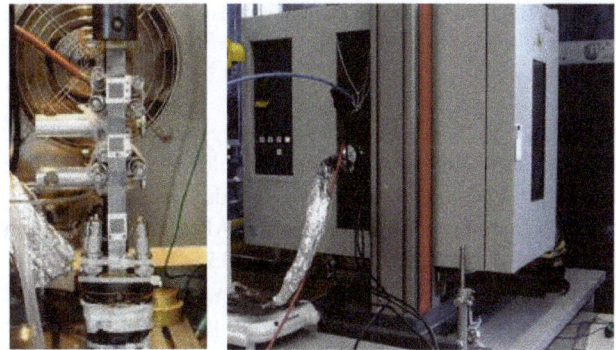

Figure 12. Left: sensors with LCP substrate mounted on a steel specimen; right: heating chamber with lead-through for water vapor, nitrogen and a humidity sensors to generate a controlled (T,RH) environment.

The tensile testing setup is integrated into a heating chamber so that it is possible to measure the influence of temperature on the capacitance and on the gauge factor of the sensor in situ. Figure 12 exhibits three single prototype sensors mounted on the steel ribbon inside this chamber.

A Digimess 300 LCR meter with a resolution of 0.001 pF was used to measure the capacitances in four-wire mode. The signal conductors lead to the outside with coaxial cables in order to shield them from electromagnetic influences. The characterization of the cross sensitivities of the capacitive sensors was a major part of this study. A mixture of nitrogen and water vapor was led into the heating chamber to generate well-defined levels of humidity. For an atmosphere with 0 % relative humidity, only nitrogen gas was used. Temperature and moisture sensors were installed in the heating chamber to control the climate with a precision of approximately 1 %. The accuracy of temperature control on the strain gauges was approximately ±2 K.

4.2 Sensor bridge characterization with an instrumentation IC

We found high sensitivities for single sensor elements on temperature and humidity. It is possible to compensate for most of the cross sensitivities with an arrangement as a

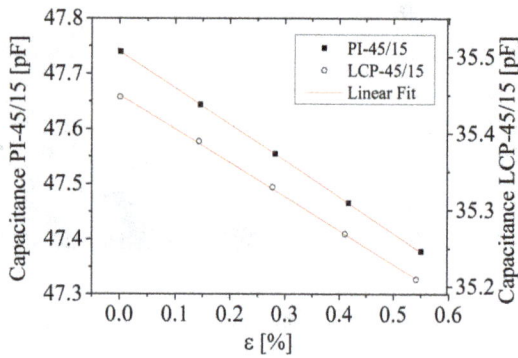

Figure 13. Measured capacitance of single sensors (design 1), PI-45/15 and LCP-45/15 as a function of strain at room temperature in a dry atmosphere.

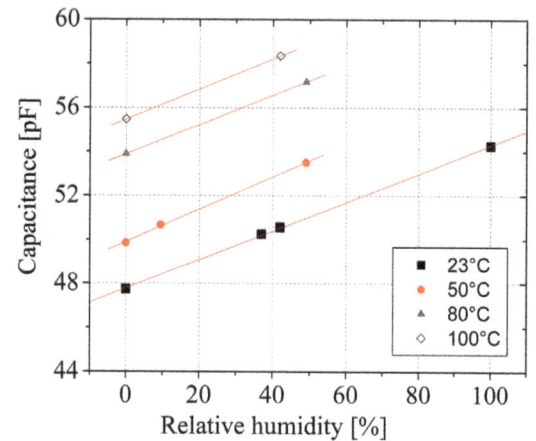

Figure 14. Measured capacitance of a single PI-45/15 sensor (design 1) as a function of relative humidity at different temperatures.

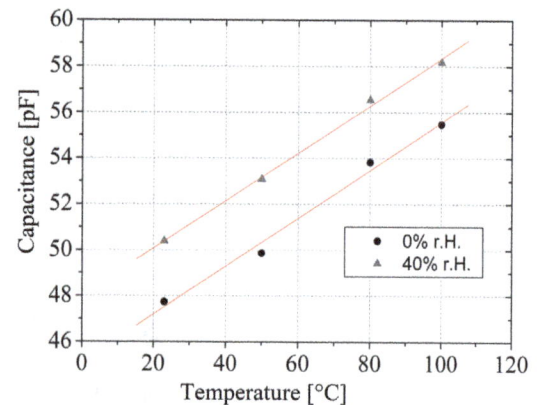

Figure 15. Capacitance of a single PI-45/15 sensor (design 1) as a function of temperature at 0 and 40 % relative humidity.

half-bridge of two capacitive structures on one substrate. When an instrumentation IC is placed close to the capacitive structure, short signal paths are provided and parasitic capacitances are kept low. A low-power AD7152 from Analog Devices with a resolution of 1 bit or 0.25 fF and with a reported power consumption of 0.2 mW was used for the measurements. Figure 13 shows the strain gauge substrate with an integrated read-out IC. The interdigital sensors of design 3 were in a 45/15 configuration on polyimide, and their sensing area was reduced to 3×3 mm^2 to obtain 5 pF per capacitance.

5   Results and discussion

5.1   Strain sensitivity

The force applied by the tensile testing machine generates a constant strain in the steel ribbon. The analyzed strain steps were 0.145, 0.282, 0.416 and 0.545 % for characterization of the strain sensitivity. Figure 13 exhibits the results for a capacitance measurement of a sensor on polyimide with an electrode width of 45 µm and a distance of 15 µm. The temperature during the measurement was 23 °C and the relative humidity 0 %. The slope of the regression line for the measured values is the strain sensitivity of the gauge with -0.658 pF %$^{-1}$. The strain coefficient is $S_C = 13\,783$ ppm %$^{-1}$ when normalized to the absolute capacitance of $C_0 = 47.74$ pF. The corresponding gauge factor is -1.38. The capacitance output signal of the gauge is linear over strain, as predicted with the analytical approach and the FEM simulations. Generally, values for the investigated capacitances obtained by FEM correspond to the measured values with a divergence of 4 %.

Figure 13 additionally depicts the results for a LCP 45/15 sensor with acrylic varnish encapsulation. The strain coefficient is $S_C = 12\,660$ ppm %$^{-1}$, normalized to the absolute capacitance of $C_0 = 35.45$ pF. The corresponding gauge factor is -1.27. The results for both sensor configurations are summarized in Table 2.

The capacitance and the gauge factor of the test sensors were obtained at different temperature and humidity levels with the described measuring setup. Figure 14 depicts the results for the capacitance of a strain gauge with polyimide substrate over the test matrix.

The sensor's capacitance shows a linear dependency on temperature at two different values of humidity (Fig. 15). The obtained coefficients for the sensitivities are $H_C = 1435$ ppm %$^{-1}$ RH and $T_C = 2168$ ppm K^{-1}.

Figure 16 exhibits the linear dependency of the gauge factor G_F on temperature. The absolute value of the factor increases with increasing temperature. The thermal coefficient of the gauge factor is obtained as -63.8 ppm K^{-1}. This can be explained by the decrease in the substrate stiffness with increasing temperature. Humidity showed no apparent influence on sensitivity.

The results for a sensor with a similar geometric ratio of the electrodes deposited on a LCP substrate are shown in Fig. 17.

Table 2. Comparison of power consumption for strain-sensing systems.

Sensor	PI 45/15 polyimide	LCP 45/15 LCP LCP encapsulated	Differential measurement polyimide
Gauge factor G_F	-1.4	-1.3	2.1
Capacitance	47.7 pF	35.5 pF	2×5 pF
Temperature coefficient	2170 ppm K^{-1}	1960 ppm K^{-1}	100 ppm K^{-1}
Humidity coefficient	1435 ppm %$^{-1}$ RH	55 ppm %$^{-1}$ RH	240 ppm %$^{-1}$ RH

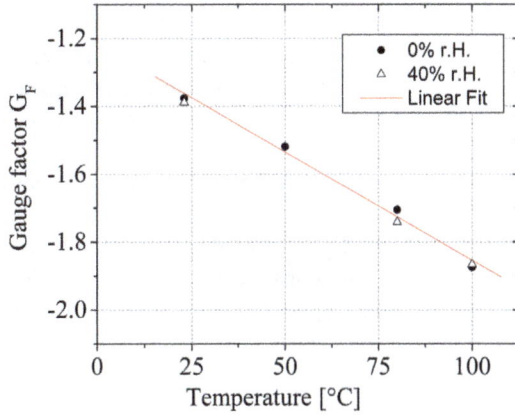

Figure 16. Gauge factor G_F of a single PI-45/15 sensor (design 1) as a function of temperature for relative humidity of 0 or 40 %.

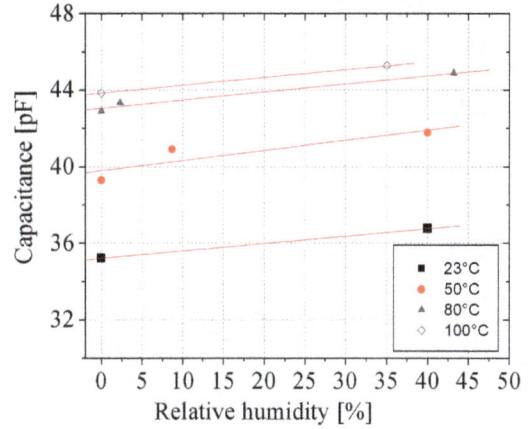

Figure 17. Capacitance of a single LCP-45/15 sensor (design 1) as a function of relative humidity at different temperatures.

For this sensor the measured coefficients are $H_C = 1149$ ppm %$^{-1}$ RH and $T_C = 2820$ ppm K^{-1}. For an LCP sensor with an encapsulation LCP foil bonded on top, the coefficients $H_C = 55$ ppm %$^{-1}$ RH and $T_C = 1958$ ppm K^{-1} are obtained. It is clearly necessary to coat the LCP-based layer with an identical material in order to fully utilize the superior properties of the substrate.

5.2 Analytical model for capacitive single sensors

Based on the experimental results for the sensors PI-45/15 and LCP-45/15, we developed a regression model for the capacitance as a function of strain, temperature and relative humidity, which is given in Eq. (9). The coefficients for the sensitivities were obtained from the measured results given in the previous section.

$$C(\varepsilon, T, H) = C_0(1 + T_C(T - T_0) + H_C(H - H_0) + S_C \varepsilon (1 + T_{C,SC}(T - T_0))) \qquad (9)$$

Coefficients for the sensor PI-45/15 are

$T_C = 2168$ ppm K^{-1} $C_0 = 47.74$ pF,
$H_C = 1435$ ppm %$^{-1}$ RH $T_0 = 23$ °C,
$S_C = -13\,783$ ppm %$^{-1}$ $H_0 = 0$ %RH,
$T_{C,SC} = -63.8$ ppm K^{-1}.

Coefficients for the sensor LCP-45/15 are

$T_C = 1958$ ppm K^{-1} $C_0 = 35.45$ pF,
$H_C = 55$ ppm %$^{-1}$ RH $T_0 = 23$ °C,
$S_C = -12\,660$ ppm %$^{-1}$ $H_0 = 0$ %RH,
$T_{C,SC} = -63.8$ ppm K^{-1} (value of PI).

With these models one can predict the value of the sensor capacitance if the environmental conditions and the level of strain are known. Furthermore, the models provide the strain of a specimen if the sensor capacitance, temperature and humidity are given.

5.3 Differential measurement with instrumentation IC

A capacitance-to-digital converter IC (Analog Devices, AD7152) measured the differential capacitances of two PI-45/15 sensors arranged perpendicularly on a flexible circuit as presented in Sect. 4. Figure 18 depicts the linear output signal during the strain measurement.

The gauge factor G_F for this polyimide-based system is $+2.1$, the temperature sensitivity is 100 ppm K^{-1} and the sensitivity against humidity is 240 ppm %$^{-1}$ RH. The capacitance disparity of the two sensors in this half-bridge is about 2 %. This fact explains the low, but not completely compensated for, temperature cross sensitivity of the system. In

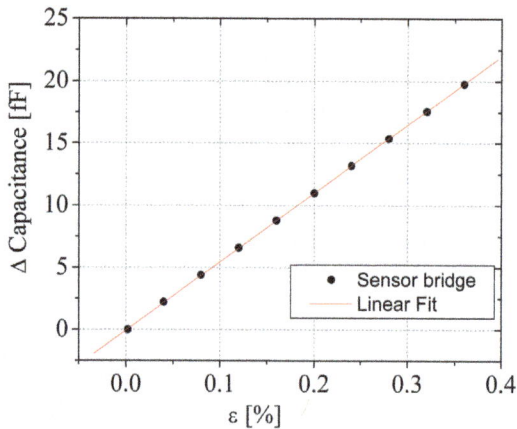

Figure 18. Differential capacitance of two orthogonal PI sensors (design 3) measured with an integrated AD7152 read-out IC as a function of strain at room temperature.

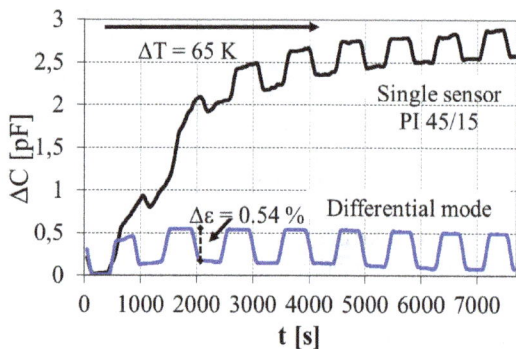

Figure 19. Comparison of the output signal of a single sensor (design 1) and a differential sensor (design 2), both on PI, during a simultaneous variation of the strain ($\Delta\varepsilon = 0.54\,\%$) and temperature ($\Delta T = 65\,\mathrm{K}$).

Fig. 19, the capacitance of a single sensor is compared with the output signal of the half bridge when the temperature is raised by 65 K during the measurement. The capacitance of the single sensor increases by 2.5 pF, whereas the differential signal increases by only 0.08 pF. Hence, the influence of temperature is nearly compensated for. The characterization results are summarized in Table 2.

5.4 Power consumption of system for wireless strain measurement

We developed two demonstrator systems on PCBs for strain measurement, one of which using the novel capacitive gauges and the other with standard resistive sensors ($R = 150$ Ω) (Joerger, 2010). In these systems, the collected data are stored in a memory-saving mode, and sleep and wake-up routines were implemented to save power. The micro-controller used in these systems already provides a wireless interface

Table 3. Comparison of power consumption for strain-sensing systems.

Power consumption	System with commercial resistive strain gauges	System with capacitive strain gauges
Electronics	< 1 mW	< 0.6 mW
Sensor-element	> 4 mW	< 0.01 mW
Total	5 mW	0.6 mW

for sending and receiving data via an antenna. The measurement circuits were assembled with low-power devices.

Both systems measured strain of a test specimen and the power consumption was monitored. In Table 3, the results for the systems are presented.

The power consumption of the instrumentation electronics for both systems was below 1 mW, the application of the low power capacitance to digital converter reduced the power by a factor of 2. The power consumption of the resistive sensor elements was 40 times higher than for the capacitive sensors.

6   Summary and conclusions

The capacitive strain gauges presented in this work were first designed and optimized based on an analytical approach in order to achieve maximum strain sensitivity and a linear output signal. FEM simulations of different sensor configurations verified the analytical model. The analyses of influencing geometry and material parameters led to the establishment of design rules for the strain gauges.

The sensors were fabricated using thin-film technology on two polymer foil substrates by means of a temporary bonding process specifically developed in this work. The characterized capacitive transducers exhibited strain sensitivities that are in the same range as those of standard resistive gauges. The characterization of single capacitive structures revealed high material-dependent cross sensitivities against temperature and humidity. Therefore, a sensor bridge design was introduced that compensates for parasitic effects from differential measurement with a low-power instrumentation IC. The gauge factor of a prototype sensor system was 65 % higher than a single strain gauge on polyimide; the sensitivity for humidity was reduced by 87 % and the influence of temperature by 98 %. Furthermore LCP proved to be a superior material when it was used for the substrate and the capping.

The conclusion of this work is that the proposed capacitive strain gauge is a suitable alternative to common resistive strain gauges. The detailed characterization of the sensitivities on strain, temperature and humidity and the linearity of the sensor capacitance on strain are aspects that have not yet systematically been taken into account for capacitive strain gauges (Matsuzaki et al., 2007; Aebersold et al., 2006; Arshak et al., 2000). The sensor structures have significant

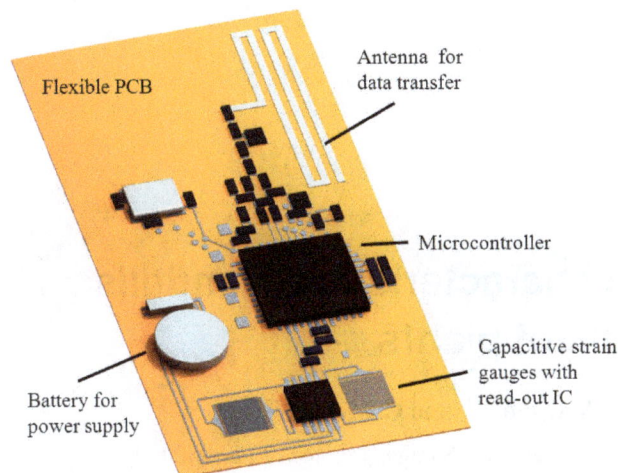

Figure 20. Concept of a wireless, intelligent sensor node with integrated capacitive strain gauges realized on a flexible substrate.

cross sensitivities against temperature and humidity. These can present limiting factors for applications in harsh environments. This work has shown that these factors can be reduced by appropriate dielectric materials. Furthermore, a very effective compensation is possible by differential measurement of orthogonal structures. An integrated system consisting of sensor elements and instrumentation electronics, including a micro-controller for data storage on a flexible substrate, was developed. The power consumption of the systems could be decreased by a factor of 10 by applying capacitive strain gauges instead of standard resistive sensors.

7 Outlook

Further steps of development will be the hybrid integration of the capacitive gauges in a wireless, intelligent sensor system realized on a flexible PCB. Figure 20 depicts a scheme of such a system with integrated capacitive strain gauges on a flexible substrate. The future perspective of such a system is a sensor node for detecting of temperature, humidity and strain that is applicable as an energy-harvesting smart label.

Acknowledgements. Part of this work was performed on behalf of the Forschungsvereinigung Antriebstechnik e. V. (FVA) using funding provided by the German Federal Ministry of Economics and Technology (BMWi) pursuant to resolution number 346 ZN passed by the German federal parliament. The authors are grateful for this support.

Edited by: R. Maeda
Reviewed by: two anonymous referees

References

Aebersold, J., Walsh, K., Crain, M., Voor, M., Martin, M., Hnat, W., Lin, J., Jackson, D., and Naber, J.: Design, modeling, fabrication and testing of a MEMS capacitive bending strain sensor, J. Phys. Conf. Ser., 34, 124–129, 2006.

Arshak, K. I., McDonagh, D., Durcan, and M. A.: Development of new capacitive strain sensors based on thick film polymer and cermet technologies, Sensor. Actuator., 79, 102–114, 2000.

DuPont: Kapton polyimide film – summary of properties, DuPont, 24, p. 5, 2008.

Igreja, R. and Dias, C. J.: Analytical evaluation of the interdigital electrodes capacitance for a multi-layered structure, Sensor. Actuator. A-Phys., 112, 291–301, 2004.

Joerger, T.: Wireless sensor-systems with energy-optimized strain measurement, B.Sc. thesis, Dept. Micro-systems Eng., Univ. of Freiburg, Freiburg, Germany, 2010.

Mamishev, A. V., Sundara-Rajan, K., Yang, F., Du, Y., and Zahn, M.: Interdigital sensors and transducers, Proc. IEEE, 92, 808–845, 2004.

Matsuzaki, R. and Todoroki, A.: Wireless flexible capacitive sensor based on ultra-flexible epoxy resin for strain measurement of automobile tires, Sensor. Actuator., 140, 32–42, 2007.

Melcher, J., Deben, Y., and Arlt, G.: Dielectric effects of moisture in polyimide, IEEE T. Electr. Insul., 24, 31–38, 1989.

Rogers Corporation: Data sheet Ultralam 3000 – liquid crystalline polymer circuit material, Rogers Corporation, 6, p. 3, 2008.

Zhang, X.: Development of SOP module technology based on LCP substrate for high frequency electronics applications, Proc. Electronics Systemintegration Technology Conference, 118–125, 2006.

13

Sensing of gaseous malodors characteristic of landfills and waste treatment plants

B. Fabbri[1,3], **S. Gherardi**[1], **A. Giberti**[1,2], **V. Guidi**[1,2,3], **and C. Malagù**[1,3]

[1]Department of Physics and Earth Science, University of Ferrara, Via Saragat 1/c, 44122 Ferrara, Italy
[2]MIST E-R S.C.R.L., Via P. Gobetti 101, 40129 Bologna, Italy
[3]CNR-INO – Istituto Nazionale di Ottica, Largo Enrico Fermi 6, 50124 Firenze, Italy

Correspondence to: B. Fabbri (bfabbri@fe.infn.it)

Abstract. We approached the problem of sensing gaseous pollutants and malodors originating as a result of decomposition of organic compounds via chemoresistive sensors. A set of four screen-printed films based on two types of mixed tin and titanium oxides, mixed tungsten and tin oxides, and zinc oxide has been tested vs. the main gaseous components of malodors. N-butanol was also considered because of its importance as a reference gas in the odorimetric intensity scale. We found that, under proper working conditions, the films can sensitively detect such gases either in dry or in wet environments, within the range of concentrations of interest for their monitoring. We also demonstrated that the array is robust under solicitation by harmful interference gases such as CO, C_6H_6, NO_2 and NO.

1 Introduction

For certain contexts, there is a need for control of the concentration of gases that result from decomposition of organic compounds. Pollutants and malodors often accompany decomposition processes, giving rise to serious health risks and/or discomfort to the human agglomerates neighboring the plants where such gases are generated. Landfills and waste treatment plants are typical cases in which several types of emission gases or vapors must be carefully taken into account. In particular, landfill emissions are characterized by very complex composition, while plants such as incinerators are potential sources of some kinds of pollutants, deriving from waste combustion such as carbon monoxide, benzene and nitrogen oxides. Among the wide variety of decomposition gases we targeted ammonia, ethyl mercaptan and hydrogen sulfide, because they are tracers of decomposition with stinging odor to our noses even down to very low concentrations in the order of ppb. Concerning the gases deriving from waste combustion, we addressed the interference of carbon monoxide, benzene, nitrogen dioxide and oxide. They can be present if the landfill or the waste treatment plant is close to engines that burn fossil fuels or incinerators or be-

cause of circulation of lorries for garbage collection, depositing their loads in the landfills. Since the odorimetric aspect of the application is highly relevant, we also took into account the responses of all the selected sensors to n-butanol, because the latter is the reference gas for the odorimetric intensity scale, useful for fixing the threshold for the human perception of an odor (Van Harreveld and Heeres, 1995; Capelli et al., 2010). The study of the decomposition and waste combustion gases is of interest in several applications, such as monitoring in green-agriculture fertilization, livestock holdings with the decomposition process of manure, landfills, food quality, and chemical industries. Chemoresistive gas sensors based on metal oxides are extremely sensitive devices, appreciated for their low cost, compactness and full compatibility with standard electronics (Capone et al., 2003; Wang et al., 2010; Fine et al., 2010). A preliminary study to monitor landfill gas emissions was done in Guidi et al. (2012), Comini et al. (2004, 2005), Micone and Guy (2007), Romain et al. (2008), and Wetchakun et al. (2011). However, lack of selectivity is an issue and for that special care is needed while selecting the sensing units. In this work, we addressed the problem of detecting typical gaseous products of waste storage, by means of gas sensors based on thick films of metal oxides and mixed solid solutions of them, also taking into account the possible interference of typical combustion products. The response at various temperatures to identify the best detecting temperature for each type of sensing material tested by the different gases will be shown. Afterwards the results of an increasing concentration in the range of interest using the same gases will be discussed. After the tests on the typical gases responsible for malodors in landfills, an analysis of possible interfering gases such as CO, $C_6H_{6,}$ NO_2 and NO was performed to determine the robustness of the array of sensors for in-field applications.

2 Experimental

2.1 Selection of the sensors and preliminary operations

A set of four metal oxide semiconducting films has been selected for the purpose. The materials chosen are ZnO, two solid solutions of SnO_2 and TiO_2, in proportions of 30–70 % (named ST30) and 90–10 %, respectively (named ST90), and a solid solution of WO_3 and SnO_2 in proportions of 30–70 % (named WS30).

ZnO powders were prepared by dissolving a proper amount of $Zn(NO_3)_2 \cdot 4H_2O$ in doubly distilled water (Carotta et al., 2009a). The reaction mixture was stirred for 1 h and kept at room temperature for 24 h; then the product was washed, filtered and dried at 80 °C, and finally calcined at 450 °C for 2 h. WS30 is a solid solution of W and Sn oxides (with Sn : W = 30 : 70) produced via hydrolysis of a WCl6 and Sn(II)ethyl hexanoate solution prior to calcination at 550 °C for 2 h under airflow. Preparation and characterization of nanostructured powders are described in Chiorino et al. (2001) and Shouli et al. (2010). Nanostructured powders of the solid solutions of Sn and Ti mixed oxide were produced via symplectic gel coprecipitation of stoichiometric Sn(4+) and Ti(4+) hydroalcoholic solutions, after calcination of the resulting xerogel at 550 °C for 2 h under airflow. The solid solutions of $Ti_xSn_{1-x}O_2$ at two values of x ($x = 0.3$, 0.9) will be hereinafter labeled as ST30 and ST90 (Carotta et al., 2008a, b, 2009b). The crystalline phase of the powders was investigated by X-ray diffraction (XRD) (Philips PW 1820/00 Cu K radiation with $\lambda = 1.54$ Å) performed at room temperature.

The powders were used to screen-print the sensing layers onto miniaturized alumina substrates. The layers were fired for 1 h at a selected temperature from 650 to 850 °C. Measurements were performed with the flow-through technique in a sealed test chamber. Air and target gases were supplied by certified bottles; humidity was provided by means of a bubbler filled with distilled water. The gases chosen for gaseous malodor application were ethyl mercaptan, ammonia, hydrogen sulfide and n-butanol, while CO, C_6H_6 and NO_2 were probed as interferents. The first three gases are primarily responsible for malodors in decomposition products, the fourth is a reference gas for the odorimetric inten-

sity scale (Van Harreveld et al., 1995; Capelli et al., 2010; Lee et al., 2013) and the last two are carcinogen combustion products. After the best operational temperature was determined, a dedicated experiment was performed to assess the effect of CO, C_6H_6, NO_2 and NO on the array of sensors. Humidity was monitored through a HIH-3610 Series Honeywell sensor. We investigated the responses of the sensors under dry (RH < 2 % at 35 °C) and wet (RH = 23 % at 35 °C) conditions. For a reducing gas the response R is defined as $(G_{gas} - G_{air})/G_{air}$, where G_{gas} and G_{air} are the conductance values in gas and in air, respectively, while for an oxidizing gas the response is defined as $\Delta R/R$.

2.2 N-butanol: reference gas for the odorimetric intensity scale

Generally, odoring compounds can be classified according to their structure, dimensions, compound family, and functional group, and they can be characterized on the basis of their concentration, intensity and hedonic tone. The majority of odoring molecules exhibit a low olfactory threshold; then odor is well detected even if the gas concentration in air is relatively low. In the literature, it is well known that the odor intensity is not proportional to the gas concentration (ASTM International E679-04, 2005; UNI EN 13725, 2004; Segura and Feddes, 2005; Brattoli et al., 2011; Lee et al., 2013; Lewis et al., 2005). In fact, due to the synergic and/or inhibiting effects between odorants, the relationships between the odor concentration and the chemical concentrations of the compounds are difficult to realize and reproduce (Micone and Guy, 2007). The odor concentration measurements can be carried out by three methods: analytical determination through mass spectroscopy or marker substance individuations, electronic nose and olfactometry (Capelli et al., 2008; Muñoz et al., 2010). The first is not a completely efficient method for the scope of this work because it does not give information about human olfactometric threshold and it is too expensive for the purpose of designing a portable device (Davoli et al., 2003; Dincer et al., 2006; Fang et al., 2012; Zarra et al., 2009). The electronic nose consists of a network of non-specific chemical sensors, the sensor array, twinned with a data-processing treatment unit in order to recognize or measure the concentration of a gas or an odor. This device allows one to get rid, firstly, of the day-to-day human subjectivity of the olfactometry, secondly, of its expensive cost and, thirdly, of elaborate data analysis used by other analytical instruments (Micone and Guy, 2007). Electronic noses are able to classify and to measure odor intensity but not the concentration of gases dangerous for health (Gardner and Bartlett, 1994; Dentoni et al., 2012; Giuliani, 2010). Lastly, olfactometry is the technique currently used worldwide in odor quantification because it gives more useful information than the other two methods (UNI EN 13725, 2004; Segura and Feddes, 2005). A test panel formed by selected and practiced persons sniffs odorous gas samples

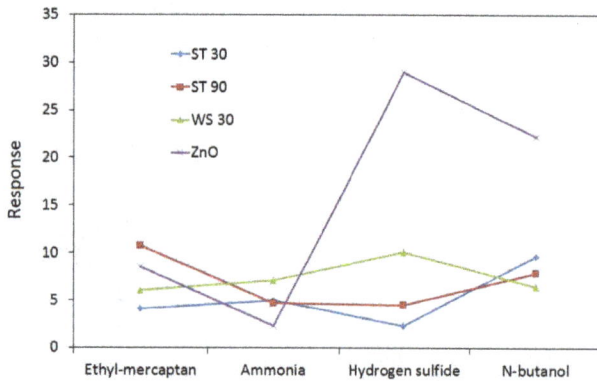

Figure 1. Best responses for each sensor at own working temperature both to gases of interest and to reference gas n-butanol, under dry conditions.

conveniently diluted by odorless air on the basis of determined relations so that each sample is presented to the panel following a decreasing dilution series. In this way the panel identifies the "odor threshold" and it is possible to digitize a sensation in order to create an odor intensity scale (odor units) (Nicolas et al., 2006; Sironi et al., 2010; Sarkara and Hobbs, 2002). Relying essentially on human expertise, it is considered a time-consuming and expensive method when used frequently (Micone and Guy, 2007). In this work we needed a reference gas for the odorimetric intensity scale in order to compare the responses obtained with gaseous malodors. Therefore, we chose n-butanol and tested the sensor array with a concentration of 5 ppm, which corresponds to a weak odor (about 12 OU – odor units) (Van Harreveld and Heeres, 1995; Capelli et al., 2010; Lee et al., 2013).

2.3   Response to decomposition gases and n-butanol

The response of the set of sensors was investigated vs. working temperature in order to determine the optimal detecting condition for ethyl mercaptan, ammonia, hydrogen sulfide and n-butanol (Fig. 1). The concentrations of the target gases were chosen in order to be comparable with the recommended exposure limits (REL) by NIOSH (United States National Institute for Occupational Safety and Health). Indeed, for ethyl mercaptan we chose the REL concentration of 0.5 ppm, 10 ppm for ammonia (REL = 25 ppm), 2 ppm for hydrogen sulfide (REL = 10 ppm) and 5 ppm for n-butanol (REL = 50 ppm).

The experimental results show diversification in the sensor responses to the first three gases and the values obtained are significant. Instead, for n-butanol the temperature test has shown homogeneous behavior and the response at best working temperature was significant for all sensing films. This is an indication of the suitability of this set of sensors to both the selective detection of the three target gases and to generic odor applications.

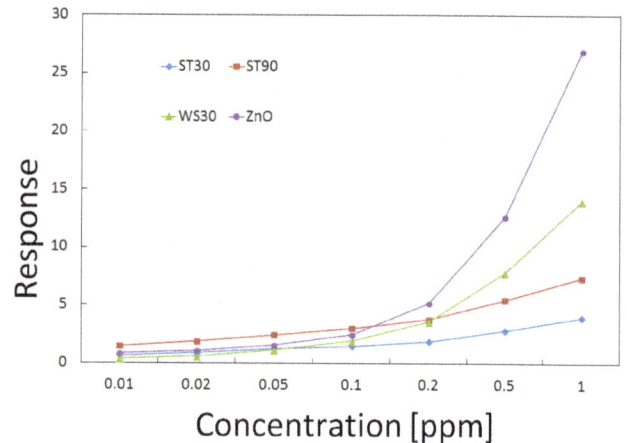

Figure 2. Responses vs. ethyl mercaptan from 0.01 to 1 ppm dispersed in a dry carrier.

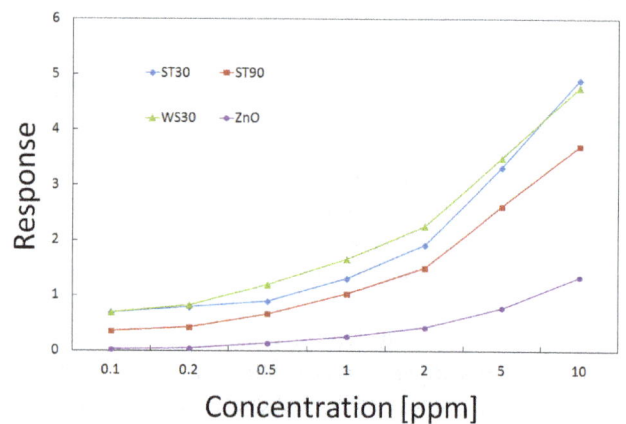

Figure 3. Responses vs. ammonia from 0.01 to 1 ppm dispersed in a dry carrier.

The optimal working temperature and the relative response for each gas are reported in Table 1.

At each best temperature for each gas, we measured the response to several concentrations of interest, highlighted in Figs. 2–5.

Regarding ethyl mercaptan, ammonia and hydrogen sulfide, the array encompasses sensors that are very unresponsive to at least one gas, while they show significant responses to the others. This feature is a very positive indication of implementation of a high-selectivity array. In particular, ZnO at 450 °C is capable of detecting hydrogen sulfide with little interference by ethyl mercaptan. ST30 at 600 °C can selectively detect ammonia; ST90 at 500 °C is capable of selectively detecting ethyl mercaptan. WS30 is sensitive to ethyl mercaptan and hydrogen sulfide at low temperatures, while it can selectively detect ammonia at high temperatures.

Since the sensors were capable of detecting concentrations of n-butanol even lower than 1 ppm, which corresponds to about 2 OU, one can conclude that sensing units are suitable

Table 1. The optimal working temperatures T_{best} and the corresponding responses R for all analyzed films tested with gases of interest and reference gas n-butanol.

Sensor	ETHYL MERCAPTAN 0.5 ppm REL 0.5 ppm		AMMONIA 10 ppm REL 25 ppm		HYDROGEN SULFIDE 2 ppm REL 10 ppm		N-BUTANOL 5 ppm REL 50 ppm	
	T [°C]	R	T [°C]	R	T [°C]	R	T [°C]	R
ST30	500	4.08	600	5	550	2.33	550	9.65
ST90	500	10.77	550	4.69	550	4.47	450	7.81
WS30	400	6.05	550	7.08	400	10.06	600	6.45
ZnO	400	8.54	400	2.34	400	29	400	22.23

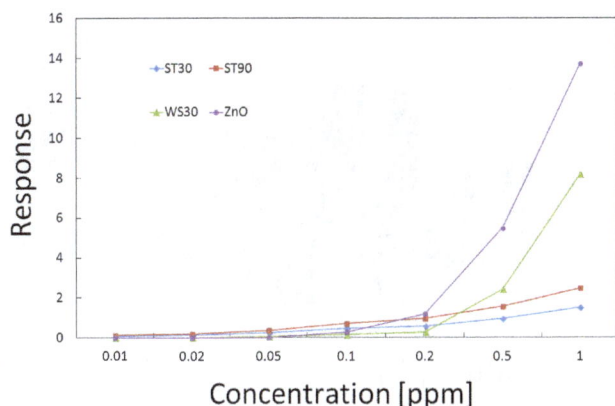

Figure 4. Responses vs. hydrogen sulfide from 0.01 to 1 ppm dispersed in a dry carrier.

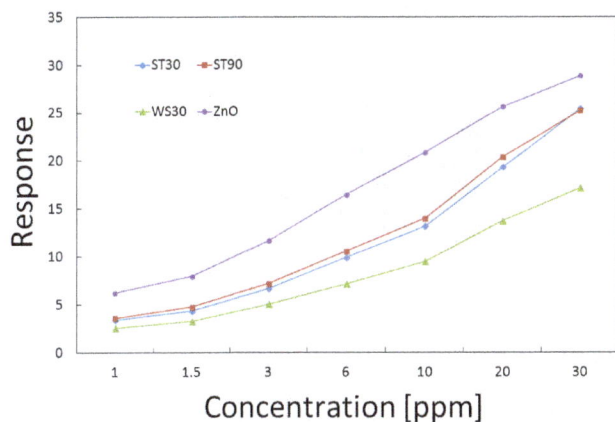

Figure 5. Responses vs. n-butanol from 1 to 30 ppm dispersed in a dry carrier.

not only for environmental monitoring but also for odorimetric applications. About this, since the human perception threshold is about 10 OU, it can be affirmed that the odorimetric performance of the sensors vs. n-butanol appears to be far better than the human olfact.

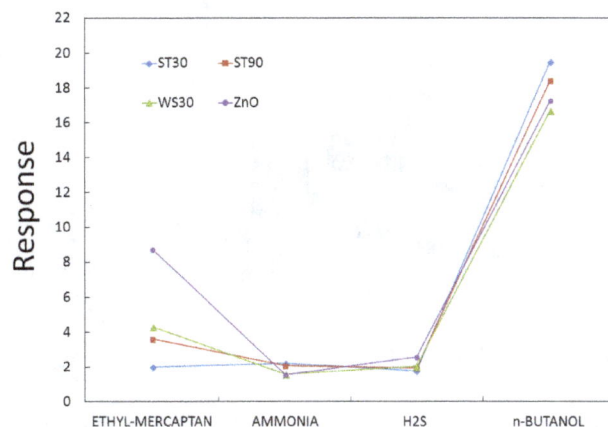

Figure 6. Responses to 0.5 ppm of ethyl mercaptan, 10 ppm of ammonia, 1 ppm of hydrogen sulfide and 5 ppm of n-butanol dispersed in a wet carrier.

In Table 2 the responses at the best working temperature for a concentration of 1 ppm for each gas are presented in order to compare the four sensors more clearly.

Consequently, an array based on ST90 for ethyl mercaptan, ST30 for ammonia and ZnO for hydrogen sulfide can be employed to detect the gases responsible for malodors in waste storage locations. The sensitivities of ZnO and WS30, defined as the derivative of the response vs. gas concentration, strongly increase after 0.2 ppm of H_2S. This is a useful feature since the sensors' lowest detection limit proves to be much lower than the concentration of interest for health risk (10 ppm). For the other sensors vs. the other gases, the response is quite high to concentrations below the exposure limits.

The response of the sensors was also tested vs. the highest concentration of each gas dispersed in a wet carrier (RH = 23 % at 35 °C). Significant decrease in the response was recorded, though it does not invalidate the capability of detection of the target gases (Fig. 6).

Table 2. The optimal working temperatures T_{best} and the corresponding responses for all analyzed films tested with gases of interest and reference gas n-butanol at the same concentration of 1 ppm.

Sensor	ETHYL MERCAPTAN 1 ppm REL 0.5 ppm		AMMONIA 1 ppm REL 25 ppm		HYDROGEN SULFIDE 1 ppm REL 10 ppm		N-BUTANOL 1 ppm REL 50 ppm	
	T [°C]	R	T [°C]	R	T [°C]	R	T [°C]	R
ST30	500	4.0	600	1.31	550	1.6	550	3.42
ST90	500	7.4	550	1.03	550	2.47	450	3.60
WS30	400	14	550	1.66	400	8.18	600	2.55
ZnO	400	27	400	0.26	400	13.72	400	6.22

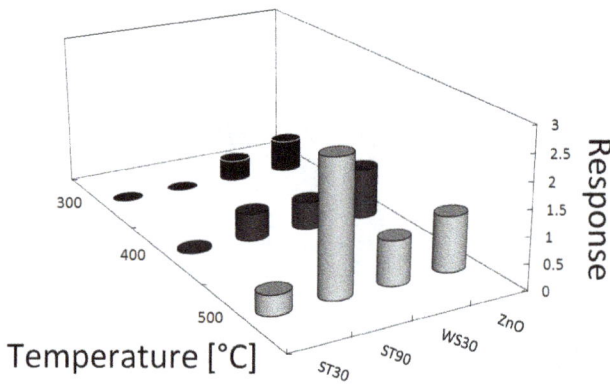

Figure 7. Responses of the sensors to 10 ppm of CO at three different temperatures.

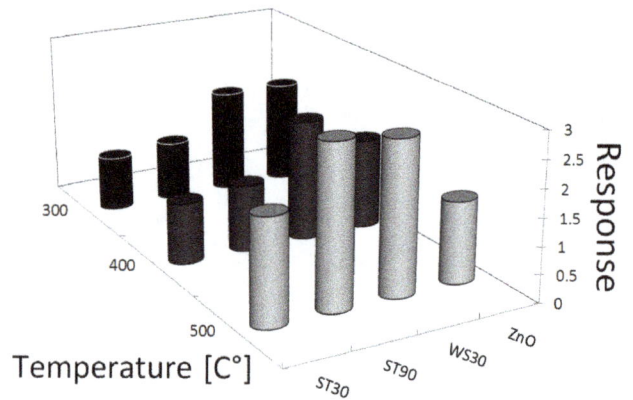

Figure 8. Responses of the sensors to 2 ppm of C_6H_6 at three different temperatures.

2.4 Response to harmful interfering gases

The responses of the sensors were also tested vs. CO, C_6H_6, NO_2 and NO since they are harmful interfering gases in zones where waste treatment plants are located. The concentrations were chosen as follows: 10 ppm for CO, 2 ppm for C_6H_6, 4 ppm for NO_2 and 4 ppm for NO, these being concentrations high enough to simulate strong interference from combustion plants near the landfills. The results are summarized in Figs. 7–10, where the responses of the sensors vs. CO, C_6H_6, NO_2 and NO at several working temperatures are reported. For a reducing gas the response R is defined as $(G_{\text{gas}} - G_{\text{air}})/G_{\text{air}}$, where G_{gas} and G_{air} are the conductance values in gas and in air, respectively, while for an oxidizing gas the response is defined as $\Delta R/R$.

As can be seen, all sensors show very low responses to these relatively high amounts of interfering gases. In Table 3 the maximum response to interfering gases are presented.

The majority of the responses are lower than 1 and in the worst case the response exceeds 2 in Fig. 7 as well as in Fig. 8, these latter being values low enough to be considered negligible in comparison with the responses to ethyl mercaptan, ammonia and hydrogen sulfide.

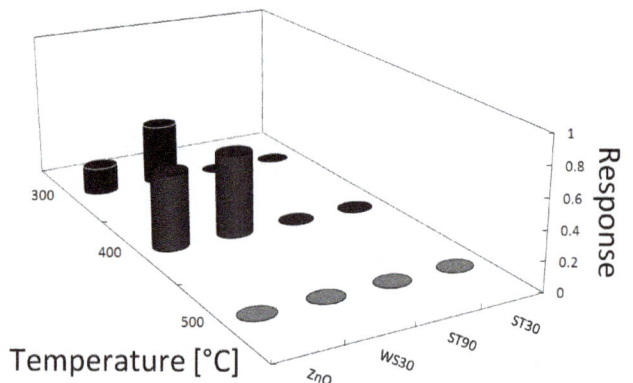

Figure 9. Responses of the sensors to 4 ppm of NO_2 at three different temperatures.

Therefore, the selected array of sensors could be employed in an array capable of selectively detecting the main gaseous malodors in landfills, without the interference of combustion gases that can harmfully interfere.

Table 3. The optimal working temperatures T_{best} and the corresponding responses for all analyzed films tested with interfering gases.

Sensor	CARBON MONOXIDE 10 ppm REL 35 ppm		BENZENE 2 ppm REL 100 ppm		NITROGEN DIOXIDE 4 ppm REL 1 ppm		NITROGEN OXIDE 4 ppm REL 25 ppm	
	T [°C]	R	T [°C]	R	T [°C]	R	T [°C]	R
ST30	500	0.38	500	1.92	$300 \div 500$	0	$300 \div 500$	0
ST90	500	2.58	500	2.94	$300 \div 500$	0	500	0.24
WS30	500	0.85	500	2.79	400	0.56	400	0.53
ZnO	500	1.07	300	1.90	400	0.51	400	0.85

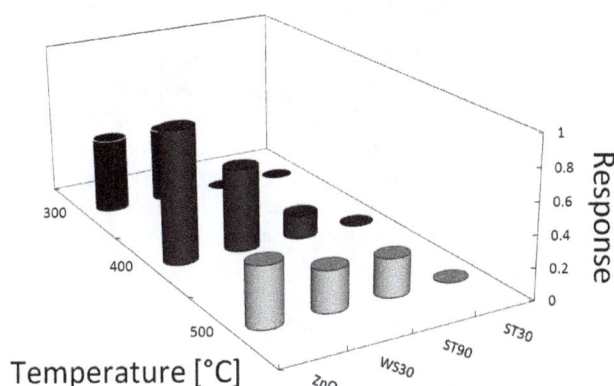

Figure 10. Responses of the sensors to 4 ppm of NO at three different temperatures.

3 Conclusions

The problem of sensing harmful gas pollutants and malodors typical of decomposition of organic substances has been addressed via chemoresistive gas sensors. A set of four sensing materials (ZnO, two mixed solutions of Sn and Ti oxides and a mixed solution of W and Sn oxides) has been tested in dry and wet conditions. It resulted that with a proper choice of the working temperatures the array can sensitively detect the target gases and quantify both the health risk and the odor intensity. The interference of typical gaseous species that can be produced in waste treatment plants has been evaluated, and it resulted in a negligible response to CO, C_6H_6, NO_2 and NO by all the sensors.

Acknowledgements. The results presented are part of the activities carried out by the MIST E-R Tecnopolo AMBIMAT laboratory of Bologna and Tecnopolo Terra&Acqua Tech of Ferrara. This work is financed by the FESR 2007-2013 operational program of Emilia-Romagna regional activities I.1.1. B. Fabbri acknowledges financial support by Spinner Regione Emilia-Romagna.

Edited by: A. Schütze
Reviewed by: two anonymous referees

References

ASTM International, E679-04: Standard Practice for Determination of Odor and Taste Thresholds By a Forced-Choice Ascending Concentration Series Method of Limits, USA, 2005.

Brattoli, M., De Gennaro, G., De Pinto, V., Demarinis Loiotile, A., Lovascio, S., and Penza, M.: Odour Detection Methods: Olfactometry and Chemical Sensors, Sensors, 11, 5290–5322, 2011.

Capelli, L., Sironi, S., Del Rosso, R., Céntola, P., and Il Grande, M.: A comparative and critical evaluation of odour assessment methods on a landfill site, Atmos. Environ., 42, 7050–7058, 2008.

Capelli, L., Sironi, S., Del Rosso, R., Centole, P., and Bonati, S.: Improvement of olfactometric measurement accuracy and repeatibility by optimization of panel selection procedures, Water Sci. Technol. – WST, 61, 1267–1278, 2010.

Capone, S., Forleo, A., Francioso, L., Rella, R., Siciliano, P., Spadavecchia, J., Presicce, D. S., and Taurino, A. M.: Solid state gas sensors: State of the art and future activities, J. Optoelectron. Adv. Mat., 5, 1335–1348, 2003.

Carotta, M. C., Gherardi, S., Guidi, V., Malagù, C., Martinelli, G., Vendemiati, B., Sacerdoti, M., Ghiotti, G., Morandi, S., Bismuto, A., Maddalena, P., and Setaro, A.: (Ti, Sn)O$_2$ binary solid solutions for gas sensing: Spectroscopic, optical and transport properties, Sensor. Actuat. B-Chem., 130, 38–45, doi:10.1016/j.snb.2007.07.112, 2008a.

Carotta, M. C., Cervi, A., Giberti, A., Guidi, V., Malagù, C., Martinelli, G., and Puzzovio, D.: Metal-oxide solid solutions for light alkane sensing, Sensor. Actuat. B-Chem., 133, 516–520, doi:10.1016/j.snb.2008.03.012, 2008b.

Carotta, M. C., Cervi, A., di Natale, V., Gherardi, S., Giberti, A., Guidi, V., Puzzovio, D., Vendemiati, B., Martinelli, G., Sacerdoti, M., Calestani, D., Zappettini, A., Zha, M., and Zanotti, L.: ZnO gas sensors: A comparison between nanoparticles and nanotetrapods – based thick films, Sensor. Actuat. B-Chem., 137, 164–169, doi:10.1016/j.snb.2008.11.007, 2009a.

Carotta, M. C., Cervi, A., Giberti, A., Guidi, V., Malagù, C., Martinelli, G., and Puzziovio, D.: Ethanol interference in light alkane sensing by metal-oxide solid solutions, Sensor. Actuat. B-Chem., 136, 405–409, doi:10.1016/j.snb.2008.12.052, 2009b.

Chiorino, A., Ghiotti, G., Prineto, F., Carotta, M. C., Malagù, C., and Martinelli, G.: Preparation and characterization of SnO$_2$ and WO$_x$-SnO$_2$ nanosized powders and thick films for gas sensing, Sensor. Actuat. B-Chem., 78, 89–97, 2001.

Comini, E., Guidi, V., Ferroni, M., and Sberveglieri, G.: TiO_2:Mo, MoO_3:Ti, TiO + WO_3 and TiO:W layer for landfill produced gases sensing, Sensor. Actuat. B-Chem., 100, 41–46, 2004.

Comini, E., Guidi, V., Ferroni, M., and Sberveglieri, G.: Detection of landfill gases by chemoresistive sensors based on titanium, molybdenum, tungsten oxides, IEEE Sens. J., 5, 4–11, 2005.

Davoli, E., Gangai, M. L., Morselli, L., and Tonelli, D.: Characterisation of odorants emissions from landfills by SPME and GC/MS, Chemosphere, 51, 357–368, 2003.

Dentoni, L., Capelli, L., Sironi, S., Del Rosso, R., Zanetti, S., and Della Torre, M.: Development of an Electronic Nose for Environmental Odour Monitoring, Sensors, 12, 14363–14381; doi:10.3390/s121114363, 2012.

Dincer, F., Odabasi, M., and Muezzinoglu, A.: Chemical characterization of odorous gases at a landfilll site by gas chromatography – mass spectrometry, J. Chromatogr., 1122, 222–229, 2006.

Fang, J.-J., Yang, N., Cen, D.-Y., Shao, L.-M., and He, P.-J.: Odor compounds from different sources of landfill: Characterization and source identification, Waste Manage., 32, 1401–1410, 2012.

Fine, G. F., Cavanagh, L. M., Afonja, A., and Binions, R.: Metal Oxide Semi-Conductor Gas Sensors in Environmental Monitoring, Sensors, 10, 5469–5502; doi:10.3390/s100605469, 2010.

Gardner, J. W. and Bartlett, P. N.: A brief history of electronic noses, Sensor. Actuat. B-Chem., 18, 211–220, 1994.

Giuliani, S.: Multisensor array system for continuous environmental odour assessment, PhD Thesis in civil engineering of Eng., X Cycle, 2010.

Guidi, V., Carotta, M. C., Fabbri, B., Gherardi, S., Giberti, A., and Malagù, C.: Array of sensors for detection of gaseous malodors in organic decomposition products, Sensor. Actuat. B-Chem., 174, 349–354, 2012.

Lee, Hyung-Don, Jeon, Soo-Bin, Choi, Won-Joon, Lee, Sang-Sup, Lee, Min-Ho, and Oh, Kwang-Joong: A novel assessment of odor sources using instrumental analysis combined with resident monitoring records for an industrial area in Korea, Atmos. Environ., 74, 277–290, 2013.

Lewis, R. S., Magers, K. D., Khoury, J., Mathison, S., Gallardo, L., and Lee, J. Measurement of Absolute Odor Intensities of Emissions from a Wastewater Treatment Plant, Proceedings of A&WMA's 98th Annual Meeting & Exhibition, Minneapolis, MN, June 21–24, Paper 05-613, 2005.

Micone, P. G. and Guy, C.: Odour quantification by a sensor array: An application to landfill gas odours from two different municipal waste treatment works, Sensor. Actuat. B-Chem., 120, 628–637, 2007.

Muñoz, R., Sivret, E. C., Parcsi, G., Lebrero, R., Wang, X., Suffet, I. H., and Stuetz, R. M.: Monitoring techniques for odour abatement assessment, Water Res., 44, 5129–5149, 2010.

Nicolas, J., Craffe, F., and Romain, A. C.: Estimation of odor emission rate from landfill areas using the sniffing team method, Waste Manage., 26, 1259–1269, 2006.

Romain, A.-C., Delva, J., and Nicolas, J.: Complementary approaches to measure environmental odours emitted by landfill areas, Sensor. Actuat. B-Chem., 131, 18–23, 2008.

Sarkara, U. and Hobbs, S. E.: Odour from municipal solid waste (MSW) landfills: A study on the analysis of perception, Environment International 27, 655–662, 2002.

Segura, J. C. and Feddes, J. J. R.: Relationship Between Odour Intensity and Concentration of n-Butanol, CSAE/SCGR 2005 Meeting Winnipeg, Manitoba June 26–29, Paper No. 05–020, 2005.

Shouli, B., Dianqing, L., Dongmei, H., Ruixian, L., Aifan, C., and Liu, C. C.: Preparation, characterization of WO_3-SnO_2 nanocomposites and their sensing properties for NO_2, Sens. Actuat. B-Chem., 150, 749–755, 2010.

Sironi, S., Capelli, L., Centola, P., Del Rosso, R., and Pierucci, S.: Odour impact assessment by means of dynamic olfactometry, dispersion modelling and social participation, Atmos. Environ., 44, 354–360, 2010.

UNI EN 13725, European Committee for Standardization: Air Quality – Determination of Odour Concentration by Dynamic Olfactom, 2004.

Van Harreveld, A. P. and Heeres, P.: Quality control and optimizaion of dynamic olfactometry using n-butanol as a standard reference odorant, Staub Reinhalt. Luft, 55, 45–50, 1995.

Wang, C., Yin, L., Zhang, L., Xiang, D., and Gao, R.: Metal oxide gas sensors: Sensitivity and influencing factors, Sensors, 10, 2088–2106, doi:10.3390/s100302088, 2010.

Wetchakun, K., Samerjai, T., Tamaekong, N., Liewhiran, C., Siriwong, C., Kruefu, V., Wisitsoraat, A., Tuantranont, A., and Phanichphant, S.: Semiconducting metal oxides as sensors for environmentally hazardous gases, Sensor. Actuat. B-Chem., 160, 580–591, 2011.

Zarra, T., Naddeo, V., and Belgiorno, V.: A novel tool for estimating the odour emissions of composting plants in air pollution management, Global Nest J., 11, 477–486, 2009.

14

Effect of thermocouple time constant on sensing of temperature fluctuations in a fast reactor subassembly

P. Sharma[1], N. Murali[1], and T. Jayakumar[2]

[1]Instrumentation & Control Group, Indira Gandhi Centre for Atomic Research, Kalpakkam-603102, India
[2]Metallurgy & Materials Group, Indira Gandhi Centre for Atomic Research, Kalpakkam-603102, India

Correspondence to: P. Sharma (paawan.sharma@gmail.com)

Abstract. Knowledge of temperature fluctuations in fast reactor subassembly is very important from a safety point of view. The time constant of thermocouples which are used for measuring coolant temperature in a fast reactor varies owing to various factors. Hence, it becomes necessary to investigate the effect of change in the time constant on sensed fluctuations. This paper investigates the dependence of temperature fluctuations on thermocouple time constants. A Scilab model consisting of source temperature profile, second-order thermocouple and histogram calculation is designed. Simulation is performed for various levels of fluctuations, fixed and variable thermocouple time constants. Kurtosis for each condition is calculated with the help of a histogram. It is found that the effect of true source fluctuations on sensor output is very large compared to that of a similar percentage of time-constant variations. Hence in systems like fast reactors, where the degree of source fluctuations (fluid enthalpy) is large in comparison to that of time-constant variations, the overall effect can be considered with great confidence to be the outcome of coolant temperature rather than thermocouple time-constant variations.

1 Introduction

A subassembly Srinivasan et al. (2006) consists of closely packed thin cylindrical tubes of fuel pellets. The coolant flows through a subassembly and transfers the heat generated due to fission. Thermocouple is placed at the subassembly outlet. The temperature reading so obtained fluctuates in nature. The analysis of such fluctuations is very important from a safety point of view. These fluctuations originate due to turbulent mixing of different temperature sodium streams across several fuel pins inside each fuel subassembly Vaidyanathan et al. (2006). The streams are at different temperatures because the fast neutron spectrum varies throughout the core, with a maximum value at the centre of the core and then gradually decreasing radially outwards. Even with arrangements for proper mixing of sodium, there exists a minimal amount of fluctuation for a particular geometry Gajapathy et al. (2007). The fluctuation level increases with overall reactor power as temperature also increases. The thermocouple time-constant value plays an important role in detecting these fluctuations, as smaller time constants are more efficient in

recording them Donaldson et al. (2008). In order to study the factors affecting the fluctuations, a proper analysis is performed considering thermocouple parameters, source temperature and fluctuating time constants. The rest of the paper is organized as follows. Section 2 briefly describes the principle of thermocouple, the core arrangement, and subassembly structure. Recorded data from Fast Breeder Test Reactor (FBTR) at different power levels are discussed in Sect. 3. The simulation setup and methodology are explained in Sect. 4 followed by analysis and discussion in Sect. 5.

2 Sensor and setup

The details about the sensor and its typical arrangement in a fast reactor subassembly are discussed here.

2.1 Thermocouple

Thermocouples are widely used in numerous industrial applications including nuclear reactors due to their range, ruggedness and accuracy. A second-order model of thermocouple

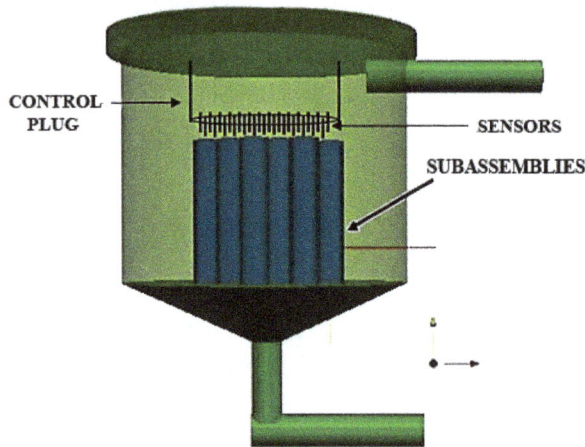

Figure 1. Basic structure of a fast reactor.

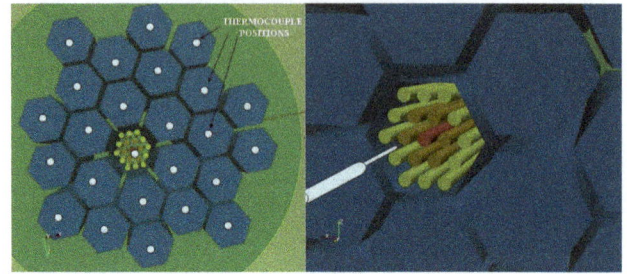

Figure 2. Subassembly arrangement.

A proper understanding of the whole arrangement is required for analysing the time constant in the case of a fast reactor subassembly.

2.2 Fast reactor subassembly

A very basic arrangement of a typical loop type fast reactor is shown in Fig. 1. The main feature of a loop type reactor is that the coolant pump is located outside the main vessel.

A group of cylindrical fuel pins is placed inside a hexagonal tube called subassembly as shown in Fig. 2.

The temperature profile inside each subassembly as well as amongst all subassemblies is not uniform since the neutron spectrum of the core peaks at the core centre and decreases radially outwards Waltar and Reynolds (1981). The fuel pins in a subassembly are at different temperatures. Coolant streams of different temperature coming out of a subassembly get mixed and form a fluctuating profile. For proper mixing of coolant streams, each fuel pin is covered with a spacer wire which also acts as a separator between different fuel pins. However, with all the provisions for better mixing, there exists a minimum amount of fluctuation level in temperature profiles. The fluctuation level depends on the overall reactor power. The centre-most subassembly is known as the central subassembly(CSA). The CSA contributes maximally to the reactor power. Thermocouples are located above each subassembly outlet and submerged in coolant. For CSA, sheathed thermocouples are used for faster response (around 100 ms), whereas they are used with thermowell in all other subassemblies (5–6 s) for a longer life time of the sensor. Figure 2 shows the top view of hexagonal subassemblies arranged compactly. Each subassembly consists of a fixed number of cylindrical fuel pins. The actual structure is however complex with more features, and only related details are shown here for a simple understanding.

3 Reactor data

Experimental data from the FBTR are used as an aid for developing a signal analysis methodology. The various states of a reactor are *shut down, reactor startup, reactor under operation (ROP), fuel handling startup* and *fuel handling*. The

can be given as

$$H(s) = \frac{1}{\tau s^2 + s + 1},\tag{1}$$

where τ is the thermocouple time constant (time taken to reach 63.2 % of the final value). Thermocouple response time varies with various configurations such as directly exposed (few ms), grounded sheathed ($\approx$ 100 ms), ungrounded sheathed (few 100 ms), and inside thermowell (5–7 s). The Time constant depends on the mass of the thermocouple (m), specific heat of thermocouple wire material (c), convective heat transfer coefficient (h) of the local medium and the surface area (A) Bentley (1998) and is given as

$$\tau = \frac{mc}{hA}.\tag{2}$$

The bandwidth of a thermocouple is given as Hung et al. (2004)

$$\omega_B = Kd^{m-2}v^m,\tag{3}$$

where K is an invariant constant, m is a constant ($0.3 \le m \le 0.7$), d is wire diameter and v is coolant velocity. In terms of wire diameter, τ is represented as Tagawa and Ohta (1997)

$$\tau = \frac{\rho cd}{4h} = \frac{\rho cd^2}{4Nu\lambda_g},\tag{4}$$

where ρ is wire material density, Nu is Nusselt's number, and λ_g is thermal conductivity of fluid around thermocouple. Nu is a function of Reynolds number, Grashoff number and Prandtl number.

$$Nu = f(Re, Gr, Pr)\tag{5}$$

It is clear from Eqs. (2), (4) and (5) that τ is dependent on many parameters involving the construction and geometry of the sensor as well as the operating conditions. Equation (3) shows that effective bandwidth offered by a thermocouple depends on coolant flow, which in turn affects τ.

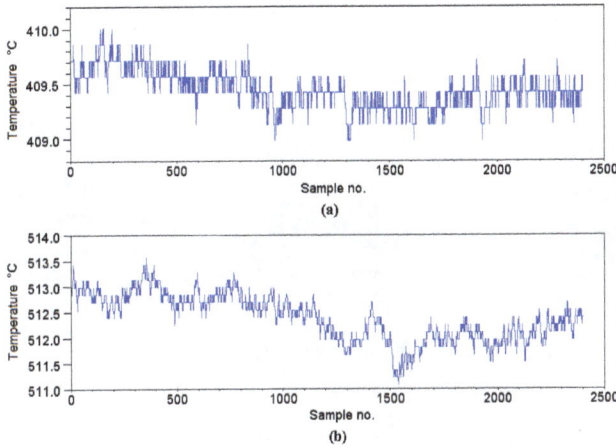

Figure 3. CSA temperature profile at stable power levels **(a)** 11 MWTh and **(b)** 18 MWTh.

CSA temperature profile, along with other important parameters, was collected for all discrete power levels in *ROP* state. Figure 3 shows the CSA temperature profile for two stable power levels of 11 MWTh and 18 MWTh (thermal).

Figure 4 shows the CSA temperature profile for gradually increasing power level.

It is clear from these figures that the amount of fluctuation increases with thermal power for nearly constant coolant flow (328–334 m^3 h^{-1}).

The parameters which are used for modelling and simulation are mean temperature value (μ) and corresponding standard deviation (σ). σ represents the degree of fluctuation. Hence a higher power level would have a high values of μ and σ. It is to be noted that, though σ increases with power, it is non-linearly localized in time since the source of such fluctuations is due to mixing of highly turbulent coolant streams passing through a subassembly.

The value of (μ, σ) is $(409.44, 0.17168)$°C for 11 MWTh and $(512.39, 0.45812)$°C for 18 MWTh. Table 1 represents unique (μ, σ) for increasing reactor power. The time series is made stationary by performing first-order differencing. This differenced series is added to the overall mean of actual series. If a is the original time series of 2400 samples taken 10 samples per second, b being the differenced series (2399 samples), stationary time series c is calculated by Eqs. (6), (7) and (8).

$$c = \mu + \sigma \tag{6}$$

$$\mu = (2400)^{-1} \sum_{i=1}^{2400} a_i \tag{7}$$

$$\sigma = b_i = \text{diff}(a) \tag{8}$$

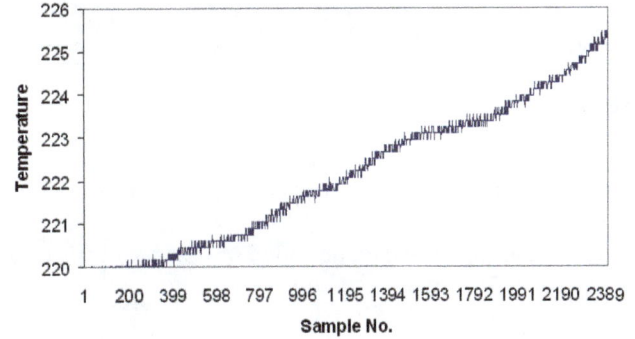

Figure 4. CSA temperature profile for increasing power.

Table 1. Reactor temperature and temperature fluctuations as a function of the reactor power.

μ	Recalculated σ
221.86	0.091
231.3	0.093
249.08	0.095
365.48	0.097
378.66	0.102
409.44	0.171
512.39	0.458

4 Simulation methodology

By using the collected reactor data, a simulation model is proposed as shown in Fig. 5.

The type K thermocouple is modelled for various values of τ ranging from 0.05 to 0.5 s using polynomial coefficients from NIST database (NIST, 2004). Based on the reactor data, the source temperature profile with fixed mean and standard deviation is obtained using the *grand* function of Scilab SCILAB (2013). The mean value is fixed at 300 °C for a closer approach towards analysis of the effect of variable fluctuations in the source (coolant) (σ) itself.

The following conditions were analysed for fixed μ to observe their relative effect:

- Various *constant* values of σ and τ

- Various *fluctuating* levels of σ and τ

Response to fluctuating (variable) values of τ is calculated by multiplexing even and odd terms of two different τ response values. The related Scilab scripts and function codes can be referenced from codes.

A histogram plot gives frequency versus variable information of the data and is performed for the calculation of mean and standard deviation from the simulated sensor output. From these two parameters, kurtosis is calculated, which gives an estimate of probability distribution (frequency distribution) of the data. So for all the test conditions, data

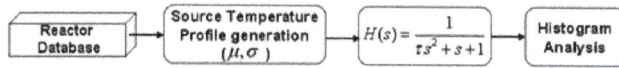

Figure 5. Simulation model.

profiles are simulated and kurtosis is calculated by using histogram.

The mean (μ_h), standard deviation (σ_h) and kurtosis (β_2) are calculated as follows.

$$\mu_h = \frac{1}{n}\sum_{i=1}^{n} f_i x_i \qquad (9)$$

$$\sigma_h = \sqrt{\frac{1}{n}\sum_{i=1}^{n} f_i(x_i-\mu)^2} \qquad (10)$$

$$\beta_2 = \frac{\sum_{i=1}^{n}(X_i-\mu)^4}{(n-1)\sigma^4} \qquad (11)$$

Maximum possible value of kurtosis is given as

$$\beta_{2,\max} = \frac{(n-1)^3-1}{n(n-1)}. \qquad (12)$$

Many works have been reported on the use of kurtosis in signal processing, such as Picard (1951), Vrabie et al. (2004) and Nopiah et al. (2009). The main characteristic of kurtosis is that it accurately depicts the peak and distribution of the data.

Any change in the distribution of observation data would reflect in a different β_2 value. The idea is to observe the relative effect on β_2 for $\pm x\,\%$ fluctuations in σ, keeping τ constant and vice versa – i.e. to observe the effect on β_2 for $\pm x\,\%$ fluctuations in τ, keeping σ constant. In this way, a relative analysis is possible to quantify the individual effect brought on by σ and τ separately on the sensed signal.

5   Results and discussion

Simulation results are obtained to study the effect of various combinations of parameters on the fluctuations. Various functions written in Scilab are mentioned below.

- *musigmakurtosis(arg1)*

  Returns data length (N), μ_h, σ_h, β_2 and $\beta_{2,\max}$ for a fixed mean value of source thermal profile (300 °C), and standard deviation (fluctuation level σ) given by *arg1*.

- *skewkurt(arg1,arg2)*

  Returns data length (N), μ_h, σ_h, β_2 and $\beta_{2,\max}$ for a source thermal profile given by *arg2* and τ given by *arg1*.

Table 2. β_2 vs. σ, $\mu = 300$ °C, $\tau = 0.1$ s.

σ	β_2
0.05	0.087
0.1	0.350
0.15	1.059
0.2	1.819
0.25	2.270
0.3	3.893

Table 3. β_2 vs. τ, $\mu = 300$ °C, $\sigma = 0.1$ °C.

τ	β_2
0.05	0.416
0.1	0.425
0.15	0.434
0.2	0.443
0.25	0.451
0.3	0.459

- *fl_sigma(arg1, arg2)*

  Returns a fluctuating source thermal profile with limits between *arg1* and *arg2*, with fixed value of μ (300 °C).

- *fl_tau(arg1, arg2,arg3)*

  Returns response to a given source *arg1* with τ fluctuating between values given by *arg2* and *arg3*.

Table 2 indicates the variation of β_2 with σ. Similarly, Table 3 gives the variation of β_2 with τ. It is clear that variations in source (coolant) temperature have greater impact on β_2.

Figure 6 shows the kurtosis variation for different values for τ and σ.

For analysing variable σ and τ (i.e. both source profile as well as thermocouple time constant tend to fluctuate due to reactor thermodynamic conditions), various levels of source and time-constant fluctuations are generated by specifying the minimum and maximum threshold values.

For example, a source profile with fluctuation levels between 0.1 and 0.2 can be generated in Scilab as

$a = fl\_sigma(0.1, 0.2).$

Similarly, thermocouple response to a source profile (either constant or fluctuating) with time constant varying between 0.1 and 0.2 s can be obtained as

$b = fl\_tau(source\ profile, 0.1, 0.2).$

Results for various combinations of source and time-constant behaviour are shown in Figs. 7 and 8.

The mean value of σ and τ are fixed at 0.1 °C and 0.1 s respectively. Fluctuation levels of 5–35 % around these mean values were simulated in different cases.

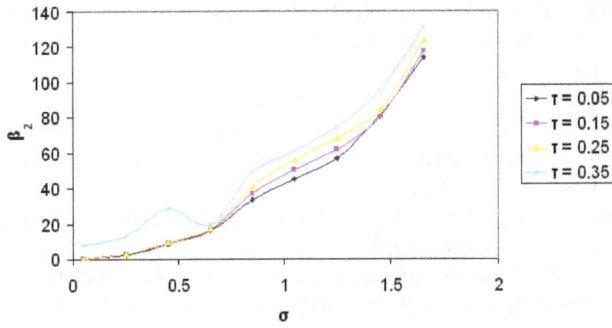

Figure 6. β_2 vs. σ.

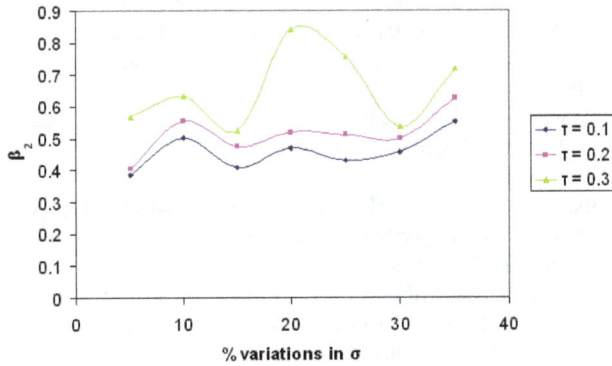

Figure 7. β_2 vs. % variations in σ.

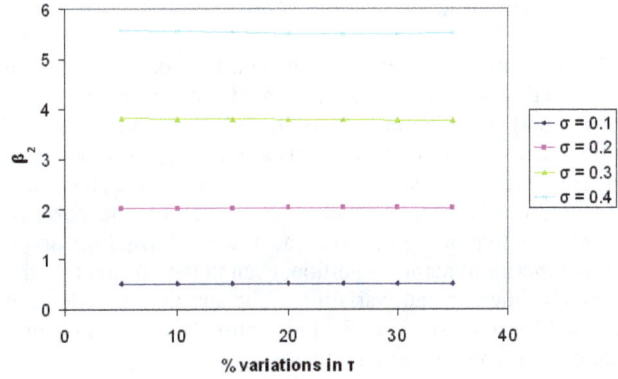

Figure 8. β_2 vs. % variations in τ.

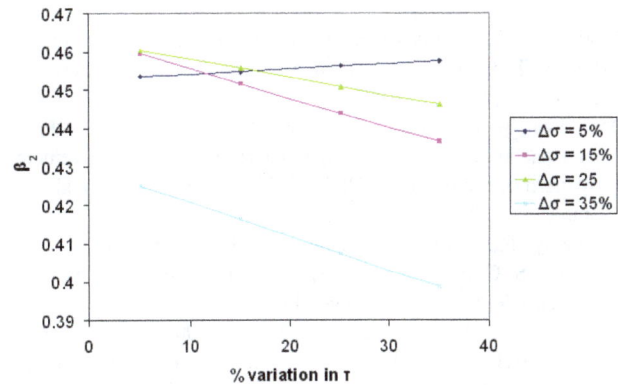

Figure 9. β_2 vs. % variations in τ and σ.

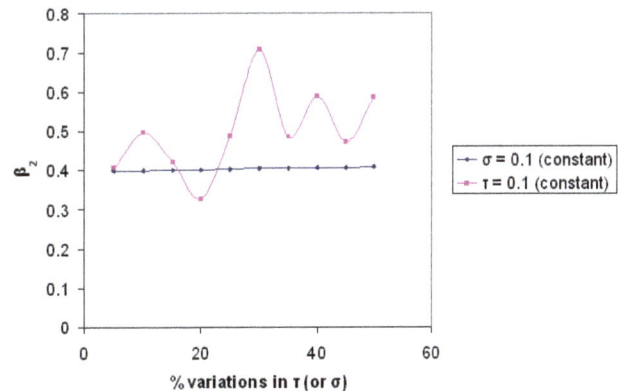

Figure 10. β_2 vs. % variations in τ or σ.

To simulate a more realistic condition, where both source and time constant tend to vary, both σ and τ are allowed to swing between fixed levels in many possible combinations – for, e.g., 5 % fluctuation in σ with $\pm(5,15,25,35)$ % fluctuations in τ and vice versa. The result is shown in Fig. 9.

The cases analysed till now are comprises of values varying around mean values of $\sigma = 0.1\,^{\circ}\mathrm{C}$ and $\tau = 0.1\,\mathrm{s}$. Result for the values of σ, τ constant and analysed with regard to fluctuating levels mutually are shown in Fig. 10. It is clear from Fig. 10 that the degree of change in kurtosis is far greater for fluctuating σ than that of fluctuating τ; i.e. fluctuations observed in thermocouple readings are largely due to the fluctuations in the source (coolant) itself and hence a good indicator of coolant condition with time.

Any change in kurtosis value denotes a shift in the signal properties by means of the frequency distribution. In a fast reactor, where the mean temperature level as well as the associated fluctuations increases with thermal power, kurtosis acts as an indicator of change in the frequency distribution of the signal. The observation that kurtosis increases with σ explains why the fluctuations at higher power level attain a sharper and narrow distribution. It means that the core temperature signal is broadly distributed with lesser peakedness at low thermal power than at higher power. These results give sufficient confidence in estimating parameters based on temperature fluctuations. Also, it supports the idea of using fluctuations in spite of its non-stationarity. There are ways by which the non-stationarity can be reduced, like first-order differencing.

6   Conclusions

Variations in thermocouple time constant over time were simulated and studied in arrangement with different power levels and fluctuations. It is found that the effect of time-constant variations on sensor readings is much less in comparison to similar variations in source (coolant) temperature. Hence, parameters calculated on the basis of source fluctuations gives very accurate information about source thermodynamic condition, even in the presence of the sensor's time-constant variations. The simulation codes can be used for analysis from the link. [https://sites.google.com/site/paawankuvera/home/kurtosisanalysis]

Edited by: G. S. Aluri
Reviewed by: two anonymous referees

References

Bentley, R. E. (Ed.): Handbook of Temperature Measurement Vol. 3: The Theory and Practice of Thermoelectric Thermometry, Vol. 3, Springer, 1998.

Donaldson, A. B., Lucero, R. E., Gill, W., and Yilmaz, N.: Problems encountered in fluctuating flame temperature measurements by thermocouple (No. SAND2008-7384J), Sandia National Laboratories, 2008.

Gajapathy, R., Velusamy, K., Selvaraj, P., Chellapandi, P., and Chetal, S. C.: CFD investigation of helical wire-wrapped 7-pin fuel bundle and the challenges in modeling full scale 217 pin bundle, Nucl. Eng. Des., 237, 2332–2342, 2007.

Hung, P., McLoone, S., Irwin, G. W., and Kee, R. J.: A total least squares approach to sensor characterisation, System Identification 2003, 321–324, 2004.

NIST: Thermocouple – NIST ITS-90 Thermocouple Database, available at: http://srdata.nist.gov/its90/download/type_k.tab (last access: 16 July 2013), 2004.

Nopiah, Z. M., Khairir, M. I., Abdullah, S., Nizwan, C. K. E., and Baharin, M. N.: Peak-valley segmentation algorithm for Kurtosis analysis and classification of fatigue time series data, Eur. J. Sci. Res., 29, 113–125, 2009.

Picard, H. C.: A note on the maximum value of kurtosis, Ann. Math. Stat., 22, 480–482, 1951.

SCILAB: Random number generator(s), SCILAB, available at: http://help.scilab.org/docs/5.3.0/en_US/grand.html (last access: 16 July 2013), 2013.

Srinivasan, G., Suresh Kumar, K. V., Rajendran, B., and Ramalingam, P. V.: The fast breeder test reactor–design and operating experiences, Nucl. Eng. Des., 236, 796–811, 2006.

Tagawa, M. and Ohta, Y.: Two-thermocouple probe for fluctuating temperature measurement in combustion – Rational estimation of mean and fluctuating time constants, Combust. Flame, 109, 549–560, 1997.

Vaidyanathan, G., Kasinathan, N., and Velusamy, K.: Dynamic model of Fast Breeder Test Reactor, Ann. Nucl. Energy, 37, 450–462, 2010.

Vrabie, V., Granjon, P., Maroni, C. S., and Leprettre, B.: Application of spectral kurtosis to bearing fault detection in induction motors, in: Proceedings of the 5th International Conference on acoustical and vibratory surveillance methods and diagnostic techniques, 11–13 October 2004, Senlis, France, 2004.

Waltar, A. E. and Reynolds, A. B.: Fast breeder reactors, Pergamon, 1981.

15

A micro optical probe for edge contour evaluation of diamond cutting tools

S. H. Jang, Y. Shimizu, S. Ito, and W. Gao

Department of Nanomechanics, Tohoku University, Sendai, Japan

Correspondence to: Y. Shimizu (yuki.shimizu@nano.mech.tohoku.ac.jp)

Abstract. This paper presents a micro optical probe, which is employed to evaluate edge contours of single point diamond tools with a size in a range of several millimetres. The micro optical probe consists of a laser source with a wavelength of 405 nm, an objective lens with a numerical aperture of 0.25, a photodiode for measurement, and a compensating optical system including another photodiode for compensation of laser intensity. A collimated laser beam, which is divided by a beam splitter in the compensating optical system, is focused by the objective lens so that the focused spot can be used as the micro optical probe. The micro optical probe traces over an edge contour of an objective tool while the signals of both the two photodiodes are monitored. The output of the photodiode for measurement is compensated by using that of the photodiode for laser intensity compensation to eliminate the influence of the laser instability. The signal of the photodiode for measurement is used to define the deviation of edge contour within the diameter of the micro optical probe. To verify the feasibility of the developed optical probe, the optical system was mounted on a diamond turning machine, and some experiments were carried out. Two types of edge contours of the diamond tools having a straight cutting edge and a round cutting edge were measured on the machine.

1 Introduction

Ultra-precision machining technology has enabled us to generate mirror surfaces by direct cutting (Moriwaki and Okuda, 1989; Evans and Bryan, 1999; Brinksmeier et al., 2010). The applications of ultra-precision machining have also been expanded by enhancements of the machining accuracy with error compensation methods (e.g. Kong et al., 2008). For ultra-precision machining, single point diamond cutting tools with a nose radius on the order of several 10 nm are often employed (Zong et al., 2010). In the fabrication of precision components such as optics and functional surfaces with micro patterns, the form error of the tool's cutting edge is one of the critical factors that will influence the quality of the machined surfaces (Brehm et al., 1979; Li et al., 2005; Lane et al., 2010). The form of the tool's cutting edge should therefore be evaluated periodically to assure the machining accuracy (Jiang, 2011; Goel, et al., 2013). In addition, on-machine measurement of the tool's cutting edge is strongly desired because a removal of the cutting tool from the ma-

chine would induce a certain amount of tool misalignment when carrying out the tool installation again. Even if such an installation error is small, it should be taken seriously in cases of ultra-precision machining (e.g. Chen et al., 2010; Fang et al., 2013).

One of the conventional methods to measure the form of a cutting tool is by scanning electron microscopes (SEMs) (Drescher, 1993; Asai et al., 1990). However, quantitative analysis of the tool's form by using the acquired SEM images, which are qualitative results, is expensive and time-consuming task (Lane, et al., 2013). In addition, the SEMs cannot be applied to on-machine tool measurement because the tool should be unmounted from the machine so that the evaluation can be carried out in a vacuum condition.

Atomic force microscopes (AFMs) are appropiate instruments to measure the tool's cutting edge (Lucca and Seo, 1993; Krulewich Born and Goodman, 2001; Zong et al., 2007). The AFMs have several advantages such as three-dimensional (3-D) imaging, sub-nanometric measurement resolution and low measuring force. An AFM-based measuring instrument, which is appropriate for on-machine measurement, has been proposed to evaluate the form of the tool's cutting edge (Gao et al., 2006). In the proposed instrument, an optical alignment system has been employed to align the AFM probe to the tool's cutting edge (Jang et al., 2011) instead of using optical microscopes on the commercial AFMs. The proposed instrument has realized on-machine quantitative measurement of the tool's cutting edge (Gao et al., 2009). However, due to the limited measurement range of the AFM system, it takes a long measuring time for form measurements of a large-scale cutting tool.

In this paper, the optical alignment system developed in the AFM-based measuring instrument is proposed to be used as a micro optical probe for rapid and quantitative measurement of large-scale cutting edge of tools with a size in a range of several millimetres. In order to evaluate the tool's edge contours, a focused beam spot is employed as the micro optical probe. In the proposed measurement method, deviation of the intensity of the light, which is a part of the optical probe passed through the tool's edge, would be converted into the information of the tool's edge contour. The principle of the proposed method and an attempt to improve the measurement resolution of the micro optical probe are described. In addition, the edge contours of both a straight-shaped cutting tool and a round-shaped cutting tool are measured by the developed on-machine measurement system.

2   Tool contour measurement with a micro optical probe

2.1   Proposed measurement method

A micro optical probe has been developed to measure edge contours of single point diamond tools. A schematic of the proposed micro optical probe is shown in Fig. 1. A rake face of the diamond tool with a large-scale edge contour is set perpendicular to the optical axis of the probe. In this paper, a focused laser beam by the objective lens is called a micro optical probe and is used for measurement of edge contours. The intensity of the laser beam, which is a part of the optical probe passed through the tool edge, is collected by a photodiode (PD). The voltage signal of a current-to-voltage circuit, which converts photoelectric current output of the PD to a voltage signal, is referred to as the PD output. A relationship between the PD output and the y-directional position of the optical probe with respect to the tool's edge is shown in Fig. 1b. By referring the relationship curve in Fig. 1b, the PD output can be converted to the relative position of the probe with respect to the tool's edge contour. When the information on the absolute probe position is available from the tool positioning system, the edge contour can therefore be acquired from the deviation of the PD output and the information on the absolute probe position; that is the core principle of the proposed method.

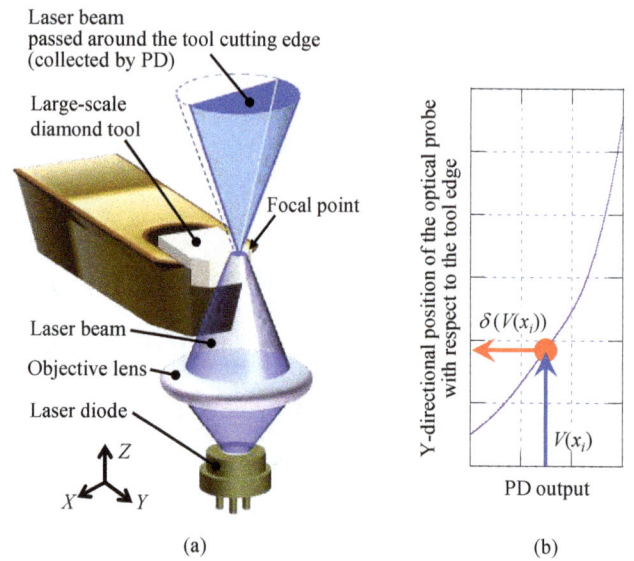

Figure 1. Schematic of the proposed micro optical probe. **(a)** Configuration of the optical setup. **(b)** Relationship between the PD output and the y-directional position of the optical probe with respect to the tool's edge.

Figure 2 shows a schematic of how the PD output can be converted to the relative position of the probe with respect to the tool's edge. In Fig. 2, $h(x_i)(= \delta(V(x_i)))$ is the relative position of the centre point of the optical probe with respect to the tool's edge, $g(x)$ is a scanning path of the micro optical probe. The tool's edge contour $f(x)$ can be expressed as

$$f(x) = g(x) + h(x). \tag{1}$$

While the optical probe is tracing the scanning path $g(x)$, both the PD output $V(x_i)$ and the probe position $g(x_i)$ at each x position (x_i, i: sampling number) are recorded simultaneously. After that, by referring the relation curve shown in Fig. 1b, the PD output $V(x_i)$ can be converted to the relative position $h(x_i)(= \delta(V(x_i)))$.

Figure 3 shows a schematic of the alignment of the tool's edge with respect to the optical probe. In the proposed method, the tool's edge should be aligned to the optical stylus precisely in the XY plane. In addition, the tool should also be positioned at the beam's waist as shown in Fig. 3a. The variation of the PD output with respect to the tool displacement along the z direction is shown in Fig. 3b. The PD output varies as the change of the sectional size of the beam, and has minimum value when the tool is positioned at the beam's waist of the optical probe. In the proposed method, the z-directional alignment can thus be carried out by monitoring the PD output before the probe scanning in the XY plane. The measurement resolution of the proposed method is mainly determined by the accuracy of the scanning motions and spot size of the optical probe, and S/N (signal-to-noise) ratio of the PD output. It should be noted that the beam

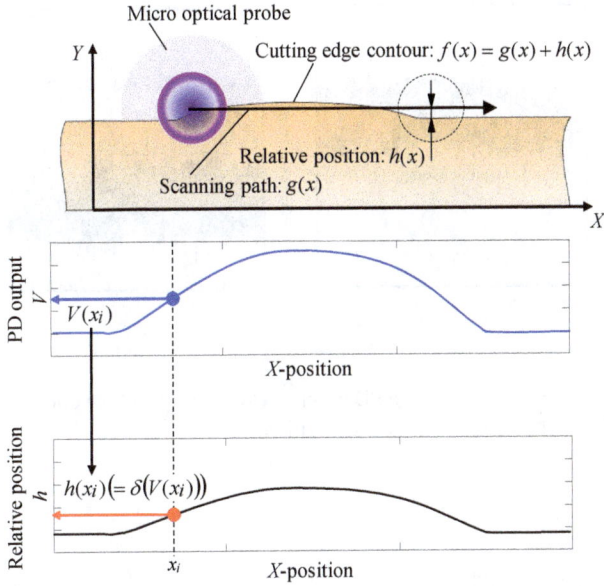

Figure 2. A schematic of how the PD output can be converted to the relative position of the probe with respect to the tool's edge.

spot is small enough to be treated as a "point" on the cutting edge in terms of the tool's nose radius.

2.2 Measurement system

In order to measure form errors and radii of tool edge contours, a measurement system including an optical system for the micro optical probe has been developed. Figure 4 shows a schematic of the developed system. The optical probe consists of a laser diode (LD), a beam splitter (BS), three lenses and two PDs. The optical system is mounted on a diamond turning machine, which has a precision positioning system along the z, y, and z directions. A laser beam, which has a wavelength of 405 nm, is emitted from the LD along the z direction, and is collimated by a collimator lens. The collimated laser beam is divided into a reference beam and a measurement beam by using the BS. Deviation of the light intensity of the reference beam is monitored by the PD for reference (PD_R) so that the influence of the LD power drift and some noises due to electric field around the LD can be compensated. Meanwhile, the measurement beam is focused on the tool's edge by an objective lens. Both the collimator lens and the objective lens are aspherized achromatic lenses with a numeric aperture (NA) of 0.25. The laser beam spot has a diameter of 2 μm at the beam waist (defined by a full width at half maximum). The laser beam's spot is employed as the micro optical probe in this paper. The size of the focused beam is mainly governed by the NA of the objective lens and the laser wavelength λ. The spot radius r_0 at the diffraction limit edge is expressed as follows (Grosjean and Courjon, 2006):

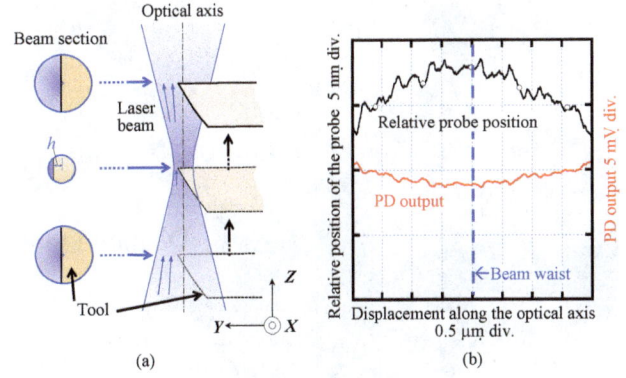

Figure 3. A schematic of the alignment of the tool's edge with respect to the optical probe. **(a)** Schematic of the tool positioning. **(b)** Relation curves for PD output, displacement along the z direction and the relative probe position converted from PD output.

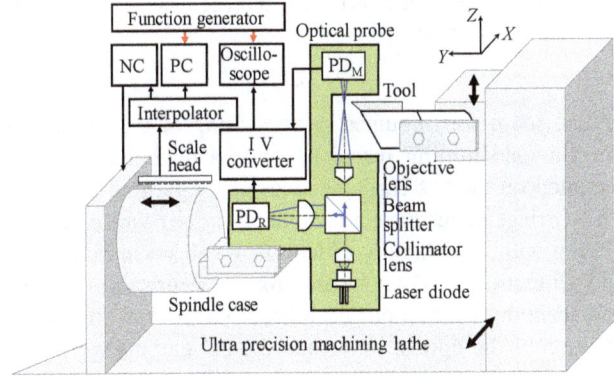

Figure 4. A schematic of the on-machine tool's edge contour measurement system.

$$r_0 \propto \frac{\lambda}{\text{NA}}. \tag{2}$$

According to Eq. (2), an objective lens with a higher NA generates a smaller laser beam spot that can achieve a higher resolution. However, attention should be paid for the light-collecting part of the micro optical probe when applying the objective lens with a high NA because the light passed through the tool's edge would propagate rapidly.

A portion of the direct beam that passed through the tool's edge is captured by the PD for measurement (PD_M) during the scanning of the micro optical probe. The output of PD_M is used as a measurement signal. In this paper, the I–V circuit for both PD_M and PD_R are arranged on the same electric circuit board so that influences of circuit power instability and noises due to electric field, which would affect signal quality of the PDs, can be minimized.

Figure 5 shows a photograph of the developed optical setup for the micro optical probe mounted on the diamond turning machine. To carry out on-machine measurement, the

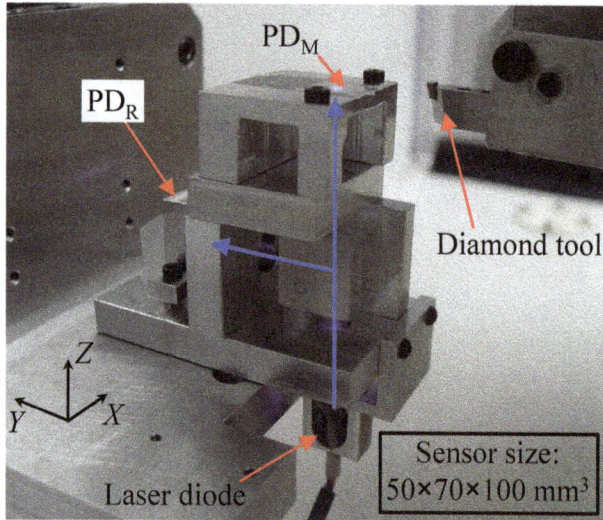

Figure 5. The developed micro optical probe mounted on the diamond cutting machine.

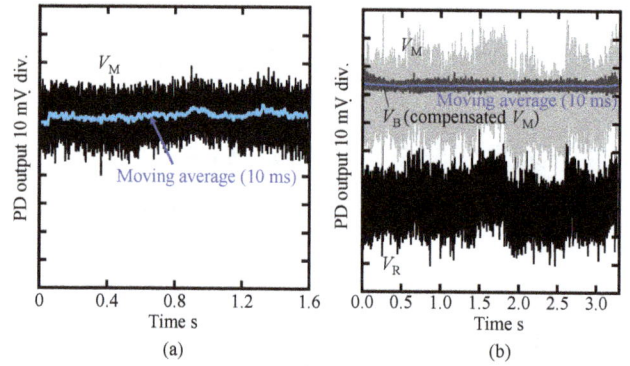

Figure 6. Stability of the PD outputs: **(a)** stability without compensation, **(b)** stability with compensation.

optical setup was mounted on the casing of the spindle of the diamond turning machine, while a diamond tool was mounted on the xyz carriage slides of the turning machine. The carriage slides have a nanometre-order positioning resolution with repeatability of less than ± 50 nm in both the x and y directions, and can be used for the accurate positioning and scanning of the optical probe. Both the xyz-directional position of the optical probe and the PD outputs were captured simultaneously during the tracing of the optical probe over the tool's edge.

3 Experiments

3.1 Stability of the developed optical probe

Stability of the developed micro optical probe was investigated in experiments. The developed optical system was mounted on the diamond turning machine as shown in Fig. 5, and the PD outputs were captured while the optical probe was kept stationary on the tool's edge. At first, the BS was removed from the optical setup, and the variation of PD_M output (V_M) was monitored. Figure 6a shows the measured output waveform of the PD_M. The amplitude of the PD_M output was found to be about 40 mV while the one with a simple moving average (window size: 10 ms) was about 10 mV. After that, the BS was integrated into the optical setup again, and variations of both the PD_M output and PD_R output (V_R) were monitored. Figure 6b shows the measured PD outputs. The amplitudes of both the PD_M and PD_R outputs were about 50 mV, which was almost the same as the one shown in Fig. 6a. In Fig. 6b, unpredictable variations of the PD outputs, which were mainly due to a drift of LD power, were also observed. In the developed optical system, the influence of the LD instability was minimized by using the following

equation:

$$V_B = V_M \times \frac{V_{R_1}}{V_R}. \tag{3}$$

In Eq. (3), V_B is the compensated PD_M output, and V_{R_1} is an initial PD_R output. The compensated PD_M output waveform had the amplitude of about 6 mV, in which a noise component with a frequency of 100 Hz due to the machine tool vibration was found. The amplitude of the V_B with a simple moving average (window size: 10 ms) was less than 1.5 mV, corresponding to the tool's edge contour of less than 10 nm. From these results, the feasibility of the developed optical system on the improvement of the optical probe stability was verified.

3.2 Contour evaluation of a straight cutting edge

A straight cutting edge of a diamond tool was measured to investigate the feasibility of the developed measurement system. The straight cutting edge was aligned to be parallel with the x axis of the measurement system within a tolerance of 1 μm–2 mm since the positioning error of the micro optical probe along its optical axis would influence the accuracy of the edge contour measurement as shown in Fig. 3b. A scanning speed of the optical probe was set to be 0.267 mm s^{-1}. It took about 13 s to scan the tool's edge with a length of 3.5 mm. The sampling frequency of the PD output was set to be 4 kHz. All the measurements were carried out on the machine, under constant temperature conditions of $24.2 \pm 0.1\,°C$.

Prior to the edge contour measurement, the relationship between the PD output and the relative position was investigated in experiments. The optical micro probe was moved along the y direction while the PD output was monitored. Figure 7 shows the acquired relationship between the PD output V and the relative position δ. The encoder readout of the y axis slide of the diamond turning machine was used to determine the relative position δ. The acquired relation curve in Fig. 7 was fitted by a polynomial function.

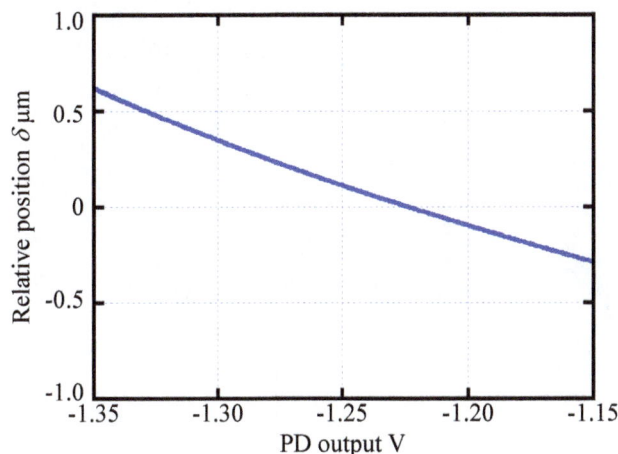

Figure 7. Relationship between the PD output and the y-directional relative position of the optical probe with respect to the flat-shaped cutting edge.

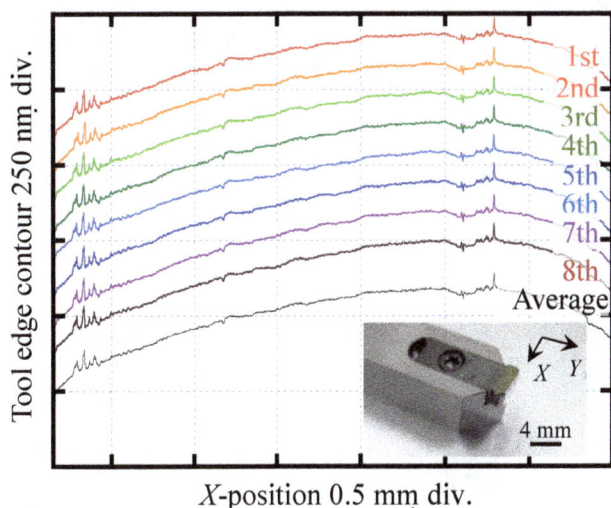

Figure 8. Measured cutting edge contour of the straight-shaped diamond tool.

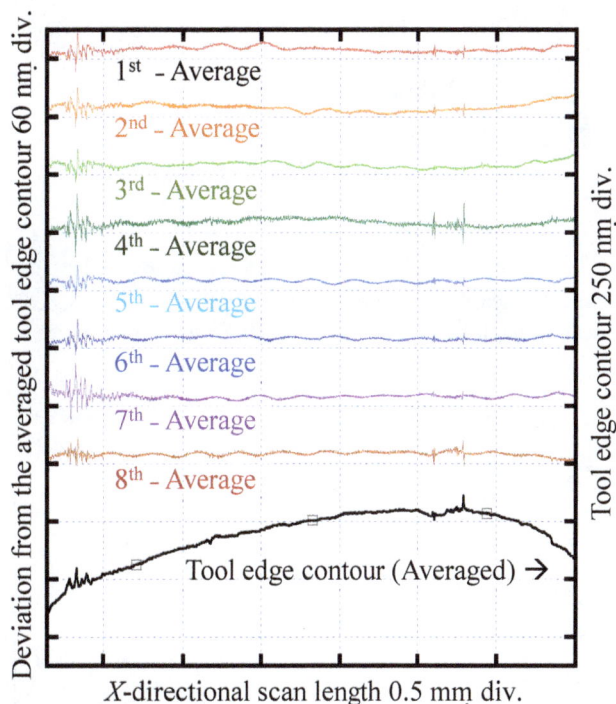

Figure 9. Averaged measurement result of a flat-shaped cutting edge from eight measurements and their deviations.

Figure 8 shows the edge contour measured by the developed system. The PD output was converted into the edge contour by using the relation curve shown in Fig. 7. In Fig. 8, the averaged edge contour acquired by averaging the results of eight time measurements is also plotted. The measurement results were found to agree with each other. A large deviation of 400 nm from the straight shape over the whole measurement range was observed. Measurement repeatability was investigated by plotting the deviations of each measured edge contour from the averaged one as shown in Fig. 9. At the areas of smooth edge contour, measurement repeatability was found to be below 10 nm. The influence of the mechanical vibration of the system, a frequency which was about 1 Hz, was observed. The measurement repeatability became worse at the area of steeply deviated edge contour. The main reason

of this is due to the slight difference of the x-directional positioning of the micro optical probe at each measurement. As shown in Figs. 8 and 9, the straight cutting edge having submicron form accuracy was successfully evaluated over the measurement range of 3.5 mm by the developed measurement system.

3.3 Contour evaluation of a round cutting edge

Another diamond tool having a round cutting edge was also measured by using the developed on-machine measurement system. Figure 10 shows schematics of two measurement methods with different scanning paths of the optical probe used in this paper.

One method used a raster-like scanning path, in which each scan line was set perpendicular to the tool's edge contour (we call this method "raster-scan method" in this paper) as shown in Fig. 10a. In the method, a priori knowledge on the form of the edge contour would be required to define the scanning path. In this paper, a "pseudo" edge contour was generated by detecting several points on the tool's edge and fitting a circle on them as shown in Fig. 10a.

The other method used a direct scanning path, in which the probe smoothly traced the tool's edge contour (we call this method "direct-scan method" in this paper). Figure 10b shows a schematic of the direct-scan path. The pseudo edge contour calculated in the raster-scan method was also employed in the direct-scan method. The direct-scan method

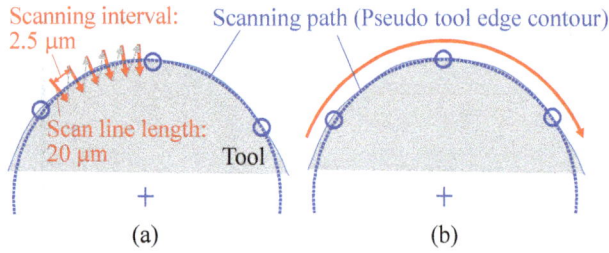

Figure 10. Schematics of the scanning paths for round cutting edge measurement. **(a)** Raster-scan method. **(b)** Direct-scan method.

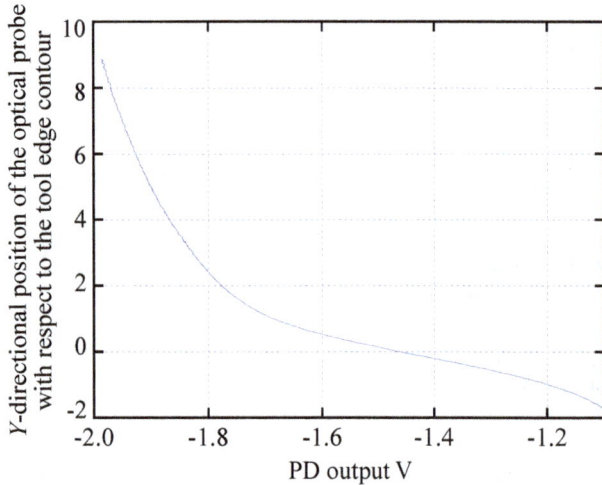

Figure 12. Matrix data of the PD output acquired at each line scan.

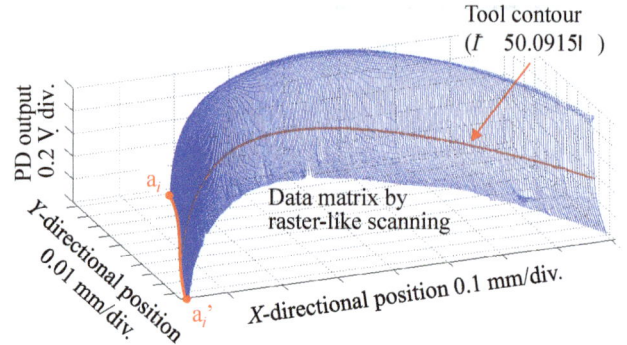

Figure 11. Relationship between the PD output and the y-directional relative position of the optical probe with respect to the round cutting edge (nose radius: 0.9 mm).

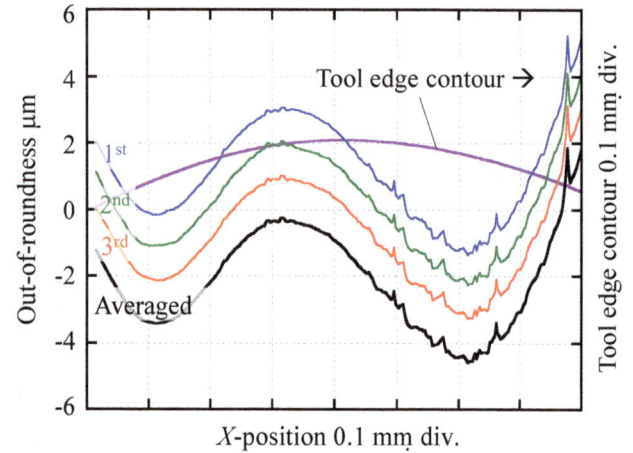

Figure 13. Edge contour and out-of-roundness measured by the raster-scan method.

could reduce measurement time. Meanwhile, the optical probe could not measure the tool's form error larger than its diameter when the direct-scan method was applied.

The tool's cutting edge with a nose radius of 0.9 mm was measured by using the two types of the scanning methods. Measurement range was set to be 0.8 mm. The relationship between the PD output V and the relative position δ was acquired in advance of the edge contour measurement as shown in Fig. 11. It should be noted that the relation curve in Fig. 11 was different from the one shown in Fig. 7 since the tool's contour of the measurement target is different.

At first, the tool was measured by using the raster-scan method. Figure 12 shows the PD output acquired at each line scan (line $a_i - a_i'$ shown in Fig. 9). The length and number of the sampling points for each scan line were set to be 20 μm and 320 points, respectively. The scanning interval along the pseudo edge contour was set to be 2.5 μm.

Figure 13 shows the measured edge contour, which was calculated from the matrix data of the PD output shown in Fig. 12. When the centre of the micro optical probe is positioned at the edge top of the tool, the intensity I of the optical probe which passes through the tool's edge would be

50.04 % of that of the whole optical probe, according to the kinematic relationship between the tool's edge (nose radius: 0.9 mm) and the micro optical probe (diameter: 2 μm). The tool's edge contour was therefore subtracted from the matrix data in Fig. 12 by using that value. In Fig. 13, out-of-roundness, which is the deviation of the edge contour with respect to a fitted circle, acquired at each measurement is also plotted (each curve was plotted with some offset just for clarity). A good agreement was found among the tool's measured edge contours. Deviations of the out-of-roundness were also shown in Fig. 14. Measurement repeatability was found to be below 0.20 μm, which was worse than that of the straight cutting edge measurement. One of the main reasons of these results is the long measurement time; it took about 7 min to carry out measurements with the raster-scan method, and the thermal drift was considered to affect the measurement repeatability.

After that, the edge contour was also measured by using the direct-scan method. Figure 15 shows the edge contour measured by the direct-scan method. In Fig. 15, the scanning path $g(x)$ set for measurement, the deviation of the PD output ($V(x_i)$, which will be converted into the relative position

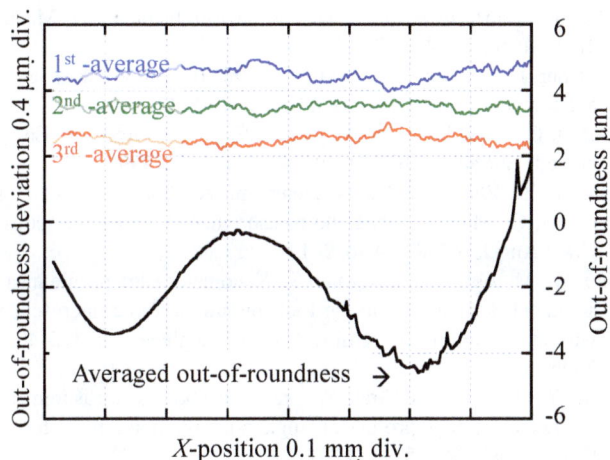

Figure 14. Deviation of the out-of-roundness measured by the raster-scan method.

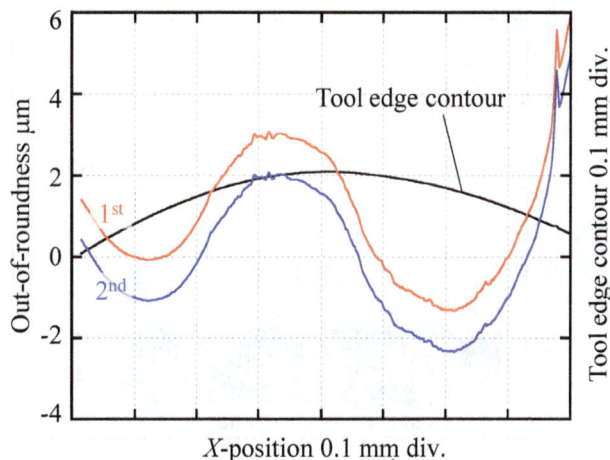

Figure 16. Out-of-roundness measured by the direct-scan method.

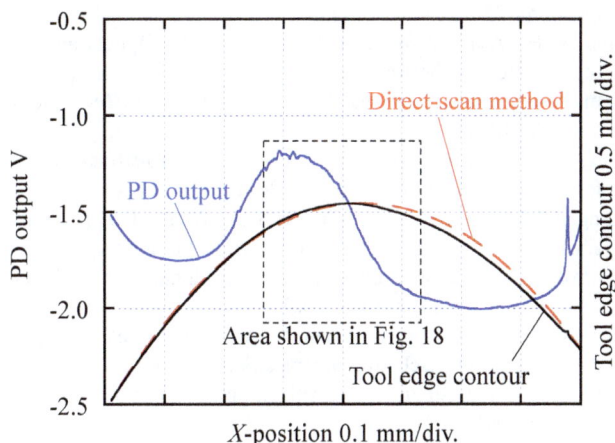

Figure 15. Edge contour of the round cutting edge measured by the direct-scan method.

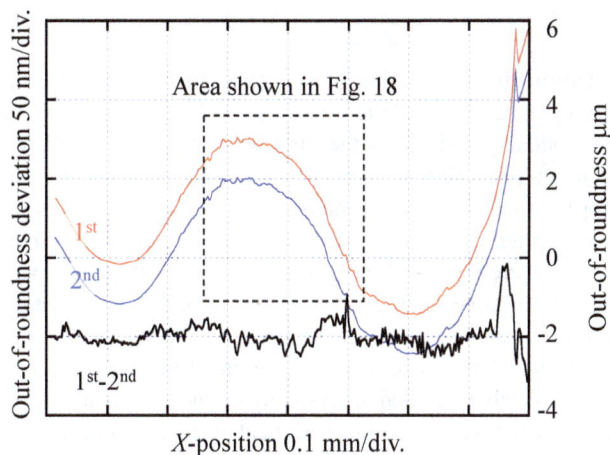

Figure 17. Deviation of the out-of-roundness measured by the direct-scan method.

$(\delta x_i))$ during the scan, and the tool's edge contour $f(x)$ calculated by using Eq. (1) are plotted. Figure 16 shows the out-of-roundness calculated from the measured edge contour shown in Fig. 15. It was also found that the tool's edge contour measured in the two time measurements agreed well with each other. Figure 17 shows the deviation of the out-of-roundness. The measurement repeatability in the two time measurements was below 80 nm, which was better than that of the raster-scan method because the required time for the direct-scanning method was relatively shorter (6 s) than that of the raster-scan method. Figure 18 shows the comparison of the out-of-roundness acquired by the two types of the scanning method. An optical microscopic image of the tool's contour is also shown in Fig. 18. The results by the direct-scanning method agreed well with the results by the raster-scan method in the region "A" in Fig. 18. Meanwhile, the direct-scanning method could not detect the steeply deviated edge contour in region "B", which was well detected by the

raster-scan method. The deviation of the tool's edge contour from the scanning path of the optical probe was large and its measurement sensitivity was considered to be degraded, resulting in a spatially averaged form in region "B" with the direct-scanning method.

4 Conclusions

An optical method for edge contour measurement by using a micro optical probe was proposed to evaluate the large-scale cutting edge of tools over the measurement range of several millimetres in a short time. An optical setup for the optical micro probe was developed, and its feasibility on the edge contour measurement was investigated in experiments. The stability of the optical micro probe was successfully improved by using the compensation optical system, and the measurement resolution of 10 nm was achieved. The on-machine measurement system was established by

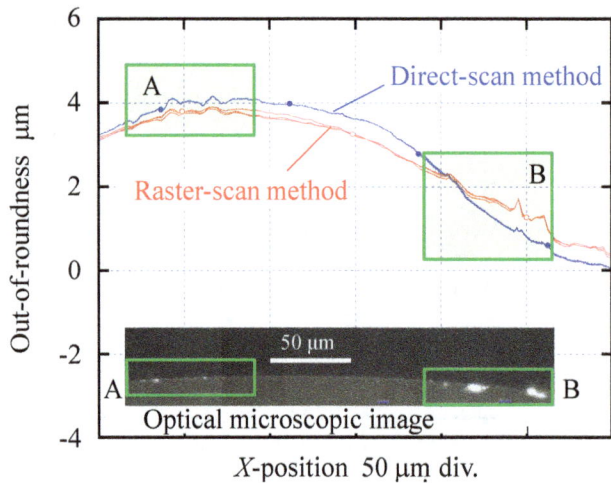

Figure 18. Comparison between the raster-scan method and the direct-scan method.

synchronizing the motion of the xyz slides on the diamond turning machine and the acquisition of PD outputs from the optical probe. Both the straight cutting edge and the round cutting edge of diamond tools were evaluated by the proposed measurement system. For the round cutting edge measurement, both the raster-scan method and direct-scan method were proposed, and the direct-scan method achieved high-speed edge contour measurements in 6 s for the measurement range of 3.5 mm. Throughout the experiments, the feasibility of the developed measurement system was confirmed. Further detailed analysis on the measurement uncertainty of the developed measurement system should be carried out as a future work.

Acknowledgements. This research was supported by JSPS KAKENHI grant number K2243843 and JST (A-STEP).

Edited by: R. Schmitt
Reviewed by: two anonymous referees

References

Asai, S., Taguchi, Y., Horio, K., Kasai, T., and Kobayashi, A.: Measuring the very small cutting-edge radius for a diamond tool using a new kind of SEM having two detectors, CIRP Annals, 39, 85–88, 1990.

Brehm, R., van Dun, K., Teunissen, J. C. G., and Haisma, J.: Transparent single-point turning of optical glass, Precision Engineering, 1, 207–213, 1979.

Brinksmeier, E., Riemer, O., Gläbe, R., Lünemann, B., Kopylow, C. V., Dankwart, C., and Meier, A.: Submicron functional surfaces generated by diamond machining, CIRP Annals, 59, 535–538, 2010.

Chen, F. J., Yin, S. H., Huang, H., Ohmori, H., Wang, Y., Fan, Y. F., and Zhu, Y. J.: Profile error compensation in ultra-precision of aspheric surfaces with on-machine measurement, Int. J. Mach. Tool. Manu., 50, 480–486, 2010.

Drescher, J.: Scanning electron microscopic technique for imaging a diamond tool edge, Precision Engineering, 15, 112–114, 1993.

Evansa, C. J. and Bryan, J. B.: "Structured", "Textured" or "Engineered" Surfaces, CIRP Annals, 48, 541–566, 1999.

Fang, F. Z., Zhang, X. D., Weckenmann, A., Zhang, G. X., and Evans, C.: Manufacturing and measurement of freeform optics, CIRP Annals, 62, 823–846, 2013.

Gao, W., Motoki, T., and Kiyono, S.: Nanometer edge profile measurement of diamond cutting tools by atomic force microscope with optical alignment sensor, Precision Engineer, 30, 396–405, 2006.

Gao, W., Asai, T., and Arai, Y.: Precision and fast measurement of 3D cutting edge profiles of single point diamond micro-tools, CIRP Annals, 58, 451–454, 2009.

Goel, S., Luo, X., Comley, P., Reuben, R. L., and Cox, A.: Brittle-ductile transition during diamond turning of single crystal silicon carbide, Int. J. Mach. Tool. Manu., 65, 15–21, 2013.

Grosjean, T. and Courjon, D.: Smallest focal spots, Optics Communications, 272, 314–319, 2006.

Jang, S. H., Asai, T., Shimizu, Y., and Gao, W.: Optical analysis of an optical probe for three-dimensional position detection of micro-objects, International Journal of Automation Technology, 5–6, 862–865, 2011.

Jiang, X.: In situ real-time measurement for micro-structured surfaces, CIRP Annals, 60, 536–566, 2011.

Kong, L. B., Cheung, C. F., To, S., Lee, W. B., Du, J. J., and Zhang, Z. J.: A kinematics and experimental analysis of form error compensation in ultra-precision machining, Int. J. Mach. Tool. Manu., 48, 1408–1419, 2008.

Krulewich Born, D. and Goodman, W. A.: An empirical survey on the influence of machining parameters on tool wear in diamond turning of large single-crystal silicon optics, Precision Eng., 25, 24–257, 2001.

Lane, B. M., Shi, M., Dow, T. A., and Scattergood, R.: Diamond tool wear when machining Al6061 and 1215 steel, Wear, 268, 1434–1441, 2010.

Lane, B. M., Dow, T. A., and Scattergood, R.: Thermo-chemical wear model and worn tool shapes for single-crystal diamond tools cutting steel, Wear, 300, 216–224, 2013.

Li, X. P., He, T., and Rahman, M.: Tool wear characteristics and their effects on nanoscale ductile mode cutting of silicon wafer, Wear, 259, 1207–1214, 2005.

Lucca, D. A. and Seo, Y. W.: Effect of tool edge geometry on energy dissipation in ultraprecision machining, CIRP Annals, 42, 83–86, 1993.

Moriwaki, T. and Okuda, K.: Machinability of Copper in Ultra-Precision Micro Diamond Cutting, CIRP Annals, 38, 115–118, 1989.

Zong, W. J., Li, D., Sun, T., Cheng, K., and Liang, Y. C.: The ultimate sharpness of single-crystal diamond cutting tools-Part II: A novel efficient lapping process, Int. J. Mach. Tool. Manu., 47, 864–871, 2007.

Zong, W. J., Li, Z. Q., Sun, T., Cheng, K., Li, D., and Dong, S.: The basic issues in design and fabrication of diamond-cutting tools for ultra-precision and nanometric machining, Int. J. Mach. Tool. Manu., 50, 411–419, 2010.

16

Ultrasound-based density determination via buffer rod techniques: a review

S. Hoche, M. A. Hussein, and T. Becker

Chair of Brewing and Beverage, Bio-PAT (Bio-Process Analysis Technology), Freising, Germany

Correspondence to: S. Hoche (s.hoche@wzw.tum.de)

Abstract. The review presents the fundamental ideas, assumptions and methods of non-invasive density measurements via ultrasound at solid–liquid interface. Since the first investigations in the 1970s there has been steady progress with regard to both the technological and methodical aspects. In particular, the technology in electronics has reached such a high level that industrial applications come within reach. In contrast, the accuracies have increased slowly from 1–2 % to 0.15 % for constant temperatures and to 0.4 % for dynamic temperature changes. The actual work reviews all methodical aspects, and highlights the lack of clarity in major parts of the measurement principle: simplifications in the physical basics, signal generation and signal processing. With respect to process application the accuracy of the temperature measurement and the presence of temperature gradients have been identified as a major source of uncertainty. In terms of analytics the main source of uncertainty is the reflection coefficient, and as a consequence of this, the amplitude accuracy in time or frequency domain.

1 Introduction

The medium density is a key parameter for most known processes in chemical, petrochemical, pharmaceutical, food and beverage, biotechnology, water and waste-water industries. The potential to determine online the quantity and quality of the process medium by means of density enables new options of process control and management. There are methods based on direct physical relations or based on the determination of parameters that can be correlated to the density for a specific chemical reaction or a characteristic process course. But most established methods, like coriolis mass flow or vibrating U-tube, have system-inherent limitations that often result in application restrictions in sensor implementation (limits in pipe diameter, limited to bypass application, limited to a certain flow range). Based on the specifications of the process, additional limitations might be sensitivity to bubbles, particles or fouling. In the case of food processing, hygienic design is a dominant constraint. The actual paper reviews ultrasound-based techniques as alternative methods which may be used where standard methods are not applicable.

The easiest way to determine the real-time density is to monitor the ultrasound velocity. According to the Newton–Laplace equation

$$\kappa_S = \frac{1}{\rho_l c_l^2},\tag{1}$$

the density ρ_l of a liquid medium can be determined knowing the isentropic (adiabatic) compressibility κ_S and the sound velocity c_l. Unfortunately, the adiabatic compressibility is usually determined from sound velocity and density measurements at atmospheric pressure (Kaatze et al., 2008). In 1967 Davis and Gordon (Davis and Gordon, 1967) developed an exact method to measure the adiabatic compressibility by determining volume and sound velocity changes under varying pressure and temperature. Davis and Gordon's research work was followed by extensive investigations to determine thermophysical properties of different materials (Bolotnikov et al., 2005; Daridon et al., 1998a, b; Esperança et al., 2006; Kell, 1975; Żak et al., 2000). Since all three parameters – density, sound velocity and compressibility – are highly temperature dependent, and since the compressibility measurement is limited to laborious methods, the application of sonic velocimetry at constant frequencies is limited to density determination of binary systems (Asher, 1987; Van Sint Jan et al., 2008). The velocimetric approach is based on temperature and, in some cases, pressure-dependent calibration measurements of sufficiently pure and well-defined liquids (Rychagov et al., 2002) and results in applications such as electrolyte measurements in accumulators or density determination of pure liquids (Swoboda et al., 1983; Vray et al., 1992; Wang et al., 2011; Kuo, 1971; Marks, 1976; Wang and Nur, 1991). The accuracy of such methods generally depends on the type of liquid and its purity (Rychagov et al., 2002; Matson et al., 2002; Wang and Nur, 1991).

Further methods to determine the density via ultrasound are waveguide and interferometric approaches. The waveguide approach generally uses propagation time variations of torsional ultrasonic waves in a transmission line immersed in the sample liquid. Besides torsional waves, the use of flexural or Rayleigh waves is also possible. Even though waveguide sensors have been used by several research groups over the last decades (Kim and Bau, 1989), it is reported (Lynnworth, 1994) that the method suffers from viscosity effects and has to be specifically designed to fulfil certain wavelength aspects.

The interferometric approaches use the effects of overlapping waves. While Pope et al. (1992, 1994) used peak FFT values of the resonance response spectrum over a certain frequency range, Sinha and Kaduchak (Sinha and Kaduchak, 2001; Kaduchak and Sinha, 2001; Sinha, 1998) used swept-frequency acoustic interferometry (SFAI) based on characteristics of standing-wave patterns. Pope's method relies on calibration measurements, and therefore is limited in the same way as the velocimetric methods. The method presented by Sinha and Kaduchak was not developed for highly accurate acoustic measurements. They reported a relative uncertainty of 0.5 % for sound speed and 5 % for the density measurement.

In conclusion to the text above, one can allege that the enormous calibration effort of most ultrasound-based methods may be the reason that, in the past decades, several research groups have focused on reflection-coefficient-based density determination methods via buffer rod systems. The plane wave propagation across one or more interface is the basis of buffer rod techniques. The history of single pulses is described with respect to the excitation amplitude considering reflection, transmission and attenuation terms. Calculating the ratios of feasible pulses results in amplitude-based representation of the reflection coefficient. Further parameters like attenuation and density can be calculated based on the knowledge of the buffer material's properties.

Sachse (1974) and Hale (1988) first reported on this method and presented validation results. Sachse analysed the amplitudes of pulses, scattered by a fluid-filled inclusion in an aluminium block to determine the reflection coefficient (RC), r of the pulse incident on the inclusion. Finally, the measured RC and the known impedance of the matrix material were used to calculate the density of the inclusion fluid.

In contrast, Hale used a transmitter–receiver configuration. From the amplitude changes of received signals, he determined the sample density with a bias of less than 2 %.

McClements and Fairly (1991, 1992) first paid attention to attenuation and temperature effects for their validation trials. The developed ultrasonic pulse echo reflectometer consists of a perspex buffer rod and an aluminium reflector plate. The reflectometer has been immersed in a water bath to stabilize the temperature to $\pm 0.1\,°C$. According to Eq. (2) the RC, $r_{\text{buffer-sample}}$ of the interface perspex buffer–sample-fluid was calculated by the use of reference signals, for which the reference medium was air. Assuming total reflection ($Z_{\text{air}} \ll Z_{\text{perspex}}$; $r \approx 1$) and constant incident pulse amplitudes A_i the ratio of the first echo's amplitudes leads to an attenuation independent term:

$$r_{\text{buffer-sample}} = A_{1\text{sample}}/A_{1\text{air}}, \qquad (2)$$

where $A_{1\text{sample}}$ is the pulse amplitude of the first pulse that is reflected from buffer–sample-fluid interface and $A_{1\text{air}}$ is the pulse amplitude of the first pulse that is reflected from buffer-air interface of the reference measurement. Knowing the RC $r_{\text{buffer-sample}}$, the specific acoustic impedance of the actual sample can be determined. McClements and Fairly achieved remarkable accuracy of $\pm 0.01 \times 10^6\,\text{kg}\,\text{m}^{-2}\,\text{s}^{-1}$ for the impedance determination. A precision of approximately $\pm 0.5\,\text{m}\,\text{s}^{-1}$ was reported for the speed-of-sound measurements. Using both to calculate densities for a series of sodium chloride solutions, an accuracy of $\pm 6\,\text{kg}\,\text{m}^{-3}$ (0.5 %) could be achieved.

In general, all subsequent investigations are based upon the same basic relations, only varying in sensor design, methodology adaptions and signal analysis. The review focuses on ultrasound-based density determination via buffer rod techniques (BRT). In Sect. 2 the physical fundamentals and basic assumptions will be discussed as well as the four basic methods that have been identified. In Sect. 3 relevant design considerations will be presented. Finally, in Sect. 4, all major analytical aspects will be discussed with respect to density accuracy, uncertainties and real process application.

2 Physical fundamentals and method classification

The basis of all BRTs is the determination of the RC, which in general is based upon the physical description of plane wave propagation across an interface (see Fig. 1). Every medium is characterized by certain sound velocity c, density ρ and sound attenuation α. Any loss of energy that appears while sound wave propagates through homogeneous medium is summarized in the attenuation term. As soon as the wave arrives at an interface, the wave will be partly transmitted and partly reflected.

The relation of transmission and reflection is governed by the specific acoustic impedance Z of the medium defined as

$$Z = \frac{\omega}{k}\rho = \frac{\omega}{\omega/c - j\alpha}\rho = \frac{c}{1 - j\alpha^c/\omega}\rho, \qquad (3)$$

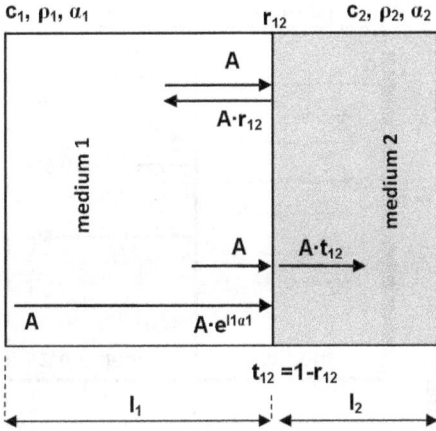

Figure 1. Schema showing the basic principles of sound propagation across an interface at normal incidence.

where k is the complex wave number and ω the angular frequency ($= 2\pi f$). For materials of sufficiently small attenuation ($\alpha \ll \omega/c$ or $\alpha c/\omega \ll 1$), Eq. (3) simplifies to

$$Z = \rho \cdot c. \tag{4}$$

The amount of a wave reflected at a plane interface is often characterized by the RC which is the ratio of the reflected (subscript r) to the incident (subscript i) wave. The RC can be expressed in terms of amplitudes A or intensities I. The intensity is proportional to the square of amplitude, which leads to the following expressions for a wave that passes from medium 1 (subscript 1) to medium 2 (subscript 2):

$$r_A = \frac{A_r}{A_i} = \frac{Z_2 - Z_1}{Z_2 + Z_1}, \tag{5}$$

$$r_I = \frac{I_r}{I_i} = \left(\frac{Z_2 - Z_1}{Z_2 + Z_1}\right)^2. \tag{6}$$

In the same way the transmission coefficient t is given as the ratio of transmitted wave (subscript t) to incident wave:

$$t_A = 1 - r_A = \frac{A_t}{A_i} = \frac{2Z_1}{Z_2 + Z_1}. \tag{7}$$

If one thinks in terms of buffer rod techniques (BRTs), medium 1 might be the buffer rod and medium 2 the sample liquid. Measuring at constant temperatures, the material properties (c and ρ) of the buffer remains constant, and any change in the RC is clearly related to a change of the specific acoustic impedance of the sample liquid. This means according to Eqs. (4)–(6), the density of the sample liquid ρ_2 can be determined via the reflection coefficient if the temperature-dependent properties of the buffer rod (ρ_1, c_1) and the sound velocity of the sample liquid (c_2) are known:

$$\rho_2 = \frac{\rho_1 c_1}{c_2} \frac{(1 + r_A)}{(1 - r_A)} = \frac{\rho_1 c_1}{c_2} \frac{(1 + r_I^2)}{(1 - r_I^2)}. \tag{8}$$

The wave propagation in its basic form is a mechanical oscillation and depends on the physical properties of the material (Saggin and Coupland, 2001; McClements, 1997; Povey and McClements, 1988):

$$\left(\frac{k}{\omega}\right)^2 = \frac{\rho}{\text{modulus of elasticity}}. \tag{9}$$

In the case of pressure waves, the appropriate modulus of elasticity is the longitudinal modulus M, which is equal to the sum of bulk modulus K and 4/3 shear modulus G. For Newtonian fluids the shear modulus can be neglected and the modulus of elasticity is assumed to be equal to the bulk modulus K ($= \kappa^{-1}$; see Eq. 1). If one considers that the wave number is complex and the attenuation in liquids is not negligible, the acoustic impedance becomes complex, expressed as the complex sum of the resistive (real) part, R_a, and the reactive (imaginary) part, X_a:

$$Z_a = \frac{P}{\xi} = R_a + jX_a, \tag{10}$$

where P is the acoustic pressure and ξ the particle displacement. Applying a BRT, the attenuation in the buffer is generally low and the simplification of Eq. (5) is valid. This may change in the case of a fluid as second phase. For high attenuation, a complex form of the RC is introduced which includes a loss angle, θ (O'Neil, 1949; Mason et al., 1949; Moore and McSkimin, 1970):

$$re^{-j\theta} = \frac{Z_2 - Z_1}{Z_2 + Z_1}, \tag{11}$$

leading to a complex acoustic impedance for the sample fluid:

$$Z_2 = R_2 + jX_2 = Z_1 \frac{1 - r^2 - j2r\sin\theta}{1 + r^2 - 2r\cos\theta}. \tag{12}$$

The resistive (real) part then becomes

$$R_2 = Z_1 \frac{(1 - r^2)}{1 + r^2 - 2r\cos\theta}, \tag{13}$$

and can be approximated as

$$R_2 \approx Z_1 \frac{1 + r}{1 - r}\left[1 - \frac{r\theta^2}{(1 - r)^2}\right] = Z_1 \frac{1 + r}{1 - r} + O(\theta^2). \tag{14}$$

Typically the acoustic impedance of liquids is less than $0.1(1 + j)$ of the buffer impedance, and therefore the loss angle was found not to exceed 5° (Mason et al., 1949). The loss angle dependent remainder can be neglected and the approximation can be used to specify the resistive component of the liquid's acoustic impedance for most buffer-liquid interfaces.

The buffer rod techniques published so far differ mainly in the way that the RC is determined, but not in the calculation of the density. Consequently, the accuracy of all BRT-density measurements basically depends on both the accuracy of the RC and the sound velocity measurement. Based upon the applied RC determination method the BRTs can be classified into multiple reflection methods (MRM), reference reflection methods (RRM), transmission methods (TM) and angular reflection methods (ARM).

2.1   Multiple reflection method (MRM)

The MRM (also known as the ABC method) was first de-
vised by Papadakis (1968). He determined the ultrasonic attenuation in a sample and the RC at the buffer–sample interface over a frequency range of 27–45 MHz. In 1972 Papadakis et al. (1973), together with Fowler and Lynnworth,
presented further results in the range 0–15 MHz and introduced a diffraction correction. Based upon the work of
Mason and Moore and McSkimin, Sachse (1974) applied
the same method to determine the density in a range up
to 10 MHz. Adamowski et al. (1998, 1995), Higuti and
Adamowski (2002a) and Bjørndal et al. (2008) used identical
principles, but enhanced some methodical aspects to overcome several error influences.

The core idea of the MRM is the use of pulse ratios. If the
correct pulses are related to each other, the unwanted attenuation, reflection and transmission terms can be neglected,
leaving a term that is only dependent on the RC of interest.
Principally the remaining term is even independent of the initially generated pulse amplitude. In general, a probe design
as shown in Fig. 2 is used for the MRM, in which medium 1
resembles the buffer (subscript 1); medium 2, the sample liquid (subscript 2); and medium 3, the reflector (subscript 3)
– all of them characterized by a certain κ, ρ and α. The reflection or transmission coefficients of the different interfaces
are indicated in terms of propagation direction and involved
mediums; for example,

RC for propagation from medium 1 to medium 2 :

$$r_{12} = \frac{Z_2 - Z_1}{Z_2 + Z_1};$$

transmission coefficient for propagation from medium 2

to medium 1 : $t_{21} = \dfrac{2Z_2}{Z_1 + Z_2}.$

Using the principles of plane wave propagation at normal
incidence, one obtains the following for A_{r1}, A_{e11} and A_{e21}:

$$A_{r1} = A_T \cdot r_{12} \cdot \exp(2l_1 \alpha_1), \tag{15}$$

$$A_{e11} = A_T \cdot t_{12} r_{23} t_{21} \cdot \exp(2l_1 \alpha_1) \cdot \exp(2l_2 \alpha_2), \tag{16}$$

$$A_{e21} = A_T \cdot t_{12} r_{23}^2 r_{21} t_{21} \cdot \exp(2l_1 \alpha_1) \cdot \exp(4l_2 \alpha_2). \tag{17}$$

The subscript r defines the captured pulse as buffer reflection
(BR) and the subscript e as an echo pulse. Furthermore in
A_{rk} and A_{ejk}, subscript k defines the pulse order (1st BR, A_{r1};
2nd BR, A_{r2}; etc.) and subscript j the echo order (e.g. pulses
of 1st echo, A_{e1k}; pulses of 2nd echo, A_{e2k}). For the ratios
A_{r1}/A_{e11} and A_{e11}/A_{e21} one obtains

$$\frac{A_{r1}}{A_{e11}} = \frac{r_{12}}{t_{12} r_{23} t_{21} \cdot \exp(2l_2 \alpha_2)}; \quad \frac{A_{e11}}{A_{e21}} = \frac{1}{r_{23} r_{21} \cdot \exp(2l_2 \alpha_2)}. \tag{18}$$

The terms of attenuation in medium 1 and the initial transmitted amplitude A_T are cancelled out. Additionally, it becomes clear that disregarding the first interface at the coupled

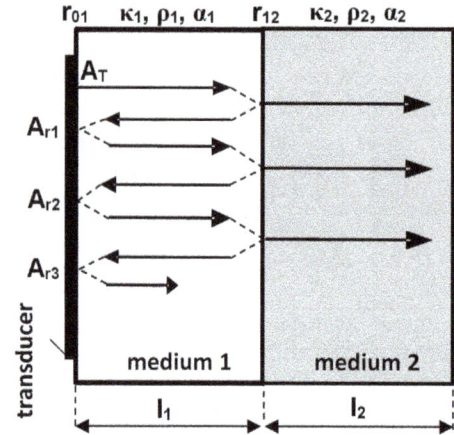

Figure 2. Schematic showing the basic principles and relevant
pulses for the MRM: buffer, medium 1; sample, medium 2; reflector,
medium 3.

sound source is a valid simplification. Every additional term
of the interface 0–1 (e.g.: $A_T = A_0 t_{01} t_{10}$)) would be added to
each of the pulses (Eqs. 15, 16 and 17) and therefore also
disappear in the ratios of (18).

Dividing now one ratio by the other, one reaches an
attenuation-independent equation, and the amplitudes A_1, A_2
and A_3 can be used to calculate the RC of interface 1–2, r_{12}:

$$r_{12} = \sqrt{\frac{x}{x-1}} x = \frac{A_{r1} A_{e21}}{A_{e11}^2}. \tag{19}$$

The resulting equation is now independent of the attenuation in medium 2. Papadakis (1968) first investigated a
glass buffer rod on a fused-silica sample. Later, in Papadakis
et al. (1973), a water buffer combined with a nickel sample was investigated; a RC of $r_{12} = 0.9435 \pm 0.0045$ was calculated, which was in good agreement with the theoretical value of 0.945. Furthermore, he introduced the so-called
A'AB method, which is more or less the first mention of the
RRM, and may be used if attenuation in medium 2 is too high
and amplitude A3 is very low. Further details about the RRM
will follow in the next section.

Instead of the normal buffer–reflector configuration,
Adamowski et al. (1995, 1998) used a double-element transducer (DET) including the buffer, a sample liquid (medium 2)
and a high-acoustic-impedance reflector (medium 3: stainless steel). The DET has a piezoceramic emitter and a
large-aperture receiver (PVDF membrane) separated by a
solid buffer rod (medium 1: PMMA) of length l_0. Another
buffer rod (medium 1: PMMA) of length l_1 is placed between receiver and sample medium. The great advantage
of Adamowski's approach is the employment of the large-
aperture receiver in the DET. The large aperture minimizes
the uncertainties if diffraction effects and the transmitted
pulse A_T can be gathered for every single excitation. That
enables calibrations due to varying excitation amplitudes as
they may occur during long-term operations. Nevertheless

applying the MRM, the use of A_T is not necessary. In Adamowski et al. (1995) a comparison of MRM and RRM is presented, and for MRM a bias of $10\,\mathrm{kg\,m^{-3}}$ is reported. The main limitation of Adamowski's DET is the PVDF's limited temperature range of application. At temperatures above 60–70 °C the piezoelectric PVDF slowly loses its imposed polarized structure. A successful application of high-temperature piezoelectric materials (PEM) in a DET has not been reported so far.

Bjørndal et al. (2008) used the MRM to verify a newly developed TM, which will be discussed later. They investigated liquids with a wide range of shear viscosities at a temperature of 27.44 ± 0.04 °C. It was reported that the systematic deviation from reference values of a calibrated pycnometer was smaller for the MRM than for the TM, and reached an error of ± 0.15 %.

A special version of the MRM is the approach of Deventer and Delsing (1997). Although this method does not follow the typical ABC approach of Papadakis, it is classified as MRM since some specific reflections are used to calculate the RC without additional calibration measurements. Delsing and Deventer used a double buffer of two different materials. Keeping the terminology of Fig. 2, medium 2 is now the second buffer and medium 3 is the sample liquid. Eliminating A_T in Eq. (17) with the use of Eq. (16) one achieves for r_{23}

$$r_{23} = \frac{A_{e11} \cdot r_{12}}{A_{r1} \cdot t_{12} t_{21} \cdot \exp(2l_2 \alpha_2)}, \tag{20}$$

and for ρ_3

$$\rho_3 = \frac{Z_2}{c_3} \cdot \frac{4 A_{r1} Z_1 Z_2 \exp(2l_2 \alpha_2) - A_{e11}(Z_1^2 - Z_2^2)}{4 A_{r1} Z_1 Z_2 \exp(2l_2 \alpha_2) + A_{e11}(Z_1^2 - Z_2^2)}. \tag{21}$$

Since the properties of medium 1 and 2 are known, the unknown parameters that have to be measured are c_3, A_{e11} and A_{r1}. So basically no echo pulse from a reflector is necessary to calculate the RC, which is a great advantage in the case of highly absorptive liquids. The disadvantage is that not only is the exact knowledge of temperature-dependent density and sound velocity of one medium required, but that of two mediums. Additionally, the attenuation in medium 2 has to be known to calculate the RC. And the sound velocity of the sample liquid is still necessary to calculate the density. Therefore transmission or pulse-echo measurements through the liquid are still a requirement to determine the density.

Deventer and Delsing (1997) used 32-times-averaged digitized signals in order to determine the densities of water at 2, 20 and 40 °C. The measured densities have been compared with tabulated data, and a mean bias of $1\,\mathrm{kg\,m^{-3}}$ was reported. In fact, the presented graph shows standard deviations from $\pm 5\,\mathrm{kg\,m^{-3}}$ at 40 °C up to $\pm 10\,\mathrm{kg\,m^{-3}}$ at 2 °C, and it was not mentioned as to how many densities have been averaged to reach the reported results. In Deventer and Delsing (2001a) the densities of glycerin, water and alcohol were determined in a temperature range from 0 to 40 °C. A mean of 100 measurements and tabulated reference data was used

for the validation. Even though a clear separation between the results of the different sample liquids is possible, the results still show varying bias and standard deviation for varying temperatures. It was stated that sound velocity inaccuracies generated an error of approximately 1 % and that a density error of 0.4 % should be reachable.

2.2 Reference reflection method (RRM)

A first version of the RRM was presented by Papadakis et al. (1973). As with all RRM the core idea is the use of plane wave propagation principles at normal incidence in combination with a reference medium. For the so-called A'AB method, Papadakis uses the 1st buffer reflection of a reference medium A' and the same 1st buffer reflection of the sample medium A to calculate the RC. The pulse amplitude B is only used to calculate the attenuation. A similar approach was used later by Adamowski et al. (1998), McClements and Fairly (1991), Saggin and Coupland (2001) and Kulmyrzaev et al. (2000).

Similar to the MRM approach of Deventer and Delsing (1997), the RC determination via RRM does not rely on the presence of a reflector. Of course, calculating the final density via Eq. (8) still requires the sound velocity of the sample medium, and therefore needs either transmission or pulse-echo measurements through the liquid, but the schematic representation of the basic principles to determine the RC can be simplified to medium 1 and 2 (see Fig. 3). For moderate attenuation and thickness of medium 1, one can obtain the amplitudes of the multiple buffer reflections A_{rk} as follows:

$$A_{r1} = A_T \cdot r_{12} \cdot \exp(2l_1 \alpha_1); \; A_{r2} = A_T \cdot r_{10} \cdot r_{12}^2 \cdot \exp(4l_1 \alpha_1);$$
$$A_{rk} = A_T \cdot r_{10}^{k-1} \cdot r_{12}^k \cdot \exp(2kl_1 \alpha_1). \tag{22}$$

The RRM based on one pulse, as applied in McClements and Fairly (1991), Papadakis et al. (1973), Püttmer and Hauptmann (1998), Püttmer et al. (1998, 2000) and Saggin and Coupland (2001), uses the ratio of any detectable buffer reflection of a sample medium and the corresponding buffer reflection of a reference medium, e.g. $A_{r1}(\text{sample})$ and $A_{r1}(\text{reference})$:

$$\frac{A_{r1}(\text{sample})}{A_{r1}(\text{reference})} = \frac{A_T \cdot r_{12}(\text{sample}) \cdot \exp(2l_1 \alpha_1)}{A_T \cdot r_{12}(\text{reference}) \cdot \exp(2l_1 \alpha_1)}. \tag{23}$$

Assuming a constant excitation pulse A_T and a similar attenuation α_1 for sample and reference signal one obtains

$$r_{12}(\text{sample}) = r_{12}(\text{reference}) \frac{A_{r1}(\text{sample})}{A_{r1}(\text{reference})}. \tag{24}$$

The RRM based on two pulses as applied in Adamowski et al. (1998) uses the ratio of any detectable buffer reflection and its following reflection, e.g. A_T and A_{r1} or A_{r1} and A_{r1}:

$$\frac{A_{r1}(\text{sample})/A_{r2}(\text{sample})}{A_{r1}(\text{reference})/A_{r2}(\text{reference})} = \frac{r_{12}(\text{reference})}{r_{12}(\text{sample})}. \tag{25}$$

Since successive ratio buffer pulses are used, the excitation pulse A_T does not have to be assumed constant anymore. But

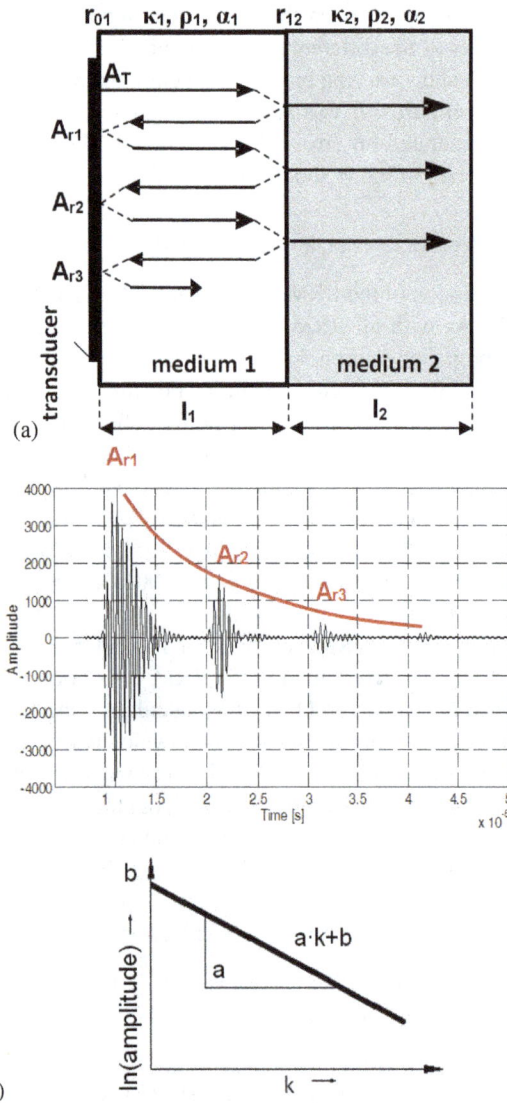

Figure 3. Basic principles and relevant pulses of the RRM: **(a)** schematic of multiple buffer reflections, **(b)** multiple buffer reflection pulses in the time domain and logarithmic decay of pulse amplitudes.

still a similar attenuation α_1 and a similar RC r_{10} have to be assumed if sample and reference measurement are compared:

$$r_{12}(\text{sample}) = r_{12}(\text{reference})\frac{A_{r1}(\text{reference})/A_{r2}(\text{reference})}{A_{r1}(\text{sample})/A_{r2}(\text{sample})}. \quad (26)$$

Since successive ratio buffer pulses are used, the excitation pulse A_T does not have to be assumed constant anymore. But still a similar attenuation α_1 and a similar RC r_{10} have to be assumed if sample and reference measurement are compared:

$$r_{12}(\text{sample}) = r_{12}(\text{reference})\frac{A_{r1}(\text{reference})/A_{r2}(\text{reference})}{A_{r1}(\text{sample})/A_{r2}(\text{sample})}. \quad (27)$$

And finally, as applied by Bamberger and Greenwood (2004a, b), the ratio of decays of multiple buffer reflections

can be used to obtain the RC via RRM. Describing the amplitude decay logarithmically:

$$\ln A_{rk} = [\ln(r_{10}) + \ln(r_{12}) \cdot 2l_1\alpha_1] \cdot k + [\ln A_T - \ln(r_{10})]$$
$$= a \cdot k + b, \quad (28)$$

and calculating the ratio $\exp[a(\text{sample})]/\exp[a(\text{reference})]$, one obtains the RC under the assumption of similar attenuation α_1 and a similar RC r_{10} for reference and sample signals:

$$r_{12}(\text{sample}) = r_{12}(\text{reference}) \cdot e^{[a_{\text{sample}} - a_{\text{reference}}]}. \quad (29)$$

McClements and Fairly (1991, 1992) applied the one-pulse RRM with air as the reference medium. They used a 2.1 MHz transducer of 10 mm diameter driven by a tone burst of 5–10 cycles. Distilled water, castor oil, olive oil, n-hexadecane and silicone fluid have been investigated at a constant temperature of 20.2 °C. For a vibrating U-tube as the reference measurement (DMA 40, Anton Paar) an error of 0.5 % is reported, which corresponds to a bias of $\pm 8\,\text{kg m}^{-3}$.

Kushibiki et al. (1995) applied a one-pulse RRM to investigate the acoustic properties of biological tissue and liquid specimen. Instead of air, water was used as the reference medium. Kushibiki et al. used a transmission line to measure velocity dispersion and attenuation. Basically the methodological assembly is comparable to Bjørndal's MRM approach. It was not mentioned why an RRM instead of an MRM was applied. Several broadband transducers (1.5 mm diameter) in combination with different gap distances have been used to cover the frequency range from 70 to 500 MHz. Different oils have been investigated and a maximum bias of 8 kg m^{-3} is reported. The temperature was reasonably constant around 23 °C, and the density validation values have been gathered via pycnometer. The investigations of Kushibiki et al. particularly show the feasibility of the method to investigate properties of very thin specimen.

Adamowski et al. (1998) applied the two-pulse RRM. Due to the special DET design it was possible to monitor the incident pulse. An unfocused 1.6 MHz broadband transducer was used, driven by a sinusoidal burst of one cycle. Distilled water, castor oil and ethanol have been investigated in a temperature range from 19 to 40 °C. The presented results have been calculated at a frequency of 1.4 MHz, and a bias of $\pm 10\,\text{kg m}^{-3}$ for reference values from the literature was reported. Furthermore, the apparatus was tested under varying flow conditions and a stable, negative bias of -3 to $-6\,\text{kg m}^{-3}$ compared to pycnometer reference measurement was reported. In Adamowski et al. (1995) similar equipment was used and results (average of 15 measurements) of RRM and MRM have been compared for constant temperatures (25 ± 0.5 °C). In the limited temperature range a bias of 1–2.5 kg m^{-3} could be reached.

Bamberger and Greenwood (2004a, b) and Greenwood and Bamberger (2004) applied the multiple-pulse RRM and used a 5 MHz transducer of 25 mm diameter. They investigated sodium compound solutions, kaolin slurries and sugar-water solutions. No information about the temperature is

Table 1. Expectable reflection coefficient difference for a defined density and sound velocity range, different buffer materials and different angles of incidence.

Material	Start value of sample medium		End value of sample medium		Longitudinal RC difference		
	ρ [kg m^{-3}]	c [m s^{-1}]	ρ [kg m^{-3}]	c [m s^{-1}]	angular incidence (45°)	angular incidence (25°)	normal incidence
PMMA					0.0095	0.0111	0.0120
quartz glass					0.0026	0.0037	0.0044
aluminium	1.055	1510	1.010	1535	0.0031	0.0038	0.0042
stainless steel					0.0013	0.0016	0.0018

given, and in terms of validation this does not matter since reference densities have been determined by weighting a known quantity. It would matter, however, if someone wants to consider applicational aspects, e.g. dynamic temperature changes. A bias of ± 10 kg m^{-3} is reported for the sodium compound solutions and ± 25 kg m^{-3} for the kaolin slurries. In Greenwood and Bamberger (2004) only the error for the acoustic impedance is given, which ranges from 1.8 % to -1.9 % for a 6.3 mm pipe wall and from -0.9 % to 8.7 % for a 3.8 mm pipe wall. The acoustic velocities have been measured by an independent system. Both the accuracy and the velocity values are not presented. In fact Bamberger and Greenwood presented a validation of the acoustic impedance and not the density. And since the velocity values are missing, an estimation of the density accuracy from the impedance validation data is not possible. There are two quite astonishing facts that are not cleared up in the publication. Table 1 in Greenwood and Bamberger (2004) indicates that only a few certain echo amplitudes are used to analyse the amplitude slope, but it is not stated why not all echoes or why exactly the presented echoes have been chosen. Furthermore, it is stated that the echo slope is a self-calibrating feature to overcome the influence of variations in the excitation voltages. But to prove the stability only the pulse width has been changed, although the published information indicates that the pulser voltage can be varied.

In summary, the following facts can be stated:

- Using the RRM to determine the RC, only buffer reflections are necessary. However, to calculate the density of the sample, the sound velocity in the medium is still necessary. Thus, aside from the angular approach (ARM), at least one echo from a reflector or some additional transmission measurements are required to determine the density.

- The RC of the used reference medium r_{12}(reference) either has to be known or, like in the case of air, can assumed to be equal to 1.

- The RRM is based on two separate measurements – of the sample and of the reference medium. The assumption of similar attenuation α_1 and RC r_{10} is only valid if a similar temperature distribution across the buffer can be guaranteed for reference and sample measurement.

- The one-pulse RRM is most susceptible to errors. The assumption of constant excitation pulses is not always valid, and has a great impact on the accuracy of the method. The excitation pulse is practically never exactly the same, and considering ageing of piezoelectric materials, the practical application would need periodic calibrations.

Besides the MRM, dual and multiple pulse RRM which are independent of the excitation amplitude, several alternative strategies have been developed to overcome the problem of varying excitation amplitudes. In Lynnworth and Pedersen (1972), Rychagov et al. (2002) and Jensen (1981) and Deventer (2004) a reference path approach is applied to monitor the excitation variations. The part of the signal that is reflected from a reference interface of constant properties can be used to standardize the received signal and negate excitation variations. Another option is the combination of reference and sample measurement as proposed by Greenwood et al. (1999, 2000) and Guilbert and Sanderson (1996). In this way the same pulse excitation can be sent to reference and sample measurement transducer. Comparable temperature distribution in both buffers can be assumed as well. But using two different transducers probably generates other systematic errors due to misalignment or differing transducer properties. A special case of this method is presented by Püttmer and Hauptmann (1998) and Püttmer et al. (1998, 2000), who used an additional delay line that is connected to the reverse side of the piezoceramic to determine signals from a reference interface. In this way a similar excitation pulse can be guaranteed for reference and sample measurement by using one transducer only. However, the advantage of similar temperature distributions is lost. A clear separation of each pulse is obtained by choosing a different length for the reference buffer and correcting the resulting difference by a calibration factor. In Fisher et al. (1995) a double buffer similar to Deventers MRM was used. However, instead of using the echo of the first buffer to calculate the RC directly, the additional reference echo was used to compensate effects such as ageing or depolarization of the piezoceramic.

2.3   Transmission methods (TM)

The TM contains all methods that use sender and receiver separately in a parallel assembly to determine the RC. Generally the TM can be classified into two approaches: the first approach is based on the work of Hale (1988), who uses only receiver signals (TMOR); the second approach as presented by Bjørndal et al. (2008) uses the signals of both transducers (TMSR).

Even though Hale's approach is not a true buffer rod technique, it is worth mentioning since it is the basis for further developments. Hale used a transmitter–receiver configuration without any additional delay line. The used configuration and terminology is given in Fig. 4, for which in Hale's approach medium 1 is the sender and medium 3 is the receiver.

Hale assumed that the attenuation does not change significantly for fluids of quite similar composition (like tap water and salty water) and that the sender impedance equals the receiver impedance ($Z_1 = Z_3$). Therefore, it was possible to state that any change in acoustic impedance of the sample liquid Z_2 is directly proportional to the measured change of amplitude A_4:

$$A_1 = \frac{(Z_1 + Z_2)^2}{4e^{-\alpha_2 l_2} Z_1 Z_2} A_4. \tag{30}$$

Considering calibration measurement for two liquids (indices c1 and c2) of known acoustic impedances Z_{c1} and Z_{c2} and constant excitation amplitude A_1, one reaches

$$\frac{(Z_1 + Z_{c1})^2}{4\exp(-\alpha_{c1} l_2) Z_1 Z_{c1}} A_{4c1} = \frac{(Z_1 + Z_{c2})^2}{4\exp(-\alpha_{c2} l_2) Z_1 Z_{c2}} A_{4c2}. \tag{31}$$

Under the assumption of similar internal losses ($\alpha_{c1} = \alpha_{c2}$) the attenuation term can be neglected, and the impedance Z_1 can be calculated:

$$Z_1 = \frac{Z_{c1} - k Z_{c2}}{1 - k} + \sqrt{\left(-\frac{Z_{c1} - k Z_{c2}}{1 - k}\right)^2 - \frac{Z_{c1}^2 - k Z_{c2}^2}{1 - k}}, \tag{32}$$

where

$$k = \frac{\exp(-\alpha_{c1} l_2) Z_{c1} A_{4c2}}{\exp(-\alpha_{c2} l_2) Z_{c2} A_{4c1}}.$$

The density results showed less than 2 % variation from the true values which have been determined via weight measurements of known volumes. McGregor (1989) discussed several possible methods to measure the density by using the same probe arrangement like Hale. He stated that a continuous wave system, with and without interference, would provide the most accurate means of determining the velocity and the characteristic impedance of the fluid under test.

Henning et al. (2000) mounted the transducers on a glass tube wall of half-wave thickness. Furthermore, the setup was

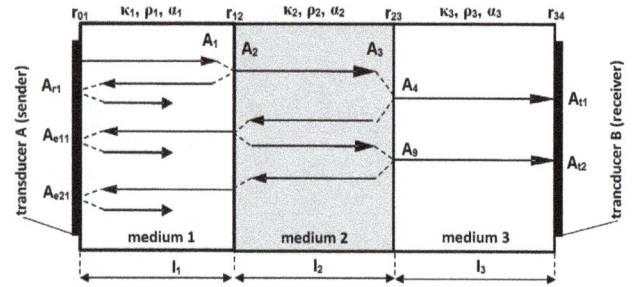

Figure 4. Schema showing the basic principles and relevant pulses for the TM and giving the terminology for Hale's, Henning's and Bjørndal's approach.

calibrated for two liquids of known acoustic impedance to determine Z_1. But in the case of Henning's setup, Z_1 is only the apparent transducer impedance. Indeed, this fictive impedance describes the combined impedance of glass wall and transducer as a result of the sound propagation through the glass wall of half-wave thickness. Furthermore the basic TMOR approach was expanded for the amplitude A_9:

$$\frac{A_9}{A_4} = \left(\frac{Z_1 - Z_2}{Z_1 + Z_2}\right)^2 \exp(-2\alpha_2 l_2). \tag{33}$$

Still the attenuation is neglected in order to calculate the transducer impedance. But now two equations can be used to calibrate the transducer impedance. Using both Eqs. (32) and (33) a mismatch between the transducer impedances was reported. In the end both impedances have been used to determine the acoustic impedance of the sample liquid. Even though the glass tube wall is of half-wave thickness, it is quite clear from theory that the amplitudes A_4 and A_9 as described by the equations are not equal to the amplitudes received by the transducer. From the physical point of view the received pulses are also influenced by the wall material and contain also information from superpositioned reflections inside the tube wall. Nevertheless, in Henning et al. (2000) both the basic and the expanded TMOR have been compared for several liquids using an aerometer measurement as reference. While the basic TMOR showed a bias of 3 to $-40\,\mathrm{kg\,m^{-3}}$, the expanded TMOR resulted in a bias of -16 to $10\,\mathrm{kg\,m^{-3}}$. Furthermore, it was reported that the absolute error increases to a few percent in the case of increasing sound absorption corresponding to the liquid properties or diffuse scattering at particles.

Additionally to the signals of the receiver (transducer B), Bjørndal et al. (2008) employs pulses received by transducer A. Comparable with the MRM, one achieves an equation that cancels the influence of the attenuation, the transducer and the electronics sensitivity. Bjørndal employs two pulses of transducer A and two pulses of transducer B (R_echo12_12 method, terminology given in Fig. 4):

$$r_{12} = \pm \left(1 - \frac{A_{e11} A_{t1}}{A_{r1} A_{t2}}\right)^{-0.5}. \tag{34}$$

It is reported that the systematic deviation from reference values was slightly higher for the TMSR compared with MRM, and it is stated that using information of both transducers, non-identical sound fields and a misalignment in the transducer configuration might be the reason for the higher deviation. In Bjørndal and Frøysa (2008) all possible pulse combinations besides Eq. (34) are discussed, even some further methods that employ transmitted pulses from both sides in which transducer A and B are used alternately as senders. After a detailed uncertainty analysis with respect to bit resolution and noise, it was outlined that the R_echo12_12 method (Eq. 34) possesses a relative uncertainty close to the optimal and case-dependent R_echo123_123 (which uses 3 pulses of receiver and transducer; details in Bjørndal and Frøysa (2008) and may be the best choice of all TMSR to be compared with the MRM).

2.4 Angular reflection method (ARM)

The ARM was presented first by Greenwood and Bamberger (2002) and Greenwood et al. (1999). Concerning the determination of the RC, the ARM is a simple one-pulse RRM (Eq. 24). But to determine the sound velocity and the density of the medium (see Eq. 5) the ARM uses measurements at two different angles.

The RC of the longitudinal wave, r_{LL} at a given angle of incidence (see Fig. 5) depends on the angle β_L, the density ρ, the longitudinal velocity c of the sample liquid and the longitudinal velocity c_L, the shear velocity c_T and the density ρ_S of the buffer material (Greenwood et al., 1999; Krautkramer and Krautkramer, 1983). The equations are generally given as

$$r_{LL} = \frac{G - H + J}{G + H + J},$$ (35)

where

$$G = \left(\frac{c_T}{c_L}\right)^2 \sin 2\beta_L \sin 2\beta_T,$$ (36)

$$H = \cos^2 2\beta_T,$$ (37)

$$J = \frac{\rho c \cos\beta_L}{\rho_S c_L \cos\beta} = \frac{Z_2 \cos\beta_L}{Z_1 \cos\beta},$$ (38)

and from Snell's law,

$$\sin\beta = \frac{c \sin\beta_L}{c_L}, \quad \sin\beta_T = \frac{c_T \sin\beta_L}{c_L}.$$ (39)

Instead of measuring the sound velocity c, the RC is determined using an RRM approach (Eq. 24) to calculate the parameter J via Eq. 35). Now Eqs. (38) and (39) can substitute the unknown angle β in

$$\sin^2\beta + \cos^2\beta = 1.$$ (40)

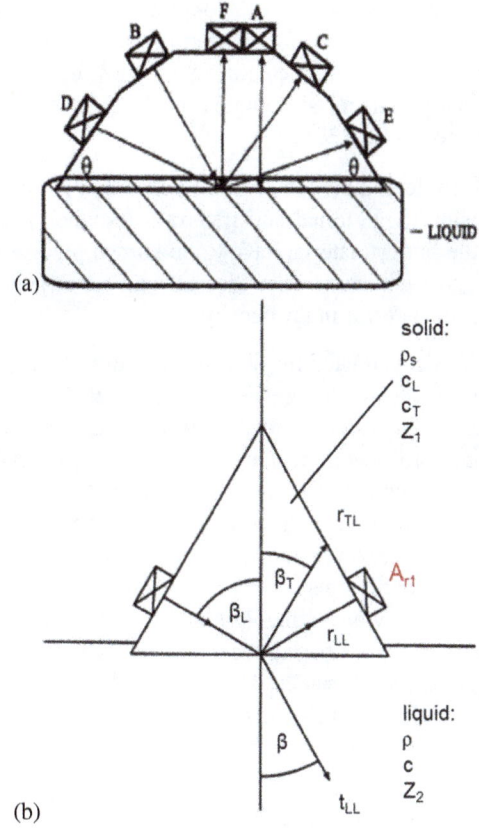

Figure 5. Schematic showing (a) the wedge design of Greenwood and Bamberger, (b) the design given by Krautkramer and the definitions of terminology.

Doing so for two different angles, equalizing both and writing the resulting equation in terms of ρ gives a term which is independent from the sound velocity in the liquid:

$$\rho = \rho_S \left(\frac{\sin^2\beta_{L1} - \sin^2\beta_{L2}}{\cos^2\beta_{L1}/J_1^2 - \cos^2\beta_{L2}/J_2^2}\right)^{0.5}.$$ (41)

Finally, the sound velocity in the liquid can be calculated with

$$c = \left(\frac{\sin^2\beta_L}{c_L^2} + \frac{\rho^2 \cos^2\beta_L}{J^2 Z_1^2}\right)^{-0.5}$$ (42)

In summary the following facts can be stated:

- The great advantage of the ARM is the determination of the sound velocity on the basis of reflection coefficient measurements at two angles. Only signal information from the interface is required, and therefore no sound propagates through the sample medium.

- The basics of the ARM reflection coefficient determination are comparable to the RRM. Consequently, all facts stated for the RRM also count for the ARM. Only the sound velocity determination is different.

– The ARM also provides the opportunity to measure the sound velocity via pulse-echo or transmission approach. Instead of measurements at two angles, one would be sufficient. The missing angle β in Eq. (38) could be calculated via Eq. (39).

– The angle and the temperature-dependent parameters – density, longitudinal and transversal sound velocity – of the buffer material have to be known precisely. The slightest deviation from the real value can generate a significant error in the density.

The ARM was validated for sugar-water solutions and surrogate slurries via weighting of known volumes. For the analysis of the sugar-water samples the wedge was submerged to reach a uniform wedge temperature. An error of 0.1–1.3 % was reported, which is a bias of 1–14 kg m^{-3}. The experiments for the surrogate slurries have been accomplished at a test loop for varying slurry flow rates, aeration flow rates and two constant temperatures (25 and 50 °C). Each density was calculated by averaging 45 signals. The validation was accomplished by comparing the average of 40 sensor densities with reference densities. The bias varied between 13 and 260 kg m^{-3}. Neglecting some extreme deviations, an overall bias of 20 kg m^{-3} could be accomplished.

3   Probe design considerations

The design of ultrasonic density probes as presented by the aforementioned authors is a complex process. In most publications, the probe's dimensions and material are simply mentioned as a given fact, not as a required necessity. In fact, an unequivocal identification of clearly unaffected pulses is one of the basic requirements for all presented methods. As soon as one of the required pulses is superpositioned by any other pulse or effect, which is not considered by the plane wave propagation theory, the resulting values will be affected by a systematic error.

3.1   Pulse excitation and separation

The best way to exemplify all interrelations clearly is to follow the design process of a buffer which might be used for an RRM approach. In its simplest version, we want to see the first reflected pulse, only affected by the reflection at the interface and the buffer material's attenuation. Neglecting all application-based boundary conditions, the only real limiting conditions are the choice of the ultrasound source and the frequency of and the type of excitation pulse. By making the right choice one can affect the pulse duration. Choosing a transducer which generates a low-damped narrowband pulse of low frequency, one achieves a relatively long pulse. Choosing a high frequency, highly damped broadband pulse, one achieves a short pulse. If a burst excitation of several cycles is used, one can specify the frequency quite accurately, but this generates a long-lasting sound pulse. Using a pulse

excitation, one can generate a shorter sound pulse, but the pulse frequency generally relies on the system's resonance frequency. In any case, often the most convenient way to investigate the resulting sound pulse duration is to test and measure the pulse length t_p of a chosen ultrasound source for varying excitation pulse amplitudes, cycles and frequencies. Knowing t_p and the temperature-dependent sound velocity c_1 of the buffer material, it is possible to calculate the minimum buffer thickness for a given temperature range to prevent superposition phenomena for the multiple buffer reflections A_{rn}.

When a reflector is used to determine the sound velocity or to adopt the MRM, further parameters besides the temporal determination of the pulse position are relevant to prevent superposition of buffer reflections and echoes. If so, the pulse amplitude and the amount of buffer reflections also have to be considered. For constant excitation amplitude those parameters only depend on the buffer materials absorption and the RC at the interphase. Combined with the pulse length t_p those parameters define the buffer reflections duration t_{br}. In order to prevent superposition between the buffer reflections A_{rn} and the echo pulses A_{ejk}, the following condition has to be fulfilled:

$$\frac{l_2}{c_2} = \text{TOF}_2 > t_{br}, \tag{43}$$

where TOF_2 is the signal's time of flight in the sample medium. Alternatively, dimensions and materials can be designed in a way that the echo pulses arrive in a time gap between two buffer reflections. This target is hard to achieve since the echo position depends on the sample mediums sound velocity, and thus such special designs are often usable only for a defined sample medium and temperature range (Bjørndal et al., 2008; Bjørndal and Frøysa, 2008). In the case of the MRM as introduced by Papadakis the superposition between the 1st pulses of the 1st and 2nd echo (A_{e11} and A_{e21}) and the reflections of those pulses inside the reflector have to be eliminated, and then the condition $l_3/c_3 = \text{TOF}_3 > t_p$ is satisfied. Bjørndal et al. (2008) presents most of those dimensional considerations. Additionally, Bjørndal and Püttmer (1998) introduce conditions for edge wave contributions with and without mode conversion. The edge wave distributions mainly depend on the buffer diameter and the ratio of transducer radius to buffer thickness and therefore also represent the near-field phenomena. The mode conversion depends on the shear wave velocity and therefore on the elastic properties of the buffer material.

3.2   The choice of material

As indicated in the previous section, most design considerations depend on the material's properties. Thus, besides the option to change the dimension of buffer or reflector, one can simply change the material to achieve a desired signal pattern. The choice of material also defines the resolution that has to be reached for a given process of defined density range. The following table shows start and end values (density and sound velocity) of a typical yeast fermentation and the resulting RC difference that can be expected for different buffer materials.

Indeed, it becomes apparent that according to Eq. (9) any buffer material can be used to determine the density using the reflection coefficient. But, as shown in Table 1, only materials of acoustic impedance comparable to the impedance of the sample medium possess an acceptable sensitivity for small density variations (Püttmer and Hauptmann, 1998; Püttmer et al., 2000; Bjørndal et al., 2008; Greenwood et al., 1999). The same holds true for the ARM; increasing angular difference to the normal incidence even decreases the RC difference.

Additional requirements for the buffer materials are good chemical resistance, reasonable temperature stability and a low sound attenuation (Püttmer and Hauptmann, 1998; Püttmer et al., 2000). If special liquids are analysed, e.g. suspensions containing abrasive materials, further criteria such as mechanical resistivity may be of importance. Concerning the mode conversion in the case of angular incidence – for example, if the ARM is applied or in the case of edge waves – the elastic properties of the buffer material may also be of interest. Materials of a high Poisson's ratio generally possess a higher conversion to shear waves.

Besides deploying the choice of material to guarantee a clear pulse separation, the pulse amplitude can be affected. Choosing a buffer material of acoustic impedance, comparable with the sample mediums impedance, results in a low reflection coefficient. The buffer reflections A_{rn} are less in quantity and lower in amplitude. Most of the energy is transferred into the sample medium. However, if an echo comes back (A_{e11}), most of the energy is transferred back into the buffer. Thus probably too little energy remains for a second detectable echo (A_{e21}). The same holds true for the reflector. Choosing a reflector material of high acoustic impedance results in high echo amplitudes. However, materials of high acoustic impedance generally possess high sound velocity, low sound attenuation and a high reflection coefficient. Therefore, resulting from extensive reflector dimensions and a considerable amount of reflections inside the reflector, this may interfere with the second echo (A_{e21}). In such cases a special reflector shape often is the most feasible alternative (Carlson et al., 2003a; Deventer and Delsing, 2001b). A reflector of low acoustic impedance may simplify the task to achieve the maximum signal purity, but also results in lower echo amplitudes.

3.3 Temperature variation, sound field and signal-to-noise ratio considerations

Regardless of the method applied or material chosen, if the temperature changes, everything changes concerning sound propagation. This fact also counts for design considerations. Every single boundary condition mentioned above has to be valid for the entire temperature range. If the temperature changes, so does the speed of sound, density, sound absorption and dimensions of all materials involved. Therefore, not only does the pulse's position change but also the pulse amplitudes. In the best-case scenario, the amplitude slightly decreases; in the worst case, whole pulses are no longer detectable, which might hamper the analysis of RC or ultrasound velocity (USV). Mak (1991) compared several MRMs concerning the influence of systematic (beam diffraction) and random errors (noise). He showed that varying attenuation and signal-to-noise ratio (SNR) affect the method's error. The higher the SNR and the less influence of diffraction, the smaller the errors in the RC. Therefore, the reference methods (ARM, RRM) might show better results, since they are independent from beam diffraction, while the accuracy of the MRM depends on the accuracy of the diffraction correction. Mak used a 50 MHz broadband transducer. Both the reference methods and the MRM showed quite low RCs at low frequencies, and both methods converged for higher frequencies near the transducer's centre frequency and showed comparable results. Adamowski et al. (1995, 1998) used a constructive solution to eliminate diffraction issues. The so-called DET technique employs a receiver of an aperture larger than the emitter that generates the sound field. As long as the beam spreading does not reach the dimensions of the receiver diameter, the principles of MRM for plane wave propagation are valid without correction.

While the correction of diffraction in the far field is discussed by several authors (Papadakis, 1959; Papadakis et al., 1973; Bjørndal et al., 2008; Kushibiki et al., 2003), the near-field problem is often not mentioned at all. Although the beam is assumed to be parallel in the near field (Povey and McClements), it is recommended to avoid it totally. The intensity varies greatly with distance, the surface's amplitudes are not constant and the whole wave front cannot be expected to be normal to the phase velocity vector. Essentially the plane wave propagation is not valid within the near field. Consequently, besides all dimensional considerations mentioned in Sects. 3.1 and 3.2, the first condition that has to be kept is the near-field distance N between the sender and first interface:

$$N = \frac{a^2}{\lambda}, \tag{44}$$

with a being the transducer radius. Table 2 shows methodic details as applied by different authors and the resulting near-field length in comparison to the chosen buffer length. Besides Greenwood, who applied the ARM, and Papadakis, who applied the MRM for attenuation measurements, the researchers used the path length of dimensions (double buffer rod length) greater than or at least in the range of the near-field distance.

Diffraction effects are generally corrected via Williams' expression (Williams, 1951; Williams and Labaw, 1945). Although Williams stated that his expression is only accurate

Table 2. Near-field relevant, methodic details of relevant publications.

Source	Transducer diameter d [mm]	Centre frequency f [MHz]	Material	Buffer rod length [mm]	Near-field distance N [mm]
Adamowski et al. (1995, 1998)	19.0	1.6	PMMA	30.0 42.0	53.48
Bjørndal et al. (2008)	12.5	5.0	aluminium	80.0	30.90
Deventer and Delsing (1997)	10.0	3.7	PEEK/PMMA	26.0 20.0	34.26
Greenwood et al. (1999)	12.5	2.25	Rexolite	6.3	37.56
McClements and Fairly (1991, 1992)	10.0	2.1	PMMA	40.0	19.44
Papadakis (1968)	12.7	10.0	fused quartz	25.4/62.2	67.66
			aluminium	25.4	63.20
			steel	18.9	68.34
Püttmer and Hauptmann (1998)	20.0	2.0	quartz glass	31.0	33.67

for $k \cdot a > 100$ and distances $z_W \geq (k \cdot a^4)^{1/3}$, the exact expression without approximations (see Williams, 1951, Eq. 17) might be usable in an extended domain. Nevertheless, so far it has not been reported whether corrections in the near field or for sound fields across an interphase within the near field can be applied successfully to reach a reflection coefficient accuracy of 1E-4 or less (see Table 5).

Knowing all these facts it becomes clear that if spatial limitations for the sensor application exist and a buffer miniaturization becomes necessary, only increasing the pulse frequency to achieve pure signals is not enough. Often the dimensions of the transducer with respect to the buffer mediums sound velocity have to be adapted.

3.4 Constructional uncertainties

The main constructional uncertainty which is occasionally discussed is the parallelism of surfaces. In ARMs, of course, the accuracy of the angles will be of similar importance. In Carlson et al. (2003b) it is reported that the misalignment of the transducer to buffer material is the main source of error causing an overestimation of attenuation and acoustic impedance. In Bjørndal et al. (2008) it is stated that effects of nonparallelism can be neglected for surfaces that are parallel within 0.01 mm. In Adamowski et al. (1995) a maximum parallelism of $0.0004\,\mathrm{mm\,mm^{-1}}$ and a change of 0.7 % in the reflection coefficient for an intentionally caused misalignment of $0.0024\,\mathrm{mm\,m^{-1}}$ was reported.

4 Discussion

While reviewing critically all published methods and validation results with regard to validation complexity, error analysis and real process relevance, several gaps and questions appeared which will be discussed in the following sections. The first point will be the analysis of relevant pulses. Further points will include the equipment used for ultrasound generation and detection, reference density and temperature

measurement, the sound velocity determination and extended uncertainty considerations.

4.1 Signal processing

Signal processing is a wide field with many fundamental details. The applied methods range from simple time domain (Greenwood and Bamberger, 2002; Greenwood et al., 1999) to extensive frequency domain methods (Bjørndal et al., 2008). The equations presented so far represent the time domain approach and refer to the signal amplitude, but do not state which pulse amplitude is used in the end. In Greenwood and Bamberger (2002), Greenwood et al. (1999), Püttmer and Hauptmann, (1998) and Püttmer et al. (1998, 2000), the maximum peak-to-peak amplitude within a certain time window has been examined:

$$A_{\mathrm{pulse}} = \mathrm{maximum}\,[A\,(t_{w1} : t_{w2})] - \mathrm{minimum}\,[A\,(t_{w1} : t_{w2})], \quad (45)$$

where A_{pulse} represents the value that is inserted in the respective equation of reflection coefficient calculation and t_{w1} and t_{w2} the time boundaries of the analysed pulse. In the following sections, $A(t)$ will represent the pulse in the time domain and $a(f)$ in the frequency domain.

Papadakis (1968) had started analysing amplitudes in the time domain for attenuation analysis, but later he changed to spectrum analysis (Papadakis et al., 1973). After correcting the frequency dependent diffraction, Papadakis et al. analysed the frequency-dependent reflection coefficient and attenuation (Papadakis et al., 1973; Sachse, 1974):

$$A_{\mathrm{pulse}}(f) = a(f). \quad (46)$$

It was found (Sachse, 1974) that the reflection coefficient and density are nearly constant over a frequency range around the centre frequency of the transducer's maximum response. That might be the reason for obtaining the amplitudes from the spectra at a particular frequency (f_1) (Adamowski et al., 1995). Higuti (Higuti and Adamowski, 2002b; Higuti et al., 2001), who followed the DET approach of Adamowski, introduced the energy method, in which the energy spectral density of each pulse is used for the reflection coefficient analysis:

$$A_{\text{pulse}} = \int_{-\infty}^{+\infty} |a(f)| \, \mathrm{d}f. \tag{47}$$

It is stated that the deployment of the energy method results in smaller variations when compared to the single-frequency method, because it averages the noise over frequency. For added Gaussian white noise of varying amplitude to simulation results, Higuti found that the energy method improves the results with smaller SNRs. By calculating the spectral density only for a small frequency band, the performance could be enhanced due to the rejection of frequencies outside the band of the transducer. Experimental results showed an error of less than 0.2 % and proved the enhanced performance of the presented new signal processing method.

In Bjørndal et al. (2008) a more detailed analysis of signal-processing methods in the time and frequency domain is presented. In the time domain the amplitude value was not determined simply as the main peak-to-peak difference per pulse; instead the peak-to-peak value was determined per period:

$$A_{\text{pulse}} = \text{maximum}\left[A(t)_{\text{pn}}\right] - \text{minimum}\left[A(t)_{\text{pn}}\right], \tag{48}$$

where $A(t)_{\text{pn}}$ represents the n-th period of the analysed time domain pulse. Depending on the amount of analysed periods (e.g. from P_1 to P_2) one can calculate a mean reflection coefficient R_{m} for each signal (Bjørndal et al., 2008):

$$R_{\text{m}} = \frac{1}{P_2 - P_1 + 1} \sum_{n=P_1}^{P_2} R_n. \tag{49}$$

It is reported (Bjørndal et al., 2008) that if the first period of the waveforms is included, there may be large errors, particularly when the amplitudes are analysed in the time domain, but also in the case of the frequency domain analysis. In the frequency domain the analysis followed the spectral density approach (Eq. 47), but the so-called l2 norm was introduced based on the mathematic basics of L^{p} spaces:

$$A_{\text{pulse}} = \sqrt{\int_{f_1}^{f_2} |a(f)|^2 \, \mathrm{d}f}. \tag{50}$$

It is stated (Bjørndal et al., 2008) that the frequency domain integration introduces a spectral-averaging approach, reducing the effect of single-frequency interference in the echo signals. The l2 norm accentuates the dominant part of the frequency spectrum, making it easier to evaluate the effect of the upper frequency limit. Equally to the periodic peak-to-peak analysis in the time domain, the frequency spectrum was analysed on a half-periodical basis. Additionally, a Hanning window function was applied to reduce the spectral leakage. The windows have been centred at the local

extreme values of each analysed peak (Bjørndal et al., 2008). The accuracy improvement compared to a frequency domain approach without window function was not reported.

Applying the different signal-processing methods to PSPICE simulation results, it was found (Bjørndal et al., 2008) that the frequency domain approach gives significantly less density deviation than the time domain analysis. The experimental results could not confirm the theoretical evaluation; in some cases the time domain analysis indicates more accurate results and less deviation. Furthermore, Bjørndal suggested a time domain integration method following Raum et al. (1998), but it was also adverted to the high sensitivity of the time integration approach to DC offsets and waveform disturbance effects:

$$A_{\text{pulse}} = \int_{t_1}^{t_2} |\text{env}(A(t))| \, \mathrm{d}t. \tag{51}$$

Besides the different signal analysis methods, the signal-processing parameters and the applied preprocessing steps are of high relevance to reach the reported accuracies. Concerning the preprocessing, most authors mentioned that a certain amount of signals have been averaged before applying the different signal analysis methods. Through signal averaging the SNR can be enhanced and the amplitude resolution can be increased beyond the AD-converter limitations (Bjørndal et al., 2008). The use of a 25 MHz low-pass filter is mentioned in Bjørndal et al. (2008); further references for filter usage have not been found. Furthermore, in Bjørndal et al. (2008) the use of least-squares-sense cubic spline approximation was reported to increase the vertical and temporal resolution.

Relevant signal-processing parameters are the pulse length in time, the amount of data points with respect to the sampling rate, the amplitude resolution and the usage of any additional processing steps to improve the frequency or magnitude accuracy, such as filtering, signal averaging, zero padding or application of window functions. Table 3 overviews the signal-processing details of several relevant authors with regard to the reached accuracies.

4.2 Signal generation and detection

Most authors used highly advanced equipment for their investigations. Generally pulse or function generators provide the electrical pulse which is converted to sound pulses by commercially available transducers. After amplification, the signal is recorded by an oscilloscope and conveyed to a personal computer for further signal analysis. Standard signal generators are generally limited to 20 V peak excitation, which is sufficient for most of the investigations. Custom signal generators for higher excitation voltages and amplifiers are available but require special circuits since the input voltage of commercial oscilloscope is often limited. To avoid noisy interferences and overloading of the oscilloscope, the

Table 3. Processing details from different literature sources with regard to density accuracies.

Source	Window size	Sampling rate (MHz)	Averaged signals	Applied method	Used domain	Density accuracy
Adamowski et al. (1995, 1998)	500 (1024, zero padding)	100	64	MRM	time/frequency	1.50 %
Bjørndal et al. (2008)	1000 (32 768, zero padding)	59	256	MRM, TRM	time/frequency	0.15 %
Deventer and Delsing (1997)	512	200	32	MRM	frequency	< 1 %
Greenwood et al. (1999)	4096	40	45	ARM	time	< 1 %
Bamberger and Greenwood (2004a, b)	–	–	–	RRM	frequency	< 1 %
McClements and Fairly (1991, 1992)	–	100	≈ 2000	RRM	time/frequency	0.50 %
Papadakis et al. (1973), Papadakis (1968)	–	–	–	MRM	time/frequency	–
Püttmer and Hauptmann (1998), Püttmer et al. (2000)	–			RRM	time	0.20 %

excitation and receiving circuit should be decoupled. Results concerning the influence of excitation voltage and voltage variations on the methods accuracy are not reported. While in Greenwood and Bamberger (2004) it is stated that the decay RRM approach is independent of changes in the pulser voltage, and although it can be assumed that the MRM is independent from the excitation voltage, it is quite doubtful that the density error is totally independent. A change of the excitation voltage or signal amplification might change the degree of interference between subsequent pulses, the SNR and the pulse appearance. The independency has definitely not been proven experimentally so far. The same counts for the excitation and transducer type. Results are reported for different excitation types (Table 4 shows an overview) ranging from peak, rectangular and sinusoidal pulses to bursts of several cycles, but a decent comparative evaluation is missing so far. Indeed, in Bjørndal et al. (2008) simulation results are reported for varying cycles, but a comparison to peak excitation and an experimental evaluation were not shown. Moreover, investigations regarding the transducers type or piezoelectric materials (PEM) have not been found so far. It is known that the very different properties of the PEM result in completely different probe types (Lach et al., 1996). Concerning the determination of the reflection coefficient, different transducers constructed with different PEM might show different sensitivities and variance.

Concerning measurements in real process environments, the use of general purpose equipment, such as oscilloscopes or function generators, is a double-edged sword. Indeed it is commercially available technology of proven accuracy, but it is often both immoderate and unfeasible for specific tasks such as reflection coefficient determination. Using the typical sampling frequency of 250 MHz to characterize a 2 MHz signal in the frequency domain is clearly oversampling – no additional information is extracted, but it might be necessary to reach high time of flight or amplitude accuracy in the time domain. In the end, the effort for signal-processing increases dramatically with increasing sampling frequency. Indeed, standard oscilloscopes can monitor the voltage-time course with a high sampling frequency but provide only a moderate vertical resolution of 8 bit. Based on simulation results it was shown (Püttmer et al., 2000; Bjørndal and Frøysa, 2008; Bjørndal et al., 2008) that a 12-bit resolution is the best choice to reach reasonable errors. Since the price of an oscilloscope is not negligible, the vertical resolution is quite low and no further usable features like amplification or variable programmable signal processing steps are provided, an oscilloscope often is replaceable. As shown in Greenwood et al. (1999, 2006), a time-to-digital converter with reasonable sampling frequency and an analogue-to-digital converter with reasonable vertical resolution also serve the purpose. Similar considerations apply to signal generation and processing. An arbitrary function generator and a personal computer might not be the best choice for measurements in real process environments, but as long as it is not clear which excitation function is the best choice for a certain method, reports about compact units that incorporate all main tasks, signal generation, signal detection and signal processing will take a while in coming.

4.3 Reference analytics, validation and uncertainty considerations

The following section reviews and discusses the measurement uncertainties in terms of density determination via BRT of all significantly involved variables: density, reflection coefficient, ultrasound velocity and temperature.

Besides the uncertainties of the simplification in Eq. (14) the reflection coefficient mainly depends on the amplitude error. According to the propagation of uncertainty the degree of dependency is defined by the equation of each method (Eqs. 19, 24, 27, 29 and 34). The amplitude error basically depends upon three main factors: the amplitude resolution, the time resolution and the SNR. The amplitude resolution dependency was discussed in Bjørndal and Frøysa (2008), Bjørndal et al. (2008) and Püttmer et al. (2000); both research groups arrived at the conclusion that a resolution of 12 bit or better is required to reach accuracies below 0.5 % error.

Table 4. Details of sound generation equipment as published by different authors.

Author/Source	Equipment		Excitation		Transducer
Adamowski et al. (1995, 1998)	function generator	oscilloscope (8 bit)	pulse/ burst	2–3 cycles	KB-Aerotech (1.6 MHz)
Bjørndal et al. (2008)	function generator	oscilloscope (8 bit)	sinusoidal burst		Panametrics (5 MHz)
Deventer and Delsing (1997)	pulse generator	oscilloscope (8 bit)	pulse	–	Panametrics (5 & 10 MHz)
Greenwood et al. (1999)	function generator	data acquisition card (PC)/digitizer (12 bit)	burst	10 cycles	–
Bamberger and Greenwood (2004a, b)	ultrasonic pulser	oscilloscope (–)	–		–
McClements and Fairly (1991, 1992)	function generator	oscilloscope (–)	burst	5–10 cycles	Karl Deutsch (0.3–1 MHz), Sonatest (1–6 MHz)
Papadakis et al. (1973); Papadakis (1968)	pulse generator	oscilloscope (–)	pulse		Y-cut quartz (30 MHz)
Püttmer and Hauptmann (1998), Püttmer et al. (2000)	analogue signal generator	time-to-digital converter (12 bit)	burst	1 cycle	lead metaniobate disk (2 MHz)

The SNR dependency was discussed in Mak (1991), Higuti et al. (2001), Bjørndal and Frøysa (2008) and Bjørndal et al. (2008). Based on theoretical uncertainty considerations it was shown that the MRM is highly sensitive to noise. The more pulses included in the reflection coefficient calculation and the lower the SNR for each included pulse, the higher the uncertainty. Particularly in the case of the MRM, the SNR of A_{e11} and A_{e21} decreases dramatically when attenuation increases. Also, the SNR of A_{e21} becomes quite low in the case of a low r_{23}. Additionally, in Mak (1991) the influence of diffraction correction uncertainties is discussed as a systematic error. Based on the fact that the RRM is independent of diffraction it was stated that the MRM is the least accurate method for calculating the reflection coefficient. Experimentally this general statement could not be proved so far; results of both MRM and RRM converged for the centre frequency of the transducer. Also the experimental results of Adamowski et al. (1995) showed similar errors for both methods. The comparison of MRM and TMSR in Bjørndal et al. (2008) showed a smaller systematic deviation from reference values for the MRM method. In Higuti et al. (2001) the statements are rested upon simulated signals with artificially added Gaussian white noise. In contrast to Bjørndal et al. (2008), who reported for a SNR of 50 an uncertainty of 25 kg m^{-3}, in Higuti et al. (2001) for a similar SNR an error of only 1–5 kg m^{-3} was presented.

So far, Bjørndal (Bjørndal et al., 2008) is one of the few to have limited the sampling frequency and investigated the time resolution uncertainty by applying cubic spline approx-

imation to synthetic 6 MHz signals. Hence, the time resolution was increased from approximately 17 ns to 1 ns via mathematical approximation. In particular, the time domain results could be improved, and it can be assumed that the effect for signals of lower time resolution is even higher.

Unfortunately, none of the authors discussed the effect of systematic errors due to interference of subsequent pulses. Indeed, most authors state that clearly unaffected pulses are required for an accurate analysis, and cite several probe design considerations based upon a defined pulse length, but the truth is that the pulses are never diminished totally (see Püttmer et al., 1998, Figs. 7 and 8). As a basic rule, a pulse is regarded as terminated when the amplitude is below the noise level. But the subsequent signal is nothing more than a systematic oscillation hidden behind noise. Analysing those effects could help in separating such systematic errors from the signal.

The USV as a source of uncertainty often seems to be ignored. Most authors do not state how the speed of sound is determined and which accuracies could be reached (see Table 6). Generally the time of flight in the sample medium is determined and related to the propagation path. But often, particularly for small distances, the propagation path cannot be determined with adequate precision. The most chosen solutions to reach a higher precision are calibration measurements with standards (Marczak, 1997; Bjørndal et al., 2008; Higuti et al., 2001; Higuti and Adamowski, 2002b; Adamowski et al., 1998), which might become quite laborious if thermal expansion of the propagation path is considered. Alternatively a material of low thermal expansion such as ZERODUR® (Bjørndal et al., 2008; Hoppe et al., 2003) could be used. In a range of ±25 K the thermal expansion can be neglected within an USV error of 0.2 m s^{-1}. Standard for the time-of-flight determination is the cross correlation which can be applied in the time domain (Adamowski et al., 1995, 1998) or frequency domain (Deventer and Delsing, 1997). The great advantage of BRTs is the provision of a stable reference pulse that can be compared to echo pulses. Therefore the time-of-flight determination in pulse echo mode is independent of electronics time jitter. The only problematic parameter is the time resolution. When a simple cross correlation is applied, the time-of-flight resolution is still dependent on the sampling rate. For example, providing sampled data of 100 MHz sampling rate leads to a 1 m s^{-1} velocity resolution for a 23 mm propagation path (Adamowski et al., 1995). That might be the reason why most researchers oversample the data. In fact, mathematical approximation is a feasible solution to achieve higher accuracies with less time resolution (Hoche et al., 2011; Hoppe et al., 2001). Apart from that, when echo detection in pulse echo mode becomes problematic (e.g. highly absorptive liquids, superposition of buffer reflections and echo pulses) often transmission measurements are necessary, which increases the uncertainties and the effort in technical equipment and analysis.

In fact, an accuracy of 0.1 m s^{-1} is reachable applying state-of-the-art technologies and methods, and the sound velocity is not actually the most critical source of uncertainty. Analysing the partial derivatives of Eq. (8) according to the propagation of uncertainties, one reaches the following: for c_1,

$$\frac{\partial \rho_2}{\partial c_1} \Delta c_1 = \frac{\rho_1 (1 + r_A)}{c_2 (1 - r_A)} \Delta c_1; \tag{52}$$

for c_2,

$$\frac{\partial \rho_2}{\partial c_2} \Delta c_2 = -\frac{\rho_1 c_1 (1 + r_A)}{c_2^2 (1 - r_A)} \Delta c_2; \tag{53}$$

for ρ_1,

$$\frac{\partial \rho_2}{\partial \rho_1} \Delta \rho_1 = \frac{c_1 (1 + r_A)}{c_2 (1 - r_A)} \Delta \rho_1; \tag{54}$$

and for r_A,

$$\frac{\partial \rho_2}{\partial r_A} \Delta r_A = \frac{2 c_1 \rho_1}{c_2 (1 - r_A)^2} \Delta r_A. \tag{55}$$

The calculated proportions of uncertainties for different assumed errors are shown in Table 5. In the first row of uncertainties a constant error of 0.1 % is assumed for all variables. The uncertainty examination shows that the contribution of reference values and measured sound velocity are comparable, while the contribution from the reflection coefficient is comparably small. Unfortunately the reachable reflection coefficient accuracies have not been reported so far. In the second row of uncertainties, realistic errors are assumed. The

Table 5. Contributed uncertainties of the relevant variables: buffer density, buffer sound velocity, sample medium sound velocity and reflection coefficient, with PMMA being the buffer and water being the sample medium.

	ρ_1 [kg m^{-3}]	c_1 [m s^{-1}]	c_2 [m s^{-1}]	r_A
value	1181.77	2764.92	1482.38	−0.3766
error 1	±0.1 %	±0.1 %	±0.1 %	±0.1 %
uncertainty 1 [kg m^{-3}]	±0.998	±0.998	±0.998	±0.438
error 2	±1 kg m^{-3}	±0.2 m s^{-1}	±0.2 m s^{-1}	±1E-04
uncertainty 2 [kg m^{-3}]	±0.085	±0.007	±0.013	±0.116

reflection coefficient of error was estimated from theoretical considerations and uncertainties. The error contribution of sound velocity and density is still small, and the reported accuracies are sufficient to reach acceptable density uncertainties. But the contribution of a realistic reflection coefficient error to the density uncertainty is comparatively high, particularly considering that the reflection coefficient can result from several amplitude errors. For the coupled PMMA–water a density uncertainty of 0.25 kg m^{-3} can be expected overall. This uncertainty is still high compared to existing reference analytics such as the vibrating U-tube (see Table 6), but seems sufficient to use the BRTs as a monitoring tool in bioprocesses of small density change (see Table 1).

The most important uncertainty contribution which controls every influencing factor discussed so far is the temperature. The temperature accuracy affects the calibration measurements of the propagation path and buffer material's properties. Moreover, the temperature error affects uncertainties of temperature-dependent reference models as provided by the literature or certified reference standards. Using, for example, Marzcak's (Marczak, 1997) model to calculate the speed of sound of water at 20 °C, a 0.1 K temperature bias results in a 0.3 m s^{-1} USV bias, but only 0.03 m s^{-1} bias for a 0.01 K temperature bias. Due to the high impact of temperature on all relevant parameters, a temperature accuracy of at least ±0.01 K is recommended. Most non-invasive temperature measurement techniques are too inaccurate and expensive (Childs et al., 2000). The standard for invasive temperature measurement is still the electrical resistance thermometry. In general, accuracies below 0.1 K can be achieved only through individual calibration regardless of the material. For highly accurate measurements, 4-wire systems, voltage reversal and low resistances are recommended.

The temperature also influences the dimensions and properties of the used materials, the characteristics of the sound field and even the properties of the PEM. So it is quite understandable that most authors have restricted their investigations to a constant temperature. In turn, the results of these works have to be evaluated with respect to the reported temperature stability. While in Bjørndal et al. (2008) a stability

Table 6. Accuracies of involved measurement principles as published by different authors.

Reference	Density reference	Reference accuracy ($kg\,m^{-3}$)	Measurement points/temperature accuracy	USV accuracy ($m\,s^{-1}$)
Adamowski et al. (1995, 1998)	pycnometer	$\pm 0.3\,kg\,m^{-3}$	$-/\pm 0.5$ K (varying)	1.0
Bjørndal et al. (2008)	literature/ standards	$\pm 0.10\,kg\,m^{-3}$	$-/\pm 0.01$ K (constant)	–
Deventer and Delsing (1997)	literature	–	2 points/± 0.01 K (varying)	–
Greenwood et al. (1999)	volume weighting	–	3 points/– (varying)	–
Bamberger and Greenwood (2004a, b)	volume weighting	–	–/–	–
McClements and Fairly (1991, 1992)	vibrating U-tube	$\pm 0.10\,kg\,m^{-3}$	$-/0.1$ (constant)	0.5
Papadakis et al. (1973); Papadakis (1968)	–	–	–/–	–
Püttmer and Hauptmann (1998), Püttmer et al. (2000)	vibrating U-tube	$\pm 0.10\,kg\,m^{-3}$	1 point/ – (constant)	–

of ± 0.04 K was reached, Adamowski et al. (1995) reported only ± 0.5 K.

Additionally, temperature gradients have to be considered. Most researchers try to avoid gradients and control not only the temperature of the sample medium but also the environmental temperature (Bjørndal et al., 2008; Higuti et al., 2007). The procedure is acceptable for highly accurate validations but of low relevance for any practical application. In real process application often the sample medium or the environmental medium temperature varies, in the worst case even both. While the temperature of the sample medium is often controlled or behaves in a predictable way, the environmental temperature does not. Depending on the time of the year, the daytime, the local weather and the location and construction of the facility, the environmental temperature can vary in a range of ± 5 to ± 20 K. The point is that, in reality, there will be temperature gradients which are generally not constant, so the gradients have to be considered. Furthermore, the temperature control of the buffer is only a solution when the sample medium is also of constant temperature.

The methods that are affected most by temperature gradients are the ARM and RRM. When reference and calibration measurements are executed at different temperatures or gradients, the error can increase enormously. As stated before, temperature control is often not an acceptable solution and often not stable enough; therefore two options remain – either the calibration for all relevant temperatures and gradients, which is extremely laborious, or an additional probe

that determines parallel, under identical conditions to the reference values (Greenwood, 2000; Greenwood et al., 1999). Indeed, the parallel reference measurement minimizes the uncertainty caused by temperature gradients, but introduces new uncertainty sources due to the use of two excitation electronics, sender, receiver, and coupling systems that might be not identical. In the case of an MRM as proposed by Deventer and Delsing (2001b), temperature differences between sample medium and buffer rod interface temperature have to be considered. Therefore both should be monitored continuously. Similar effects have to be considered for propagation path calibrations (Higuti et al., 2007) and varying dynamic behaviour due to temperature changes of different magnitude which results in hysteresis effects (Deventer and Delsing, 2001a; Higuti et al., 2007).

In fact, there is another temperature gradient that has not been considered so far – the temperature gradient in the sample medium. As long as there is a temperature difference between sample medium and environment, there will be a gradient at the buffer–liquid interface, which implies three major issues:

1. The temperature variation over the sound propagation path influences the accuracy of the sound velocity measurement. In general, the properties vary with propagation path, and so does the sound velocity. In the end, the measured velocity, USV_p represents the average of all variations. For a known temperature dependency of

the velocity, USV(T) and a known temperature gradient $T(x)$ over the propagation path x, the relation can be described as follows:

$$\text{USV}_p = \frac{1}{T(x_2) - T(x_1)} \int_{x1}^{x2} \left[USV(T(x)) \cdot \frac{\partial T(x)}{\partial x} \right] dx. \quad (56)$$

The main conclusion of this expression is that if one wants to determine the temperature that fits to the measured USV, or vice versa, one has to determine the temperature at the right position or the mean temperature over the propagation path.

2. Equation (56) only introduces the general problem. The basic problem concerning the density determination is the combination of propagation path information and interfacial information. Knowing the temperature gradient means only that the measured sound velocity is not the sound velocity as it is next to the interface which is the relevant sound velocity for the reflection coefficient.

3. Thinking in terms of real process measurements, the temperature gradient cannot be considered to be simply a function of temperature difference. As soon as the sound velocity is measured in flows the gradient becomes dependent on the flow conditions.

To summarize, it can be expected that highly accurate measurements require multiple-point temperature measurements (see Table 6: Deventer and Delsing, 1997 and Greenwood et al., 1999) to gather all relevant temperatures and to estimate the gradients. Relevant temperature-dependent validations of ultrasound-based density determination are published in Adamowski et al. (1998), Greenwood and Bamberger (2002), Higuti et al. (2007), Deventer and Delsing (1997) and Deventer and Delsing (2001a).

The only method that can be assumed to be independent of gradients in the sample medium is the ARM. The density is determined via RRMs at two different angles (Eq. 41). The sound velocity can be calculated as an additional parameter from the determined density, but is not necessary for the density determination. If Eq. (42) is used, the calculated sound velocity can be assumed to be the interfacial sound velocity of the sample medium. On the other hand, the density uncertainties of the ARM can be assumed to be even more complex than presented in Eqs. (52)–(55). And, in case the sound velocity is not determined by the TOF-distance relation but by Eq. (42), the sound velocity uncertainty becomes similar in complexity.

The last point concerning the temperature-related uncertainties will be the temperature dependency of transducers and PEM. Most transducers possess a matching layer or wear plate. The transmission through such layers clearly is temperature dependent and can be described in terms of wavelength and layer thickness. Furthermore, for quartz crystals

and piezocermic materials, it is known that the resonance behaviour changes with temperature (Hammond and Benjaminson, 1965; Yang, 2006). This effect can actually be used to measure the temperature. Once an MRM is used or the RRM and ARM are calibrated for different temperatures, those influences can be neglected in terms of attenuation or varying transmission coefficients, but the frequency behaviour might change significantly. Consequently, signal-processing methods in the frequency domain possibly have to be modified to consider temperature-dependent variations, particularly the single-frequency method (see Eq. 46).

4.4 Relevant errors for industrial conditions

This section discusses errors which are especially relevant for industrial applications. First of all, errors due to thin layers, which may represent coupling layers, matching layers or buffer surface deposits, will be discussed. Surface deposits might be applied as a protective layer or might appear as a result of fouling.

In Püttmer et al. (1999), the focus is on investigation of surface deposits by simulations via SPICE. After validation with polystyrene layers of varying thickness, the developed model was applied for materials of varying acoustic impedance and thickness. Scattering effects due to non-plane surfaces have been neglected. The results show that for layers of impedance lower than the buffer material and $\lambda/100(\lambda/50)$ thickness, the error of the sample medium's acoustic impedance can reach up to 0.5 % (2.6 %); the USV error up to 0.05 % (0.1 %). For layers of impedance higher than the buffer material, the error increases rapidly. It is stated that deposits of low acoustic impedance such as polymers can be tolerated with a thickness up to $\lambda/50$.

In Deventer (2003) also the influence of fouling deposits is investigated via a PSPICE model. Commensurate with a different probe design, the effects of deposits are simulated for a PMMA buffer instead of quartz glass (Püttmer et al., 1999). For the deposit material a density of $1500 \, \text{kg} \, \text{m}^{-3}$ and a sound velocity of $3000 \, \text{m} \, \text{s}^{-1}$ was assumed and thicknesses of 0.5, 1 and 2 µm have been investigated. It was stated that, compared to a clean surface, the amplitude difference is quite high, but changing the layer thickness results only in small changes. While comparing the results with those of Püttmer et al. (1999), it was assumed that the model might be inconsistent. But comparing the details of both publications explains the difference: (1) in Püttmer et al. (1999) layer thicknesses relative to wavelength in the deposit material are investigated, which would correspond more likely to 8 and 17 µm layer thicknesses in the case of Deventer (2003). (2) In Püttmer et al. (1999) no results of amplitude changes but errors in the determination of acoustic impedance and sound velocity are presented. (3) Checking the presented results of Püttmer et al. (1999) for impedances higher than the buffer materials, as investigated in Deventer (2003), one can assume that the amplitude difference is quite high compared to clean surface. Thus, based on the information given in Deventer (2003), no inconsistency is noticeable.

In Higuti et al. (2006) a model of acoustic or electroacoustic transmission lines was developed. The model was validated experimentally with signals from the true measurement cell, but without deposits. Metallization layers on the PVDF-receiver surface, varying thicknesses of the PVDF receiver, varying coupling layers and deposits on the buffer surface have been investigated. The thickness of the metallization layers was reported to be around 500 Å. In contrast to Deventer (2003) it was stated that layer thicknesses up to 1 μm do not introduce significant changes in the signals, and their effects can be neglected. In the case of the receiver thickness, the pulse centre frequency changes with temperature, while the bandwidth remains constant. It is shown that layer thickness variations significantly change the frequency domain information, which might result in errors > 2 % when applying the single-frequency approach. The error can be minimized by using the energy method and time delay compensation. The density error was kept within ±0.2 % for receiver thickness variations and within ±0.1 % for coupling layer variations up to 50 μm. Deposit results have been presented for varying thickness and different materials. For all presented materials the density error does not exceed 0.2 % up to 2 μm layer thickness. For higher thicknesses the error quickly reaches 6 % and more.

Actually, neither in Püttmer et al. (1999) nor in Deventer (2003) or Higuti et al. (2006) is the relevance of the assumed fouling properties and layer thicknesses discussed. For milk fouling layers, for example, a layer thickness of 500–700 μm and an impedance of 2.97 MRayl has been reported (Wallhäußer et al., 2009). Hence, concerning the impedance of biological fouling layers, the assumption of lower acoustic impedance seems to be correct for most buffer materials. Whether relevant thicknesses have been investigated so far is questionable. Generally it can be stated that not much is known about the acoustic properties of real fouling layers and that electrical analogous systems can be applied to investigate the influence of thin layer deposits under ideal conditions (Deventer, 2003; Higuti et al., 2006; Püttmer et al., 1999) and to simulate design aspects of probes with a few limitations (Deventer, 2004). In Püttmer et al. (1999) it is shown that the error due to thin layers can be reduced as long as the degree of fouling can be detected. Reference calibrations with air are proposed, while in Deventer (2003) it is recommended to detect fouling at higher frequencies via broadband transducers. Also, in Higuti et al. (2006) it is stated that a periodic calibration with a reference medium might be necessary.

Besides surface deposits, short-term variations of process variables might have an influence on the method's accuracy. The influence of temperature variations and measurement accuracy has already been discussed above. Also, the influence of varying flow condition on temperature gradients has already been indicated, but not the direct signal diversion due to a flow perpendicular to the propagation path. Generally it is assumed that the diversion can be neglected as long as the sound velocity in the medium is considerably higher than the flow velocity. Assuming a moderate flow of $5\,\mathrm{m\,s^{-1}}$ typically results in a diversion angle of $0.2°$. In consequence, each molecule is distracted approximately $0.003\,\mathrm{mm}$ per mm propagation path while the signal propagates through the sample medium. First of all, the diversion results in an offset diffraction, and furthermore the angular difference from normal incidence causes a difference of approximately 0.1 % in the reflection coefficient. Greenwood et al. (1999) investigated flow velocities up to $2.5\,\mathrm{m\,s^{-1}}$ and found that the varying flow conditions did not significantly affect the average density bias. In Adamowski et al. (1995, 1998) varying flow velocities up to $10\,\mathrm{m\,s^{-1}}$ were investigated. It was found that the experimental results are not affected by the flow rate. Indeed, changes of reflection coefficient, sound velocity and density appeared, but relative to the temperature variation, the observed deviations have been within the precision range of the method. It is reported that cavitation occurred for mean flow velocities above $10\,\mathrm{m\,s^{-1}}$, and for this reason the results became inconsistent. Further issues might occur in the case of non-homogenous suspensions or bubbly flow. As correctly stated by Schäfer et al. (2006), the measurement effect bases on reflection at interfaces. Non-homogenous distributions of solid or gaseous objects across the interface would lead to a certain error. In Greenwood and Bamberger (2002) the feasibility of the ARM for homogenous suspensions was proven. The influence of bubbly flow was also investigated, and it was reported that three of the six investigated instruments have been significantly affected by the air feed. It can be assumed that generally the bubble dependency depends on the design and placement of the probe. As long as the bubbles do not adhere to the interface, no significant effect on the reflection coefficient should be noticeable. For the ARM also, the sound velocity determination only depends on the interfacial information. In the case of the other methods the situation for the sound velocity is quite different. Depending on the amount of air inside a certain volume, the density and compressibility change:

$$\rho = \frac{(M_1 + M_2)}{(V_1 + V_2)}, \tag{57}$$

$$\kappa = \frac{(\kappa_1 V_1 + \kappa_2 V_2)}{(V_1 + V_2)}, \tag{58}$$

where M and V represent the mass and volume and the indices indicate the particular phase. According to Eq. (1) the sound velocity changes as a result. In Hoppe et al. (2002) it was stated that the bubbles operate like a high-pass filter. It was shown in Hoppe et al. (2001) that the amplitude and the zero crossing times of detected pulses decrease, but the arrival time of the signal does not change. It was further stated that the influence of gas bubbles on the speed-of-sound accuracy can be minimized by adequate signal processing.

Generally the attenuation due to bubbles is frequency dependent. The bubble size governs the resonance frequency of a bubble, and therefore the bubble size distribution with respect to the main frequency defines the degree of attenuation (Carstensen and Foldy, 1947; Silberman, 1957; Fox et al., 1995). According to Eq. (3), also the acoustic impedance could be affected for disadvantageous bubble distributions. Henning et al. noticed only a change of impedance for high bubble intensities (Hoppe et al., 2002).

5   Conclusions

In the last decades, several research groups have investigated varying methods based on BRTs. The reported methods can be classified into four main groups: MRM, TRM, RRM and ARM. Each method holds characteristic advantages and disadvantages. ARM and RRM are perfectly suited for highly sound absorbing liquids but require calibration measurements. The RRM is only suited for moderate sound absorbing liquids, but does not require calibrations. The TRM can be ranked somewhere in between, but as with the ARM, the method requires an additional receiver, which introduces additional sources of uncertainty. Although the RRM was proven theoretically to be more sensitive to SNR-caused inaccuracies than any other method, the experimental results did not confirm the theoretical evaluations. Basically all methods are sensitive to temperature gradients. While for MRM it is sufficient to determine the accurate temperature at the interface in order to determine the correct acoustic impedances, in the case of ARM and RRM it might be necessary to calibrate the probe for all relevant temperature gradients. An appropriate correction seems to be possible, but so far has not been proven to work accurately.

The main design limitations result from intentions to avoid pulse superposition. Pure pulses can be guaranteed by avoidance and suppression of radial mode vibrations and adequate dimensioning with respect to the given pulse duration and material properties. In some cases additional near-field constraints might have influenced the chosen dimension. Although angular reflections within the near field might disturb the sound field in a way that one should prevent the assumption of plane wave propagation, the ARM as well as the RRM can be assumed to be widely unaffected by those phenomena as long as all changes of the sound field are considered in the calibration. In the case of MRM and TRM, diffraction correction often is a major requirement for adequate errors. Alternatively to corrections, large-aperture receivers can be used in some applications to minimize the error.

The published results show minimum achievable density errors of 0.15 % for constant temperature and 0.4 % for varying temperatures, which is sufficient to identify liquids of significant different density. The question if the reported errors are sufficient for a suitable control of a specific process or not in the end depends on the density variation that

can be expected. Sensitive biotechnological processes such as yeast fermentation generally show a density variation of $< 60\,\mathrm{kg\,m^{-3}}$, which results in density accuracy requirements of at least $1\,\mathrm{kg\,m^{-3}}$ or 0.1 %. In the case of density-based models for concentration measurements of multicomponent mixtures, an even lower error might be necessary.

The uncertainty analysis shows that errors in the reflection coefficient contribute significantly to the overall density error but has been investigated least so far, whereas the contributions of realistic errors of the sound velocities and buffer material's density are comparably low. Indeed, most authors neither state the accuracies of the sound velocities nor the accuracy of the reflection coefficient measurement. Although the few presented USV errors are $\geq 0.5\,\mathrm{m\,s^{-1}}$, state-of-the-art technologies can provide accuracies $\leq 0.1\,\mathrm{m\,s^{-1}}$ even for low sampling frequencies. Moreover, the buffer material's density can be determined with acceptable accuracies keeping the uncertainties of the sample liquid's density within the required accuracy. Consequently, improvements in the reflection coefficient determination are the right choice to improve the density accuracy. Main improvements are reached by increasing the SNR and improving the amplitude determination. Most authors apply signal averaging, which reduces the Gaussian noise. But averaging of the whole signal is only a feasible method as long as the signal acquisition rate is much higher than changes of process parameters. In the case of fast varying sound velocity, signal averaging can cause systematic errors. We assume that it might be better not to average the whole signal but only the relevant pulses after being centred to a characteristic location. Errors due to systematic changes in the frequency domain can be minimized by applying the integration method to an adequate frequency band. The temperature measurement is identified as another main source of error. Often the temperature at a certain position is required to calculate the buffer material's properties from reference polynomials. In addition, temperature gradients may occur, particularly during dynamic process changes. Thus, for real-time process application and exact validation it is necessary to measure the temperature as accurately as possible ($\leq \pm 0.01$ K) and to observe temperature gradients as they may arise. Altogether it seems possible to reach an accuracy of $\leq 1\,\mathrm{kg\,m^{-3}}$ even for dynamic conditions. At present, the remaining uncertainty could be a result of both the assumed simplifications for the reflection coefficient at solid–liquid interfaces or the technological limitations – state of the art is a 12-bit resolution at 1 GHz sampling rate; a higher vertical resolution of 14 bit or more often results in significantly lower sampling rates.

A sensor system for real-time process application will have to be suitable to fulfil all involved task reaching, from generation of the excitation signal and sound signal capturing over temperature measurement and up to signal processing. To date, most of the basics have been investigated, but still final statements about which technology or method suits best a certain case of application are not possible. It is not known if simple peak excitations are sufficient or if bursts of a certain frequency are the best choice. It is not clear exactly if signals of a specified frequency require a certain sampling frequency in order to reach the desired density accuracy or not. Similar can be stated for the different signal-processing methods. Applying spline interpolation in the time domain might reach comparable results such as integration in the frequency domain. The big question is which one requires less computational effort. From the technological point of view it is clear that a vertical resolution of 12 bit or better is required to reach accurate results. For statements about electronic effort, computation power and the required memory, first the basic aspects of signal generation and signal processing have to be discussed in more detail. Definitely not all methodical options to determine the reflection coefficient via BRT have been investigated so far, but the basic rules are clear: minimization or correction of temperature gradients, and maximization of SNR.

Edited by: M. Jose da Silva
Reviewed by: three anonymous referees

References

Adamowski, J. C., Buiochi, C., Simon, C., Silva, E. C. N., and Sigelmann, R. A.: Ultrasonic measurement of density of liquids, J. Acoust. Soc. Am., 97, 354–361, 1995.

Adamowski, J. C., Buiochi, C., and Sigelmann, R. A.: Ultrasonic Measurement of Density of Liquids Flowing in Tubes, IEEE Transactions on Ultrasonics, Ferroelectrics, and Frequency Control, 45, 48–56, 1998.

Asher, R. C.: Ultrasonics in chemical analysis, Ultrasonics, 25 17–19, 1987.

Bamberger, J. A. and Greenwood, M. S.: Measuring fluid and slurry density and solids concentration non-invasively, Ultrasonics, 42, 563–567, 2004a.

Bamberger, J. A. and Greenwood, M. S.: Non-invasive characterization of fluid foodstuffs based on ultrasonic measurements, Food Res. Int., 37, 621–625, 2004b.

Bjørndal, E. and Frøysa, K. E.: Acoustic Methods for Obtaining the Pressure Reflection Coefficient from a Buffer Rod Based Measurement Cell, IEEE Trans UFFC, 55, 1781–1793, 2008.

Bjørndal, E., Frøysa, K. E., and Engeseth, S. A.: A Novel Approach to Acoustic Liquid Density Measurements Using a Buffer Rod Based Measuring Cell, IEEE Trans UFFC, 55, 1794–1808, 2008.

Bolotnikov, M. F., Neruchev, Y. A., Melikhov, Y. F., Verveyko, V. N., and Verveyko, M. V.: Temperature Dependence of the Speed of Sound, Densities, and Isentropic Compressibilities of Hexane + Hexadecane in the Range of (293.15 to 373.15) K, J. Chem. Eng. Data 50, 1095–1098, 2005.

Carlson, J. E., Deventer, J., and Micella, M.: Accurate temperature estimation in ultrasonic pulse-echo systems, World Congress on Ultrasonics, Paris, 2003a.

Carlson, J. E., Deventer, J., Scolan, A., and Carlander, C.: Frequency and Temperature Dependence of Acoustic Properties of Polymers Used in Pulse-Echo Systems, IEEE ULTRASONICS SYMPOSIUM, 8030032, 885–888, 2003b.

Carstensen, E. L. and Foldy, L. L.: Propagation of Sound Through a Liquid Containing Bubbles, J. Acoust. Soc. Am., 19, 481–501, 1947.

Childs, P. R. N., Greenwood, J. R., and Long, C. A.: Review of temperature measurement, Rev. Sci. Instrum., 71, 2959–2978, doi:10.1063/1.1305516, 2000.

Daridon, J. L., Lagourette, B., Xan, B., and Montel, F.: Petroleum characterization from ultrasonic measurement, J. Petrol. Sci. Eng., 19 281–293, 1998a.

Daridon, J. L., Lagrabette, A., and Lagourette, B.: Speed of sound, density, and compressibilities of heavy synthetic cuts from ultrasonic measurements under pressure, J. Chem. Thermodynam., 30, 607–623, 1998b.

Davis, L. A. and Gordon, R. B.: Compression of Mercury at High Pressure, J. Chem. Phys., 46, 2650–2660, 1967.

Deventer, J.: Detection of, and compensation for error inducing thin layer deposits on an ultrasonic densitometer for liquids, Instrumentation and Measurement Technology Conference 2003, 648–651, 2003.

Deventer, J.: One dimensional modeling of a step-down ultrasonic densitometer for liquids, Ultrasonics, 42, 309–314, 2004.

Deventer, J. and Delsing, J.: An Ultrasonic Density Probe, IEEE ULTRASONICS SYMPOSIUM, 1997.

Deventer, J. and Delsing, J.: Thermostatic and Dynamic Performance of an Ultrasonic Density Probe, IEEE Trans UFFC, 48, 675–682, 2001a.

Deventer, J. and Delsing, J.: Thermostatic and Dynamic Performance of an Ultrasonic Density Probe, IEEE Trans. UFFC, 48, 675–682, 2001b.

Esperança, J. M. S. S., Visak, Z. P., Plechkova, N. V., Seddon, K. R., Guedes, H. J. R., and Rebelo, L. P. N.: Density, Speed of Sound, and Derived Thermodynamic Properties of Ionic Liquids over an Extended Pressure Range. 4. [C3mim][NTf2] and [C5mim][NTf2], J. Chem. Eng. Data, 51, 2009–2015, 2006.

Fisher, B., Magpori, V., and von Jena, A.: Ultraschall (US)-Dichtemesser mm Messen der spezifischen Dichte eines Fluid, EP 0 483 491 81, Europe, 1995.

Fox, F. E., Curley, S. R., and Larson, G. S.: Phase Velocity and Absorption Measurements in Water Containing Air Bubbles, The J. Acoust. Soc. Am., 27, 534–539, 1995.

Greenwood, M. S.: Ultrasonic fluid densitometer having liquid/wedge and gas/wedge interfaces, 6, 082, 181, United States, 2000.

Greenwood, M. S. and Bamberger, J. A.: Ultrasonic sensor to measure the density of a liquid or slurry during pipeline transport, Ultrasonics, 40, 413–417, 2002.

Greenwood, M. S. and Bamberger, J. A.: Self-Calibrating Sensor for Measuring Density Through Stainless Steel Pipeline Wall, J. Fluid. Eng., 126, 189–192, 2004.

Greenwood, M. S., Skorpik, J. R., Bamberger, J. A., and Harris, R. V.: On-line Ultrasonic Density Sensor for Process Control of Liquids and Slurries, Ultrasonics, 37, 159–171, 1999.

Greenwood, M. S., Adamson, J. D., and Bamberger, J. A.: Long-path measurements of ultrasonic attenuation and velocity for very dilute slurries and liquids and detection of contaminates, Ultrasonics, 44, e461–e466, 2006.

Guilbert, A. R. and Sanderson, M. L.: A novel ultrasonic mass flowmeter for liquids, IEE colloquium on: Advances in Sensors for Fluid Flow Measurement, London, 1996.

Hale, J. M.: Ultrasonic density measurement for process control, Ultrasonics, 26, 356–357, 1988.

Hammond, L. D. and Benjaminson, A.: The Linear Quartz Thermometer - a New Tool for Measuring Absolute and Difference Temperatures, Hewlett-Packard Journal, 16, 1965.

Henning, B., Prange, S., Dierks, K., Daur, C., and Hauptmann, P.: In-line concentration measurement in complex liquids using ultrasonic sensors, Ultrasonics, 38, 799–803, 2000.

Higuti, R. T. and Adamowski, J. C.: Ultrasonic Densitometer Using a Multiple Reflection Technique, IEEE Trans UFFC, 49, 1260–1268, 2002a.

Higuti, R. T. and Adamowski, J. C.: Ultrasonic densitometer using a multiple reflection technique, IEEE Trans. Ultrason., Ferroelect., Freq. Contr., 49, 1260–1268, 2002b.

Higuti, R. T., Montero de Espinosa, F. R., and Adamowski, J. C.: Energy method to calculate the density of liquids using ultrasonic reflection techniques, Proc. IEEE Ultrason. Symp., 319–322, 2001.

Higuti, R. T., Buiochi, C., Adamowski, J. C., and Espinosa, F. M.: Ultrasonic density measurement cell design and simulation of non-ideal effects, Ultrasonics, 44, 302–309, 2006.

Higuti, R. T., Galindo, B. S., Kitano, C., Buiochi, C., and Adamowski, J. C.: Thermal Characterization of an Ultrasonic Density-Measurement Cell, IEEE Transactions on Instrumentation and Measurement, 56, 924–930, 2007.

Hoche, S., Hussein, W. B., Hussein, M. A., and Becker, T.: Time-off light prediction for fermentation process monitoring, Eng. Life Sci., 11, 1–12, 2011.

Hoppe, N., Schönfelder, G., Püttmer, A., and Hauptmann, P.: Ultrasonic density sensor – Higher accuracy by minimizing error influences, Proc. IEEE Ultrason. Symp., 361–364, 2001.

Hoppe, N., Schönfelder, G., and Hauptmann, P.: Ultraschall-Dichtesensor für Flüssigkeiten – Eigenschaften und Grenzen, Technisches Messen, 3, 131–137, 2002.

Hoppe, N., Püttmer, A., and Hauptmann, P.: Optimization of Buffer Rod Geometry for Ultrasonic Sensors with Reference Path, IEEE Trans UFFC, 50, 170–178, 2003.

Jensen, B. R.: Measuring equipment for acoustic determination of the specific gravity of liquids, 4, 297, 608, United States, 1981.

Kaatze, U., Eggers, F., and Lautscham, K.: Ultrasonic velocity measurements in liquids with high resolution – techniques, selected applications and perspectives, Meas. Sci. Technol., 19, 1–21, doi:10.1088/0957-0233/19/6/062001, 2008.

Kaduchak, G. and Sinha, D. N.: Apparatus and method for remote, noninvasive characterization of structures and fluids inside containers, 8, 186, 004 B1, United States, 2001.

Kell, G. S.: Density, Thermal Expansivity, and Compressibility of Liquid Water from 0° to 150 °C: Correlations and Tables for Atmospheric Pressure and Saturation Reviewed and Expressed on 1968 Temperature Scale, J. Chem. Eng. Data, 20, 97–105, 1975.

Kim, J. O. and Bau, H. H.: Instrument for simultaneous measurement of density and viscosity, Rev. Sci. Instrum., 60, 1111–1115, 1989.

Krautkramer, J. and Krautkramer, H.: Ultrasonic Testing of Materials, 3rd edn. ed., Springer-Verlag, New York, 1983.

Kulmyrzaev, A., Cancelliere, C., and McClements, D. J.: Characterization of aerated foods using ultrasonic reflectance spectroscopy, J. Food Eng., 46, 235–241, 2000.

Kuo, H. L.: Variation of Ultrasonic Velocity and Absorption with Temperature and Frequency in High Viscosity Vegetable Oils, Japanese Journal of Applied Physics, 10, 167–170, 1971.

Kushibiki, J., Akashi, N., Sannomiya, T., Chubachi, N., and Dunn, F.: VHF/UHF range bioultrasonic spectroscopy system and method, IEEE Trans. Ultrason., Ferroelec. Freq. Contr., 42, 1028–1039, 1995.

Kushibiki, J., Okabe, R., and Arakawa, M.: Precise measurements of bulk-wave ultrasonic velocity dispersion and attenuation in solid materials in the VHF range, J. Acoust. Soc. Am., 113, 3171–3178, 2003.

Lach, M., Platte, M., and Ries, A.: Piezoelectric materials for ultrasonic probes, NDTnet, 1, 1996.

Lynnworth, L. C. and Pedersen, N. E.: Ultrasonic mass flowmeter, Proc. IEEE Ultrason. Symp., 87–90, 1972.

Lynnworth, L. C.: Ultrasonic nonresonant sensors, in: Sensors – A Comprehensive Survey, edited by: Göpel, W., Hesse, J., and Zemel, J. N., Mechanical Sensors, VCH Publishers Inc., New York, 311–312, 1994.

Mak, D. K.: Comparison of various methods for the measurment of reflection coefficient and ultrasonic attenuation, British Journal of NDT, 33, 441–449, 1991.

Marczak, W.: Water as standard in the measurements of speed of sound in liquids, J. Acoust. Soc. Am., 102, 2776–2779, 1997.

Marks, G. W.: Acoustic Velocity with Relation to Chemical Constitution in Alcohols, The Journal of the Acoustical Society of America, 41, 103–117, 1976.

Mason, P., Baker, W. O., McSkimin, H. J., and Bepiss, J. H.: Measurement of Shear Elasticity and Viscosity of Liquids at Ultrasonic Frequencies, Phys. Rev., 75, 936–946, 1949.

Matson, J., Mariano, C. F., Khrakovsky, O., and Lynnworth, L. C.: Ultrasonic Mass Flowmeters Using Clamp-On or Wetted Transducers, 5th International Symposium on Fluid Flow Measurement, Arlington, Virginia, 2002,

McClements, D. J.: Ultrasonic Characterization of Foods and Drinks: Principles, Methods, and Applications, Critical Reviews in Food Science and Nutrition, 37, 1–46, 1997.

McClements, D. J. and Fairly, P.: Ultrasonic pulse echo reflectometer, Ultrasonics 29, 58–62, 1991.

McClements, D. J. and Fairly, P.: Frequency scanning ultrasonic pulse echo reflectometer, Ultrasonics, 30, 403–405, 1992.

Mc Gregor, K. W.: Methods of Ultrasonic Density Measurement, Australasian Instrumentation and Measurement Conference, Adelaide, S. Aust., 1989.

Moore, R. S. and McSkimin, H. J.: Physical Acoustics, Academic Press, New York, 167–242, 1970.

O'Neil, H. T.: Reflection and Refraction of Plane Shear Waves in Viscoelastic Media, Phys. Rev., 75, 928–935, 1949.

Papadakis, E. P.: Correction for Diffraction Losses in the Ultasonic Field of a Piston Source, J. Acoust. Soc. Am., 31, 150–152, 1959.

Papadakis, E. P.: Buffer-Rod System for Ultrasonic Attenuation Measurements, J. Acoust. Soc. Am., 44, 1437–1441, 1968.

Papadakis, E. P., Fowler, K. A., and Lynnworth, L. C.: Ultrasonic attenuation by spectrum analysis of pulses in buffer rods: Method and diffraction corrections, J. Acoust. Soc. Am., 53, 1336–1343, 1973.

Pope, N. G., Veirs, D. K., and Claytor, T. N.: Fluid Density and Concentration Measurment using noninvasive in situ ultrasound resonance interferometry, Ultrasonics Symposium, 1992, 855–858, 1992.

Pope, N. G., Veirs, D. K., and Claytor, T. N.: Fluid density and concentration measurment using noninvasive in situ ultrasonic resonance interferometry, 5, 359, 541, United States, 1994.

Povey, M. J. W. and McClements, D. J.: Ultrasonics in Food Engineering. Part I: Introduction and Experimental Methods, J. Food Eng., 8, 217–245, 1988.

Püttmer, A. and Hauptmann, P.: Ultrasonic density sensor for liquids, Proc. IEEE Ultrason. Symp., 497–500, 1998.

Püttmer, A., Lucklum, R., Henning, B., and Hauptmann, P.: Improved ultrasonic density sensor with reduced diffraction influence, Sensors Actuators A, 67, 8–12, 1998.

Püttmer, A., Hoppe, N., Henning, B., and Hauptmann, P.: Ultrasonic density sensor–analysis of errors due to thin layers of deposits on the sensor surface, Sensor. Actuator., 76, 122–126, 1999.

Püttmer, A., Hauptmann, P., and Henning, B.: Ultrasonic density sensor for liquids, IEEE Trans. Ultrason., Ferroelec. Freq. Contr., 47, 85–92, 2000.

Raum, K., Ozguler, A., Morris, S. A., and O'Brien, W. D. J.: Channel Defect Detection in Food Packages Using Integrated Backscatter ultrasound Imaging, IEEE Trans UFFC, 45, 30–40, 1998.

Rychagov, M. N., Tereshchenko, S., Masloboev, Y., Simon, M., and Lynnworth, L. C.: Mass Flowmeters for Fluids with Density Gradient, IEEE Ultrasonics Symposium, 465–470, 2002.

Sachse, W.: Density determination of a fluid inclusion in an elastic solid from ultrasonic spectroscopy measurements, Proc. IEEE Ultrason. Symp., 716–719, 1974.

Saggin, R. and Coupland, J. N.: Concentration Measurement by Acoustic Reflectance, J. Food Sci., 66, 681–685, 2001.

Schäfer, R., Carlson, J. E., and Hauptmann, P.: Ultrasonic concentration measurement of aqueous solutions using PLS regression, Ultrasonics, 44, e947–e950, 2006.

Silberman, E.: Sound Velocity and Attenuation in Bubbly Mixtures Measured in Standing Wave Tubes, J. Acoust. Soc. Am., 29, 925–933, 1957.

Sinha, D. N.: Noninvasive identification of fluids by swept frequency acoustic interferometry, 5, 767, 407, United States, 1998.

Sinha, D. N. and Kaduchak, G.: Chapter 8: Noninvasive determination of sound speed and attenuation in liquids, in: Experimental Methods in the Physical Sciences, Academic Press, 307–333, 2001.

Swoboda, C. A., Frederickson, D. R., Gabelnick, S. D., Cannon, P. H., Hornestra, F., Yao, N. P., Phan, K. A., and Singleterry, M. K.: Development of an Ultrasonic Technique to Measure Specific Gravity in Lead-Acid Battery Electrolyte, IEEE Transactions on Sonics and Ultrasonics, 30, 69–77, 1983.

Van Sint Jan, M., Guarini, M., Guesalaga, A., Ricardo Perez-Correa, J., and Vargas, Y.: Ultrasound based measurements of sugar and ethanol concentrations in hydroalcoholic solutions, Food Control, 19, 31–35, 2008.

Vray, D., Berchoux, D., Delachartre, P., and Gimenez, G.: Speed of Sound in Sulfuric Acid Solution: Application to Density Measurement, Ultrasonics Symposium, 465–470, 1992.

Wallhäußer, E., Hussein, M. A., Hinrichs, J., and Becker, T.: The acoustic impedance – an indicator for concentration in alcoholic fermentation and cleaning progress of fouled tube heat exchangers, 5th International Technical Symposium on Food Processing, Monitoring Technology in Bioprocesses and Food Quality Management, Potsdam, Germany, 1 September, 2009.

Wang, H., Cao, Y., Zhang, Y., and Chen, Z.: The design of The ultrasonic liquid density measuring instrument, Third International Conference on Measuring Technology and Mechatronics Automation, 2011.

Wang, Z. and Nur, A.: Ultrasonic velocities in pure hydrocarbons and mixtures, J. Acoust. Soc. Am., 89, 2725–2730, 1991.

Williams, A. O. J. and Labaw, L. W.: Acoustic Intensity Distribution from a "Piston" Source, J. Acoust. Soc. Am., 16, 231–236, 1945.

Williams, A. O. J.: The Piston Source at High Frequencies, J. Acoust. Soc. Am., 23, 1–6, 1951.

Yang, J.: Chapter 10: Temperature Sensors, in: Analysis of Piezoelectric Devices, World Scientific Publishing Co. Pte. Ltd., Singapore, 371–386, 2006.

Żak, A., Dzida, M., Zorbęski, M., and Ernst, S.: A high pressure device for measurements of the speed of sound in liquids, Rev. Sci. Instrum., 71, 1756–1765, 2000.

17

Principal component analysis for fast and automated thermographic inspection of internal structures in sandwich parts

D. Griefahn, J. Wollnack, and W. Hintze

Institute of Production Management and Technology (IPMT), TU Hamburg-Harburg, Denickestrasse 17, 21073 Hamburg, Germany

Correspondence to: D. Griefahn (dominik.griefahn@tuhh.de)

Abstract. Rising demand and increasing cost pressure for lightweight materials – such as sandwich structures – drives the manufacturing industry to improve automation in production and quality inspection. Quality inspection of honeycomb sandwich components with infrared (IR) thermography can be automated using image classification algorithms. This paper shows how principal component analysis (PCA) via singular value decomposition (SVD) is applied to compress data in an IR-video sequence in order to save processing time in the subsequent step of image classification. According to PCA theory, an orthogonal transformation can project data into a lower dimensional subspace with linearly uncorrelated principal components preserving all original information. The effect of data reduction is confirmed with experimental data from IR-video sequences of simple square-pulsed thermal loadings on aramid honeycomb-sandwich components with CFRP/GFRP (carbon-/glass-fiber-reinforced plastic) facings and GFRP inserts. Hence, processing time for image classification can be saved by reducing the dimension of information used by the classification algorithm without losing accuracy.

1 Introduction

Lightweight materials – such as sandwich structures – experienced and are forecasted to see a rising demand due to overall increasing transportation volumes especially in aviation. Driven by fuel efficiency requirements, the higher share of lightweight materials also in the traditional transportation industry will further augment this demand. This overall increase continuously drives the manufacturing industry to improve automation in production and quality inspection. Today, sandwich is – thanks to its excellent combination of mechanical strength but also damping properties and the low average material density – a commonly used macro- and micro-composite construction. Sandwich components with carbon- or glass-fiber-reinforced plastic (CFRP/GFRP) facings are typically deployed in rough environments with locally high loadings. In order to cope with heavy concentrated loads or to connect with other structures, components are designed with molded-in inserts, e.g., made from short glass fiber-

reinforced plastic. These inserts replace the honeycomb core to absorb stresses in a defined way (Bitzer, 1997). Quality inspection requires controlling these inserts for presence, correct type, and deviation of geometrical location inside the component after the fabrication step. Due to the mostly intransparent sandwich facings, normal visual inspection methods fail to perform the described tasks, whereas infrared (IR) thermography combined with image classification algorithms delivers promising results for facing thicknesses below half a millimeter.

Active thermography methods are typically classified by excitation method – namely optical, electromagnetic and mechanical excitation. Most commonly applied methods for composite materials use optical excitation, since these do not require electrical conductivity and are contactless. Lock-in thermography with modulated optical excitation is typically deployed for defects at high depth relative to their size, being comparatively time consuming due to the load modulation.

Pulsed thermography methods are characterized by a shorter cycle time but lower depth resolution. Both are commonly used methods and established for the testing of small lots at laboratory level (Maldague, 2001; Ibarra-Castanedo et al., 2009).

This study evaluates the potential of the square-pulsed thermography for detection of macroscopic subsurface structures in large sandwich components and shows an approach for automated inspection.

2 Background on principal component analysis and automated detection in IR sequences

As described, thermography is a very well investigated NDT (nondestructive testing) method with many different technical variants for the active testing approach (Maldague, 2001; Ibarra-Castanedo et al., 2009). All techniques aim at maximizing contrast directly in the thermal image or to apply algorithms to create or improve contrast in a second step. Principal component thermography (PCT) is a computational approach for analyzing thermal material behavior over time (Rajic, 2002), further improvement can be obtained with contrast enhancement methods and thermal behavior modeling (Omar et al., 2010; Feuillet et al., 2012). An automation of the qualitative PCT approach can be achieved by adding a supervised learning step for image classification (Marinetti et al., 2004).

2.1 Principal component analysis (PCA) using singular value decomposition (SVD)

Principal component analysis is a technique widely used in the context of machine vision (e.g., face recognition or remote sensing), but also for image and video compression. PCA applies a linear transformation to a group of correlated variables in such a way that the obtained set of transformed variables is uncorrelated (Jackson, 1991). The principal components are typically computed via a SVD.

In order to perform a PCA using SVD on infrared video sequences (spatial temperature information over time) the 3-D thermographic data need to be rearranged into a 2-D matrix. Image information (n_x-by-n_y), where n_x and n_y represent the number of photosensitive elements on the sensor in x and y direction, is reshaped into an $n_x \cdot n_y$-by-1 matrix for every time step. This operation preserves the original spatial information of temperature on the specimen surface, since the reverse transformation is unique. The subsequent transformation of all n_t time steps in the video sequence creates an $n_x \cdot n_y$-by-n_t matrix $\mathbf{A}$ in which time variations are stored column-wise and spatial variation row-wise.

According to the theory of SVD, any matrix $\mathbf{X}$ (P-by-Q, $P \leq Q$) can be factorized as follows:

$$\mathbf{X} = \mathbf{\Omega} \mathbf{\Gamma} \mathbf{V}^{\mathrm{T}}, \tag{1}$$

where $\mathbf{\Omega}$ is a Q-by-Q matrix, $\mathbf{\Gamma}$ is a P-by-Q matrix with positive or zero diagonal elements representing the singular values and $\mathbf{V}^{\mathrm{T}}$ is the transposed of a P-by-P matrix. The decomposition of IR-data in the matrix $\mathbf{A}$ (M-by-N, $M = n_x \cdot n_y$ and $N = n_t$ and therefore $M > N$) can be determined by computing and decomposing $\mathbf{A}\mathbf{A}^{\mathrm{T}}$ or using the "reduced" or "economy" SVD form to obtain

$$\mathbf{A} = \mathbf{U}\mathbf{S}\mathbf{V}^{\mathrm{T}}, \tag{2}$$

where $\mathbf{U}$ is an M-by-N matrix containing spatial information in the orthogonal space. Since spatial information in $\mathbf{A}$ are arranged vertically, the columns of $\mathbf{U}$ represent a set of orthogonal statistical modes called empirical orthogonal functions (EOF) (Emery and Thomson, 2004). The rows of $\mathbf{V}^{\mathrm{T}}$ describe the characteristic time behavior of the corresponding orthogonal function – called principal component vectors building the principal component space. The vectors can provide a measure for time behavior and characterize the defect depths in the material. The matrix $\mathbf{S}$ is an N-by-N diagonal matrix with the singular values s_j of $\mathbf{A}$. The principal components are obtained scaling the EOFs by multiplying $\mathbf{U}$ with $\mathbf{S}$ or by projecting $\mathbf{A}$ via a multiplication with $\mathbf{V}$ into the principal component space.

It can be shown that

$$\mathbf{A}\mathbf{A}^{\mathrm{T}} = \mathbf{U}\mathbf{S}^2\mathbf{U}^{\mathrm{T}} \tag{3}$$

to derive that the singular values s_j are the square roots of the positive eigenvalues of $\mathbf{A}\mathbf{A}^{\mathrm{T}}$, which is the co-variance matrix of $\mathbf{A}$ multiplied with the factor $(M-1)$. This relationship allows creating a relative measure for the share of cumulated variance included in the first i EOFs.

$$v_{\mathrm{EOF}}(i) = \frac{\sum_{j=1}^{i} s_j}{\sum_{j=1}^{n_t} s_j} \qquad i \in [1, n_t] \tag{4}$$

Earlier investigations state that more than 95 % of variance can be contained in the first three to five statistical modes and respective components (Marinetti et al., 2004).

2.2 Instance-based learning with k-nearest neighbor

Instance-based learning is used for classification when an explicit description of the target function is not available. The instance-based algorithms store training data for classification of future instances. The k-nearest neighbor algorithm is the most basic and very common kind of instance-based learning classifiers.

The k-nearest neighbor algorithm classifies points in an n-dimensional space based on the Euclidean distance to the k nearest points in the training sample. Depending on the selection of k the classification result can differ (Mitchell, 1997). The algorithm is a powerful tool for classification of multi-attribute instances with high robustness to noise for sufficiently large sets of training data.

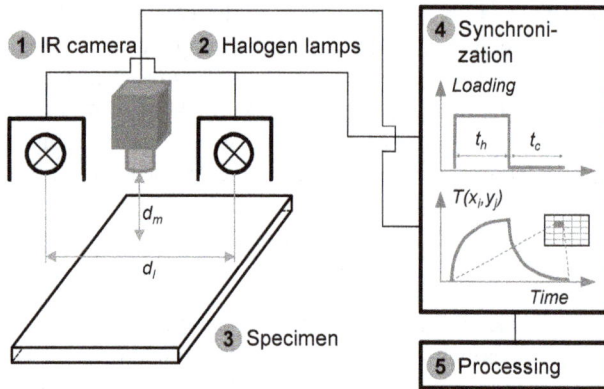

Figure 1. Test field for infrared inspection.

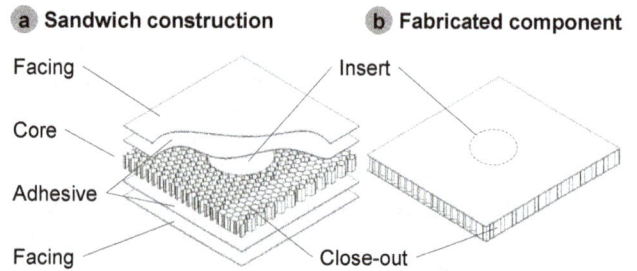

Figure 2. Schematic setup of sandwich structures with close-out and inserts (**a**) and fabricated component (**b**) according to Bitzer (1997).

The approach for supervised learning on IR sequences is to generate sets of training data from EOFs with known geometries for different materials and corresponding test settings.

3   Experimental setup for square-pulsed thermographic inspection

The following section describes the experimental setup for the square-pulsed thermographic inspection including test field configuration, the deployed sandwich specimen, and the test settings

3.1   Test field configuration

Figure 1 shows the test field setup used for the described experiments. An IR camera (1) is installed equilaterally with two 400 W halogen lamps at a lateral distance of $d_l = 250$ mm (2), which are used as heat sources. The halogen lamps as well as the IR camera have a distance of $d_m = 200$ mm to the tested sandwich specimen. The deployed IR camera is an Optris PI400 with sensor resolution of 382 pixels by 288 pixels, a thermal sensitivity of 80 mK, and maximum measurement frequency of 80 Hz. The halogen lamps are equipped with a reflector plate in order to homogenize the radiation on the specimen (3) surface. The camera captures the radiation emitted by the specimen's surface.

The synchronization unit (4), which is also linked to the IR-camera recording software, triggers the halogen lamp via a relay and applies the thermal loading during the heating phase. The camera records the heating and the cooling phase. The algorithms for SVD and k-nearest neighbor described in Sect. 2 subsequently perform the processing (5) using MATLAB (version 2012b). LabVIEW (version 2011) coupled with a digital I/O (input/output) device synchronizes the measurements.

3.2   Tested specimens

Sandwich panels (Fig. 2) are generally built from a dense and strong facing, an adhesive layer and a core. The role of the adhesive layer is to bond the facing to the top and bottom sides of the lightweight core.

Facing material can be metallic such as steel, titanium or aluminum as well as nonmetallic material such as glass fiber, Kevlar-reinforced plastic or carbon-fiber-reinforced plastic. For composite materials such as prepregs, the matrix material may substitute the effect of the adhesive layer (Bitzer, 1997). The earliest core material used for aviation purposes was balsa wood after World War I, and is still in use for some applications. Mostly for nonaerospace applications, expanded polymer foams and aluminum foams can be found as core material today. Honeycomb cores clearly dominate all other cores in aerospace. Honeycomb core structures can be produced from almost all typical lightweight materials such as aluminum, regular, and reinforced polymers or paper. Aramid-fiber paper impregnated with phenolic resin is today's most used honeycomb material (Karlsson and Åström, 1997).

Figure 2a additionally shows an example for a closeout and a high-strength insert element. Close-outs fulfill the function of mechanical protection of the component's edges and a barrier for humidity penetration. Close-outs are added cofabricated during master shaping as polymer filling of the honeycombs as shown in Fig. 2 (Bitzer, 1997).

3.2.1   Sandwich panel fabrication

Autoclave and pressing are the two main methods for bonding sandwich components at industrial scale. Autoclaves are used for curve-shaped components. Hydraulic presses are deployed for flat components and can produce large panels with sizes of up to 3000 mm $\times$ 15 000 mm (Euro-Composites$^©$, 2013). Several smaller components are typically fitted onto lager master plates hence separated and trimmed in a secondary machining step. The tested specimens are produced on a multilayer press and are machined into manageable sizes of less than $800\,\text{mm} \times 800\,\text{mm}$, but only smaller sections are inspected.

3.2.2 Specimens material

All specimens used for the experiments are fabricated from typical aircraft-grade materials and produced under condition of mass production for the aviation industry.

For test purposes, two types of pressed honeycomb sandwich modifications with intransparent facings have been selected – one with a CFRP-based facing and the other with a GFRP-based facing. The aramid honeycomb core used has a cell-size of $3.2\,\text{mm}$. The CFRP specimen, with a total thickness of $9.7\,\text{mm}$, is covered with a $0.5\,\text{mm}$ facing based on woven carbon-fiber phenolic resin and bonded to the aramid honeycomb core. The GFRP specimen, with a total thickness of $15.5\,\text{mm}$, is composed of a $0.3\,\text{mm}$ woven fiberglass facing with phenolic resin and also bonded to an aramid honeycomb. For in-service reasons, the GFRP sample is covered with a thin but intransparent polymer protection foil. Both are equipped with GFRP inserts and in the potting step locally filled with thermoplastic polymer for edge close-out. The diameter of the inserts, locally replacing the honeycomb core of the specimens, is $18\,\text{mm}$ for the CFRP specimen and $45\,\text{mm}$ for the GFRP specimen. The dimensions of edge close-outs range from approximately $10\,\text{mm}$ for the CFRP specimen to $20\,\text{mm}$ for the GFRP specimen; given the accuracy of the potting and the filling behavior of the honeycomb cells the width varies a few millimeters.

3.3 Test settings

Heating time t_h and cooling time t_c have to be selected depending on the material of the sandwich facing (see Sect. 4.1 for the exact settings) and the facing layer thickness. CFRP facings require increased heating time or higher power of halogen lamps. This is due to the high heat flow transversal to the test direction given the higher conductivity of carbon compared to glass fibers. The required spatial resolution for the purpose of quality assurance defines the distance between camera and specimen resulting from the field of view.

4 Experimental results

4.1 Contrast improvement via PCA

Video sequences are acquired with heating time $t_{h,\text{GFRP}}$ and cooling time $t_{c,\text{GFRP}}$, each of $10\,\text{s}$, on a GFRP sample resulting in a sequence length $t_{m,\text{GFRP}}$ of $20\,\text{s}$ for the total measurement. The CFRP sample was tested at a heating time $t_{h,\text{CFRP}}$ and cooling time $t_{c,\text{CFRP}}$, each of $20\,\text{s}$, resulting in a sequence length $t_{m,\text{CFRP}}$ of $40\,\text{s}$ for the total measurement. In order to investigate the influence of the amount of provided input data on the PCA, measurement frequency f_m is varied with the steps 0.5, 1, 2, and 3 Hz. The number of resulting images or

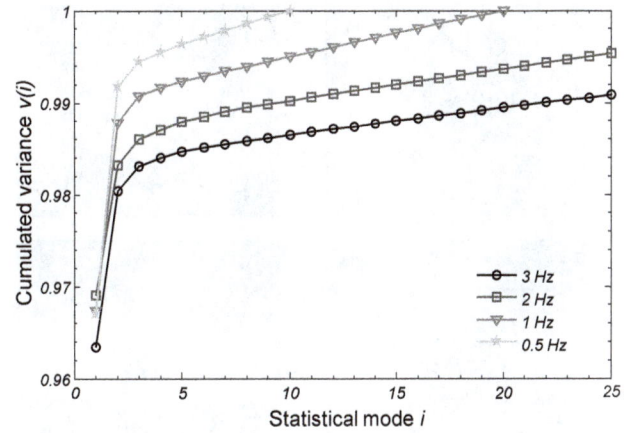

Figure 3. Cumulated and normalized variance v_{EOF} for the first 25 statistical modes.

dimension n_t of the matrix $\mathbf{A}$ is given by the following equation:

$$n_t = (t_h + t_c) \cdot f_m = t_m \cdot f_m. \tag{5}$$

Figure 3 shows the normalized variance v_{EOF} cumulated in the first i so-called statistical modes and corresponding EOFs for the GFRP sample. According to Eq. (3), this measure cumulates the first i singular values in $\mathbf{S}$ corresponding to the first i spatial components or EOFs in columns of the matrix $\mathbf{U}$. The value is normalized with the total variance.

The first dimension of the video matrix $\mathbf{A}$ contains the total number of $n_x \cdot n_y = 382 \cdot 288 = 110\,016$ elements. At the maximum frequency of $3\,\text{Hz}$ and the given recording time, the number of time steps and second dimension n_t of the video matrix $\mathbf{A}$ equals to 60 elements.

The analysis shows that the relatively slow process of heat conduction through materials with partially very low thermal conductivity does not require measurement frequencies above $1\,\text{Hz}$ to cover more than 99 % of the time behavior of sandwich material in the first three EOFs. At minimum, the Shannon theorem in the time domain must be fulfilled.

For a measurement frequency of $1\,\text{Hz}$ raw, thermal data without emissivity correction are shown on a grayscale in Fig. 4 (GFRP specimen (a) and CFRP specimen (b)) for three selected and representative instances. The first image of the heating phase as well as the first and the last image of the cooling phase are displayed. The SVD is performed in a subsequent step to obtain the EOFs from the thermal data.

Figures 5 and 6 show the effect of decreasing variance on the specimen with GFRP and CFRP facings respectively. It visualizes that the information from an IR-video sequence of square-pulsed thermal images are compressed into three to four EOFs. The retransformation of the columns of matrix $\mathbf{U}$ as described in Sect. 2.1 delivers the spatial information that are scaled to a grayscale image. In contrast to the images from the GFRP sample, the second statistical mode from the

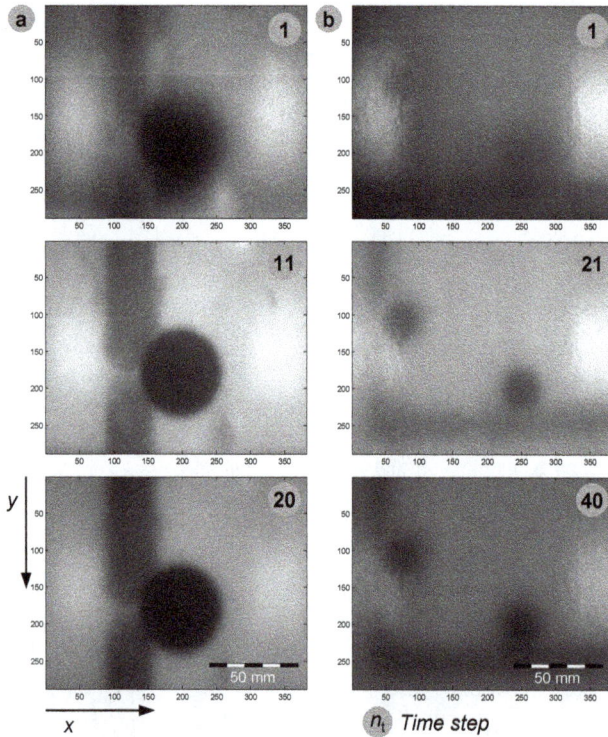

Figure 4. Thermal images at the time steps n_t for the GFRP specimen (a) and the CFRP specimen (b) (first image in heating, first and last image in cooling phase).

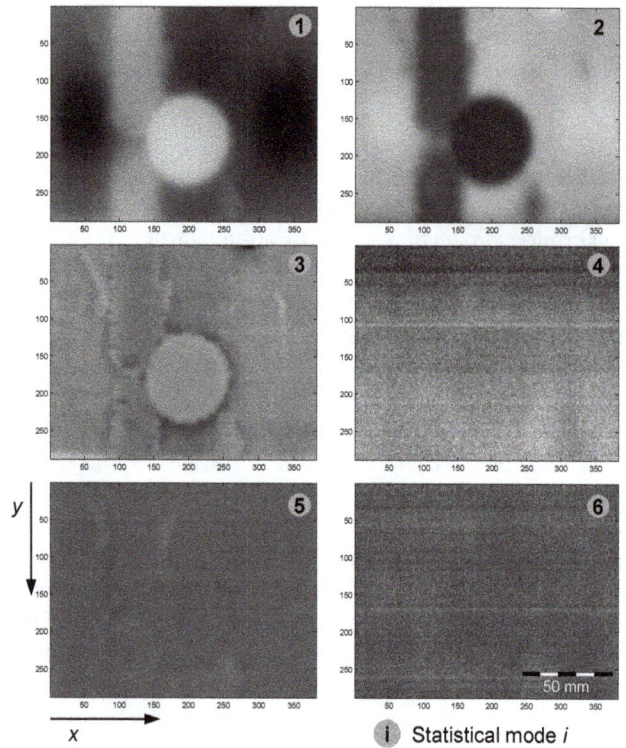

Figure 5. First 6 EOFs of the GFRP specimen.

4.2 Automated detection

CFRP sample contains the direct reflections from the heat sources occurring at the shiny specimen surface.

This section demonstrated that PCA can be applied to square-pulsed thermography for inspection of sandwich components by delivering compressed spatial information with improved contrast between subsurface elements in sandwich structures by preserving time-variation data.

4.2 Automated detection

The aim of the automated detection is to obtain segmented images for the purpose of further inspection. The k-nearest neighbor algorithm requires an amount of preclassified data. These so-called training data are generated from the sandwich samples from known geometric locations (e.g., center of the insert or close-out, plain honeycomb) that are manually classified.

Figure 7 visualized such set of training data for the first three EOFs of the GFRP specimen for approximately 5000 preclassified pixels, which are a subset in the dimension M ($n_x \cdot n_y$). Each data point corresponds to the intensity values from the first three images in Fig. 5 for the same selected pixel. The k-nearest neighbor algorithm uses the training data to classify a "new" instance (in this case the nonclassified data from the sequence) of data – pixel by pixel – based on the Euclidian distance to the k-nearest neighbors. For the ex-

periments, k varies from three to five depending on the other test settings. In theory, the algorithm can perform the classification tasks in real numbers $\mathbb{R}^n$, whereas only $i = n_t$ features for classification are available from the IR sequence. Based on the result that 99 % or more of the variance is retained in the first few EOFs, even a reduction of the dimension of the feature space has to be considered.

Two types of definitions for classification errors are used to assess the performance of the algorithm depending on the dimension of the feature space in terms of classification accuracy and computation speed. An algorithm implements the definitions to obtain repeatable and automated results. All inspected parts contain subsurface elements that can be assumed as closed contours on the level of pixel size. If all (or all but one) neighboring pixels in a classified image differ from the class of the selected pixel, it is obviously falsely classified. Figure 8 illustrates the definition of classification errors type I and type II in a segmented image.

Figure 9 shows the results from the classification performance analysis. All data are normalized to 100 % for $i = 1$ to evaluate the relative performance to the smallest possible dimension of feature space. The experiment varies the dimension of the feature space used for classification from one to the maximum possible n_t and evaluates the number of errors in the classified image as well as meters the elapsed computation time. Using the ten first instead of only the first statistical mode as feature space for classification increases

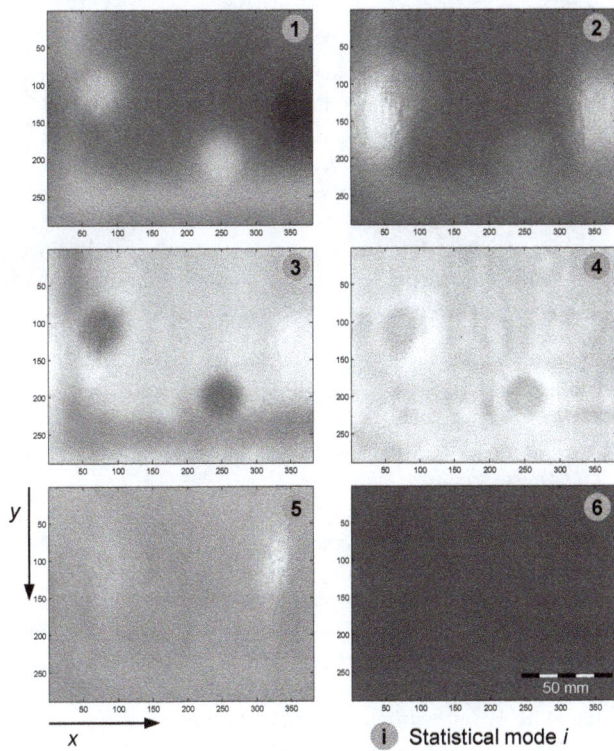

Figure 6. First 6 EOFs of the CFRP specimen.

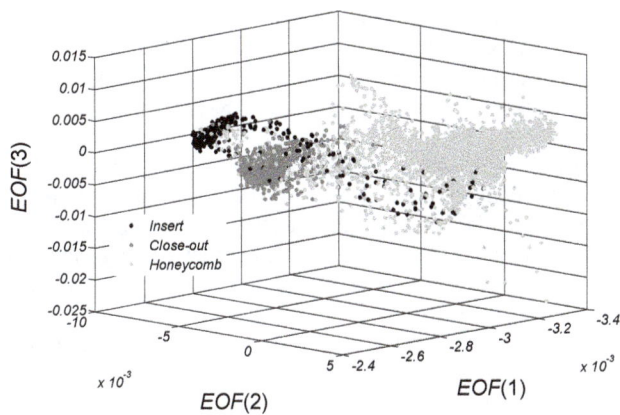

Figure 7. Visualization of first three EOFs of the GFRP-training sample.

the computation effort by a factor of 100, while significantly increasing the computation effort, the classification accuracy does not improve but worsens by 15 % for error type I. Using only the first EOF as feature space for the k-nearest neighbor classification is comparable to applying a histogram-based approach with multiple thresholds. Feature space dimensions between two and five deliver up to 30 % improved results for classification accuracy regardless of the type of error definition and show the advantage of reduced ambiguity. Results from Fig. 3 explain the described effect of falling accuracy when adding statistical modes beyond the fifth one.

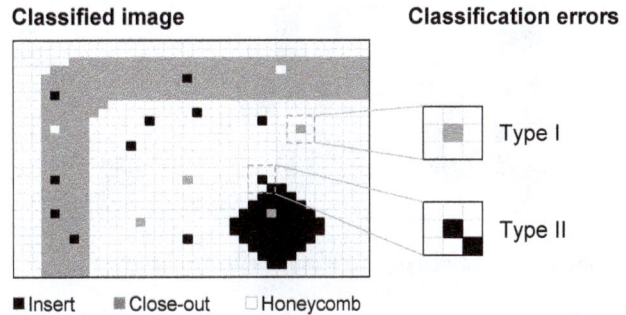

Figure 8. Definition of detectable classification errors by type.

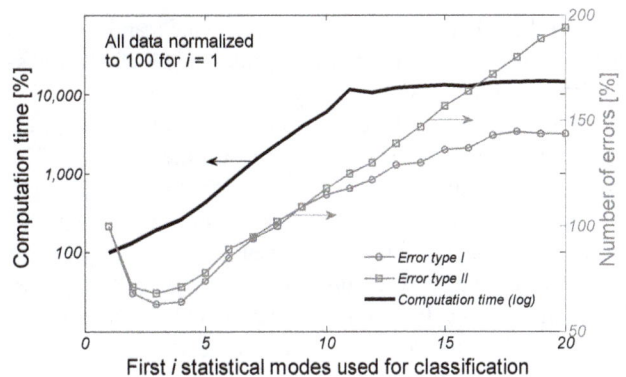

Figure 9. Normalized plot of computation time and number of occurring errors for classification for the GFRP sample based on the first i statistical modes.

Every additional mode contains only a very small amount of additional information useful for classification. It mostly increases noise in the image due to the high rate of data compression.

Figure 10 shows the final results for the classification of the GFRP specimen (a) and the CFRP specimen (b). The GFRP specimen – tested at the setting specified above – is processed using the three first EOFs and based on the three nearest neighbors relationship. Heating and cooling time for the CFRP sample are both increased by 5 s to obtain sufficient contrast. Processing requires including the first five EOFs and using the five nearest neighbors to improve classification results. The color code in the images for the different subsurface structures reflects the classification result. Both tests show that subsurface structures in sandwich components with fully intransparent facing materials are detectable. The images show a specimen of surface of approximately $100 \, \text{cm}^2$, which is tested in less than 30 s including processing time for classification. Decreasing the spatial resolution by pixel in the test setup further improves this ratio.

5 Conclusions

The described experiments demonstrate that PCA on simple square-pulsed thermography in combination with an

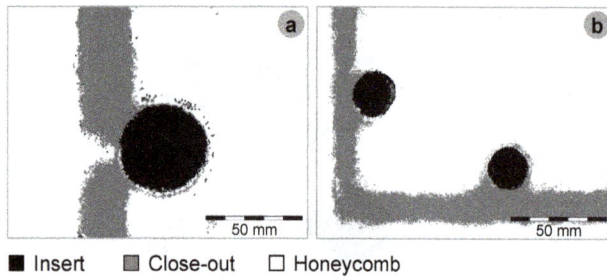

■ Insert　■ Close-out　□ Honeycomb

Figure 10. Classification result for the GFRP specimen (**a**) and the CFRP specimen (**b**).

instance-based classification algorithm detects and separates subsurface structures in sandwich components with intransparent facing material.

All investigations aimed at identifying suitable test settings to ensure fast and reliable results for the detection. The experiments confirm the effect of data reduction via PCA into the first three to five statistical modes from previous investigations and suggest limiting the measurement frequency to avoid noise from oversampling of a slow thermal process. The numerical evaluation of the cumulated variance in the transformed sequences fortifies the result. Principal component thermography on sandwich components is a very robust technique to improve contrast on square-pulsed-tested IR sequences and has a lower sensitivity to inhomogeneous lightning than e.g., simple threshold methods in image processing.

The combination with the k-nearest neighbor algorithm enhances the setup to a method for automated detection and classification of subsurface structures. Three to five statistical modes covering more than 99 % of variance deliver clearly an optimum result with respect to classification accuracy and a relatively low computation effort.

Future investigations focus on a prediction of thermal behavior of sandwich material based on numerical simulations. This will help to improve the current set of training data in the transition between different subsurface structures and will show approaches for automated population of training data.

Acknowledgements. The project on which this paper is based was funded by the German Federal Ministry of Economics Affairs and Energy under funding code 20W1115C. The authors assume all responsibility for the content of this publication.

Edited by: R. Tutsch
Reviewed by: two anonymous referees

References

Bitzer, T.: Honeycomb technology: Materials, design, manufacturing, applications and testing, 1st Edn., Chapman & Hall, London, 1997.

Emery, W. J. and Thomson, R. E.: Data analysis methods in physical oceanography, 2. and rev. ed., 3. impr., Elsevier, Amsterdam, 2004.

Euro-Composites©, Infrastructure and production technologies – Panel Production: http://www.euro-composites.com/en/technology/Seiten/panel.html (last access: 13 March 2013), 2009.

Feuillet, V., Ibos, L., Fois, M., Dumoulin, J., and Candau, Y.: Defect detection and characterization in composite materials using square pulse thermography coupled with singular value decomposition analysis and thermal quadrupole modeling, NDT&E Int., 51, 58–67, 2012.

Ibarra-Castanedo, C., Piau, J.-M., Guilbert, S., Avdelidis, N. P., Genest, M., Bendada, A., and Maldague, X. P. V.: Comparative Study of Active Thermography Techniques for the Nondestructive Evaluation of Honeycomb Structures, Res. Nondestruct. Eval., 20, 1–31, 2009.

Jackson, J. E.: A user's guide to principal components, Wiley series in probability and mathematical statistics, Wiley, 1991.

Karlsson, K. F. and Åström, T. B.: Manufacturing and applications of structural sandwich components, Compos. Part A-Appl. S., 28, 97–111, 1997.

Maldague, X. P. V.: Theory and practice of infrared technology for nondestructive testing, Wiley series in microwave and optical engineering, Wiley, New York, NY, 2001.

Marinetti, S., Grinzato, E., Bison, P. G., Bozzi, E., Chimenti, M., Pieri, G., and Salvetti, O.: Statistical analysis of IR thermographic sequences by PCA, Infrared Phys. Techn., 46, 85–91, 2004.

Mitchell, T. M.: Machine learning, International ed., McGraw-Hill series in computer science, McGraw-Hill, New York, NY, 1997.

Omar, M. A., Parvataneni, R., Zhou, Y.: A combined approach of self-referencing and Principle Component Thermography for transient, steady, and selective heating scenarios, Infrared Phys. Techn., 53, 358–362, 2010.

Rajic, N.: Principal component thermography for flaw contrast enhancement and flaw depth characterisation in composite structures, Composite Structures, 58, 521–528, 2002.

18

Chain of refined perception in self-optimizing assembly of micro-optical systems

S. Haag[1], **D. Zontar**[1], **J. Schleupen**[1], **T. Müller**[1], **and C. Brecher**[2]

[1]Fraunhofer Institute for Production Technology IPT, Aachen, Germany
[2]Chair at the Laboratory for Machine Tools and Production Engineering (WZL) at RWTH Aachen University, Aachen, Germany

Correspondence to: S. Haag (sebastian.haag@ipt.fraunhofer.de)

Abstract. Today, the assembly of laser systems requires a large share of manual operations due to its complexity regarding the optimal alignment of optics. Although the feasibility of automated alignment of laser optics has been shown in research labs, the development effort for the automation of assembly does not meet economic requirements – especially for low-volume laser production. This paper presents a model-based and sensor-integrated assembly execution approach for flexible assembly cells consisting of a macro-positioner covering a large workspace and a compact micromanipulator with camera attached to the positioner. In order to make full use of available models from computer-aided design (CAD) and optical simulation, sensor systems at different levels of accuracy are used for matching perceived information with model data. This approach is named "chain of refined perception", and it allows for automated planning of complex assembly tasks along all major phases of assembly such as collision-free path planning, part feeding, and active and passive alignment. The focus of the paper is put on the in-process image-based metrology and information extraction used for identifying and calibrating local coordinate systems as well as the exploitation of that information for a part feeding process for micro-optics. Results will be presented regarding the processes of automated calibration of the robot camera as well as the local coordinate systems of part feeding area and robot base.

1 Introduction

Optical systems and lasers belong to high-technology sectors with high technical and economic potential in the near future. Especially, the laser industry is regarded to be highly innovative with a leverage effect on other industrial branches. New and improved products are developed and brought to the market frequently. Diode laser systems have a market share of about 50 %[1], and they are characterized by good energy efficiency and small size on the one hand and a relatively large beam divergence angle on the other. The latter requires the challenging assembly of collimation optics. The scope of this paper addresses the assembly of micro-optics and especially the assembly of collimation optics in diode laser systems.

The alignment of micro-optics requires ultra-high precision in up to six degrees of freedom. For meeting the demands of the alignment task, active alignment needs to be applied, which means that relevant beam characteristics are monitored and evaluated during the alignment process of the optics. The observed values are processed cognitively by the operator or by dedicated program logic. Due to its complexity, industrial assembly of high-technology diode laser systems is dominated by manual processes, which determine the majority of overall production costs. The feasibility of automation of such assembly tasks has been proven in several research projects (Brecher, 2012; Haag and Härer, 2012; Loosen et al., 2011; Pierer et al., 2011; Miesner et al., 2009) as well as in a few industrial applications. A breakthrough of automation in this field has not yet been achieved. Mainly, economic reasons are accountable for this situation as in many business cases there are relatively small production

---

[1]LaserFocusWorld, LASER MARKETPLACE 2013, January 2013.

volumes so that automation is not profitable due to a large portion of non-operational times caused by planning, commissioning and frequent changeovers. In recent years, flexible assembly systems for optics assembly have been developed (Brecher, 2012; Haag and Härer, 2012) aiming for shorter non-operational times and hence for higher machine utilization times in scenarios with many product variants. Flexibility has mainly been achieved through modularization of tools and standardization of mechanical interfaces. Distributed multi-agent systems have been implemented in order to provide flexible architectures for assembly execution. Further flexibility can be achieved through interaction of a machine with its environment. This requires sensor integration allowing the perception of the environment (Russell et al., 2010).

Higher flexibility and sensor integration lead to higher complexity, which is a challenge regarding the efficient planning and commissioning of alignment processes for optics. The work presented in this paper is motivated by the current discrepancy between the benefits of flexibility and sensor integration and the increased complexity. Therefore, this paper shows how 2-D bin-picking of micro-optics in part feeding can efficiently be realized and embedded in a model-based control scenario using low-cost hardware.

2   Chain of refined perception

The integration of sensors allows the perception of crucial process data and its use for optimizing individual steps as well as the overall result of the assembly task. For making full use of available models such as geometric model of the product and the assembly cell from computer-aided design (CAD) or the optical setup from ray-tracing simulation, coarse information from the large workspace and high-resolution information from local regions has to be evaluated. In most cases it is inconvenient or even impossible to use high-precision sensors, which cover a large workspace at the same time.

This section introduces the architecture of flexible assembly systems and the principles of self-optimizing optics assembly on which the concept of this work is built. The final part of this section introduces the chain of refined perception, which is applied for the model-based execution of optics assembly.

2.1   Flexible assembly cell concept for micro-optical systems

Flexible assembly systems for micro-optics usually combine a macro-workspace covering a large area by a robot or gantry with a micro-workspace in which a micromanipulator locally carries out sensor-guided high-resolution motion in the sub-micrometer range (Brecher et al., 2012).

In previous research projects, modular micromanipulators with three or six degrees of freedom have been developed

to enable common robotic systems and gantries to carry out micro-optical assembly. Additional modules such as cameras can be integrated in the micromanipulator in order to monitor the grasped part or the grasping area (Brecher et al., 2013). Additionally, such mobile cameras can be used to detect local reference marks for the calibration of spatial relations between local coordinate systems as will be described in the following sections.

Schmitt et al. (2008) describe a multi-agent system for providing the required flexibility regarding the control architecture of a flexible assembly system.

2.2   Self-optimizing assembly of laser optics

One focus of the research in the Cluster of Excellence "Integrated Production Technology for High-wage Countries" at RWTH Aachen University[2] is put on self-optimizing assembly systems, which aim for the reduction of planning efforts for complex and sensor-based assembly tasks (Brecher, 2012). Self-optimizing assembly of laser optics is applied for the production of high-quality laser modules coping with finite positioning accuracy of the actuation system, noisy perception, and tolerances of laser beam sources and optics. Therefore, model-based approaches for assembly execution under the presence of uncertainties are investigated.

Conceptually, self-optimizing systems follow a three-step cycle. Firstly, the current situation is analyzed considering the objective of the task, the current state of the assembly system including the product as well as a knowledge base holding additional information provided prior to assembly or collected during assembly execution. Secondly, internal objectives such as reference values for internal closed-loop controls are adapted based on reasoning on the analysis carried out in the previous step. This step goes beyond the classical definition of closed-loop controls and adaptive closed-loop controls. In a third step, self-optimizing systems adapt their behavior either through parameter adaption or through structural changes.

Hence, key aspects of self-optimizing assembly systems are model-based control and sensor integration. Model-based control allows automatisms during the planning phase and therefore drastically reduced planning times. Yet the approach requires the use of sensors in order to identify and compensate differences between ideal models and real-world situations.

Figure 1 shows the reduced ontology of a model-based self-optimizing assembly system. Different types of models such as product models (e.g., geometry, optical function), production system models (e.g., kinematic chains) as well as process knowledge and system objectives provide information for the cognition unit to select and configure algorithms and program logic. For example, the product model might provide a certain geometrical or optical constraint to

---

[2]See http://www.production-research.de.

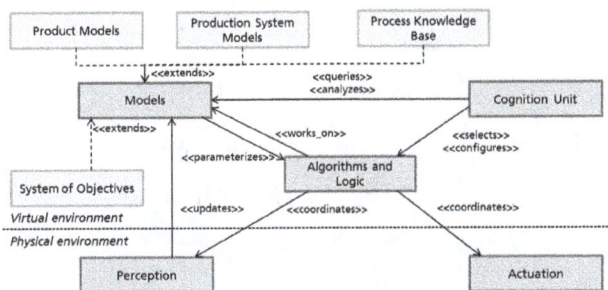

Figure 1. Reduced ontology of a self-optimizing assembly system.

3D-measurement of assembly cell (volume in the range of 1 m³) using structured light sensor (accuracy in the range of 10 mm) and CAD object matching (accuracy in the range of 5 mm)

Detection and localization of reference marks and parts in an area in the range of 50 x 50 mm² through stationary and mobile cameras (accuracy in the range of 50 μm)

Passive alignment of geometric features in an area in the range of 5 x 5 mm² using stationary cameras (accuracy in the range 5 μm)

Active alignment according to optical function through observing laser beam profile (accuracy in the range of 100 nm)

Figure 2. Chain of refined perception for self-optimizing assembly of micro-optical systems indicating roughly the volume or area covered by measurements as well as the measurement accuracy achieved.

be fulfilled by an assembly step. The cognition unit selects a certain type of mounting sequence consisting of a standardized sequence of steps (mounting template) such as part pickup, dosing of adhesives, active alignment, etc. In the context of this paper the part pickup is of special interest. The rough coordinates of the optical element can be retrieved from the geometrical model of the production system. The small size and the presence of uncertainties require the localization of the part with a precision sufficient for part pickup. This paper presents a sensor-guided approach for localizing micro-optical parts realized on a low-cost robot-based assembly station.

2.3 Chaining of process steps in micro-optical assembly

In order to overcome the gap between ideal models and uncertain reality, crucial assembly steps are implemented based on sensor guidance. Individual tasks during the assembly of micro-optics such as the pickup of parts or alignment of optics require different levels of accuracy ranging roughly from 10 mm measuring accuracy achieved by low-cost structured light sensors down to 100 nm positioning accuracy achieved through active alignment. Figure 2 shows the concept of a chain of refined perception as propagated by the work presented. For carrying out process steps, this concept uses several means of perception at different levels of granularity. The objective of one process step is to transform the assembly state to the tolerance level of the subsequent process step. The approach enables the advantages of planning complete assembly tasks based on models with the flexibility and precision of sensor-integrated systems. Figure 2 shows the chain of refined perception for the case of micro-optical assembly. In the top level it covers a large workspace in the range of one or more cubic meters for autonomously planning collision-free paths of the macro-positioner. In the bottom level, a motion resolution for optical alignment in the range of 10 nm is possible.

For the task of collision-free path planning, software tools such as MoveIt! as part of the ROS package[3] have been developed in the robotics community. The work related to this paper applies such software in combination with structured

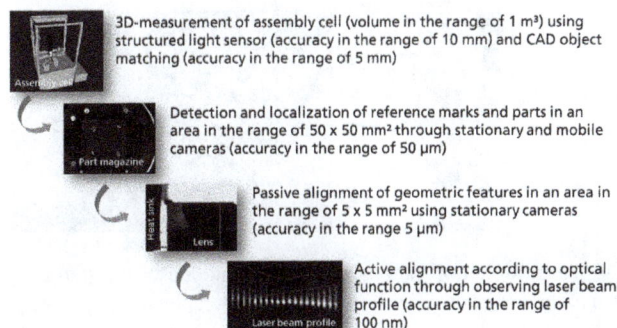

light sensors such as Microsoft Kinect. The environment can be scanned in 3-D with such sensors. Additional point cloud processing software tools[4] allow the matching of CAD models with the detected point clouds. The result is a collision model that allows the planning of collision-free paths.

For tasks such as 2-D bin-picking of micro-optical components, local coordinate systems need to be calibrated with reference to each other. Figure 3 shows a typical setup with a fixed camera and a mobile camera (the mobile camera and the micromanipulator it is attached to are carried by a positioning system such as a robot or a gantry). The fixed camera and its objective cover a large area such as a part carrier. The mobile camera covers a much smaller area intended for detecting local reference marks. The detection of defined reference marks allows the calibration of the cameras and their spatial relation.

Passive alignment is a step usually required prior to active alignment, and it is based on the detection of reference marks or geometric features using charge-coupled device (CCD) chips. During passive alignment parts are pre-positioned with reference to each other so that the initial starting point for active alignment is within a certain tolerance with high probability.

In the context of optics assembly, the task of active alignment accounts for the quality of the optical system. In the case of collimation optics, an optical measurement setup and a CCD chip are used for determination of the current state of alignment. Alignment algorithms have been and currently are subject to recent research activities (Brecher, 2012; Haag and Härer, 2012; Pierer et al., 2011; Miesner et al., 2009).

3 Calibration of stationary and mobile camera

As depicted above, two camera systems are used in the setup for the calibration of the positioner coordinate system with reference to the local coordinate system defined by local

---

[3]See http://moveit.ros.org/wiki/MoveIt!.

[4]See http://pointclouds.org/.

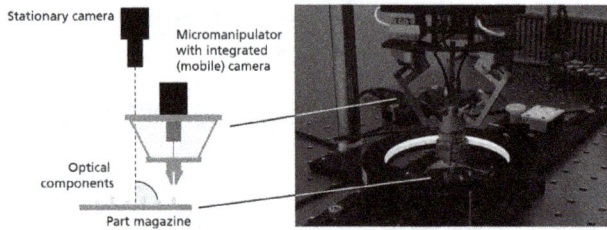

Figure 3. Model of the position and orientation of the stationary and mobile camera systems (left) and photography of the part magazine setup including dark field illumination using a ring light and the micromanipulator (right).

reference marks. The first camera is fixated perpendicular above the part carrier. The second is mounted to the micromanipulator, which is attached to the macro-positioner (cf. Fig. 3). In order to use image data as input for further calculations, both cameras have to be calibrated first. The calibration process allows compensating optical and perspective distortion and determining a scaling factor between image pixel and real-world metrics.

3.1   Calibration of stationary camera

The stationary camera system is equipped with a common entocentric optic and is appointed to monitor parts on the Gel-Pak magazine (part carrier). The predominant kinds of distortion consist of a radial barrel distortion, which is generally associated with the deployed kind of lens, and a trapezoidal distortion resulting from a misalignment of the camera with respect to its optimal perpendicular orientation. The scaling factor is calculated for the surface plane of the Gel-Pak because it depends on the object's distance to the camera. Determining and compensating camera distortion is a common task in computer vision. Hence, algorithms are widely available as frameworks in many programming languages.

The scaling factor can be obtained during the determination of the distortion or the local coordinate system simply by comparing known physical features like the calibration pattern or the distance between two reference marks with their representation in the image.

The calibration process has to be carried out when the position or orientation of the stationary camera changes.

3.2   Calibration of mobile camera

The mobile camera system is equipped with a telecentric lens to provide local image data of components during assembly tasks. Due to specific properties of telecentric lenses, there is no need to compensate any imaging deformations caused by perspective. Therefore, the calibration process only includes the identification of the relationship between the camera's local coordinate system and the robot's tool center point (TCP) (cf. Fig. 4).

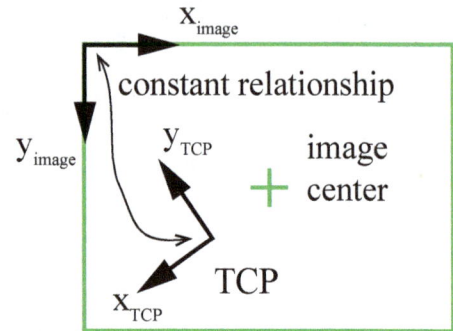

Figure 4. Relationship between the image and TCP coordinate system.

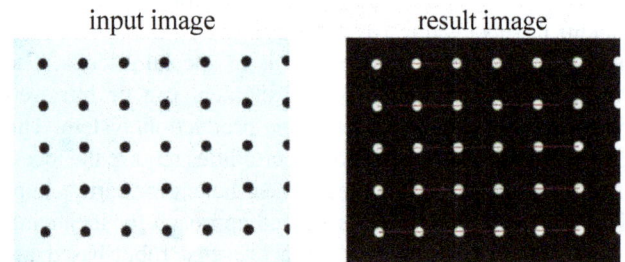

Figure 5. The scaling factor is obtained by comparing the detected pattern with known physical features (center point distances).

The telecentric lens is mounted approximately at the center of the robot's TCP while the fixed focus plane is tuned to be aligned with an attached gripper. For a mathematical coordinate transformation, four parameters have to be identified: the x and y offset of the image center from the z axis of the TCP, the camera orientation described by the angle between both x axes and finally the scaling factor to transform pixels into millimeters.

To obtain the scaling factor, the camera is positioned above a calibration pattern (dot target) with known physical features. The distance between two points in the image is then compared to its physical equivalent (cf. Fig. 5).

The angular offset is determined in a two-step approach. First, the robot camera is positioned above a reference mark. The camera image is then analyzed to determine its center point, which is stored along with the current robot coordinates. For the second step the robot is moved in a plane so that the reference mark stays in the image region, which will be analyzed again. The orientation can then be calculated by comparing the vector described by the movement of the robot with the movement of the reference mark in the image (cf. Fig. 6).

Due to the fact that the camera is positioned parallel to the z axis of the TCP, its x and y offset can be determined by stepwise rotation of the TCP and by analyzing the path of a reference mark in the image. The path is expected to describe

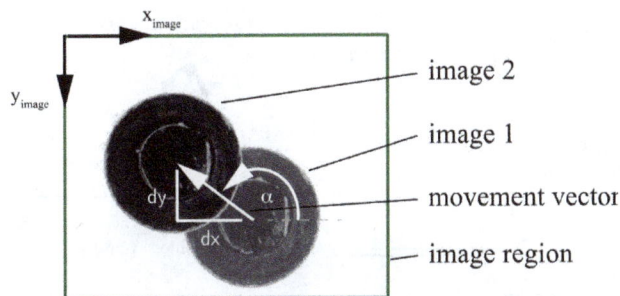

Figure 6. Images before and after a movement have been overlaid. The camera orientation is calculated by comparing the robot's movement and the vector described by the reference mark.

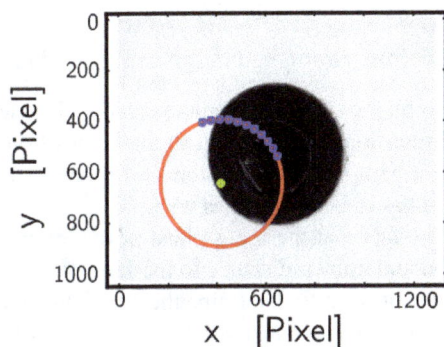

Figure 7. A reference mark describes a circle while the camera system is rotated. The pivot point identifies the z axes of the TCP.

a circle that can be fitted to measured points. Its center point depicts the origin of the x and y axes of the TCP (cf. Fig. 7).

3.3 Calibration of local coordinate systems

In order to calculate the position and orientation of components in robot coordinates, a local coordinate system has to be defined first. Therefore two reference marks have been placed alongside the Gel-Pak to identify the origin and the direction of the y axes. The z axis is defined to be perpendicular to the surface pointing upwards. Finally, the x axis is positioned to complete a right-handed coordinate system.

The reference marks must be positioned in the image area of the stationary camera, in order to be identified and used to describe parts on the Gel-Pak in a local coordinate system. However, the distance of both reference marks should be maximized in order to minimize any error on the measured y axis orientation.

The local coordinate system can be automatically measured in positioner coordinates by moving the positioner with its mobile camera above each reference mark. Image data can then be analyzed to detect the center point of each reference mark. Based on the calibration of the mobile camera, coordinates can further be transformed into TCP coordinates and finally into positioner base coordinates.

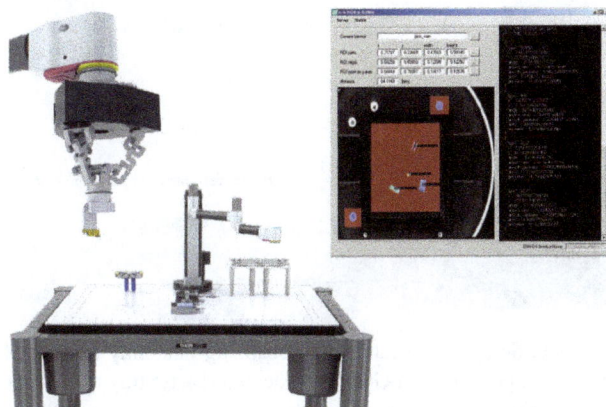

Figure 8. Hybrid assembly setup consisting of a SCARA kinematic as macro-positioner and a micromanipulator for fine alignment. The software window shows the result of image processing (detection of reference marks and localization of optics of different types).

4   2-D Bin-picking of micro-optical components

Part identification has been implemented as a stand-alone application. Through standard networking APIs, process control scripts can retrieve detailed information about every detected part on the magazine (cf. Fig. 8). OpenCV has been used for image processing. See Laganière (2011) and Bradski (2000) for reference.

Gel-Paks® provide a convenient way to handle optical components during transport. Due to a proprietary elastomeric material, parts can be placed freely on the carrier and kept in position to ensure safe transportation and storage. Hence, this kind of magazine is a standard way of presenting optical components. Pickup positions can no longer be statically defined and therefore have to be identified through sensor evaluation. In the setup presented in this paper, a camera fixated above the Gel-Pak® covers the complete lens presentation area as well as a set of reference marks in its field of view (cf. Fig. 9). Applying image processing, optics can be located in the local 2-D coordinate system. In order to carry out the robot-based pickup, the local coordinate system needs to be calibrated with respect to the positioner's base coordinate system as explained above.

4.1   2-D localization of micro-optical components

For an automated pickup process, optical components on the Gel-Pak have to be localized and identified. The localization step determines the x and y position and orientation of all parts in a local coordinate system. In a following step parts are distinguished and grouped by their type. This is accomplished by comparing visual features, which in combination allow a reliable identification of the investigated optical components. These features include without limitation the length and width, the visible area and its perimeter, as well as different ratios of these parameters. The grayscale histogram is

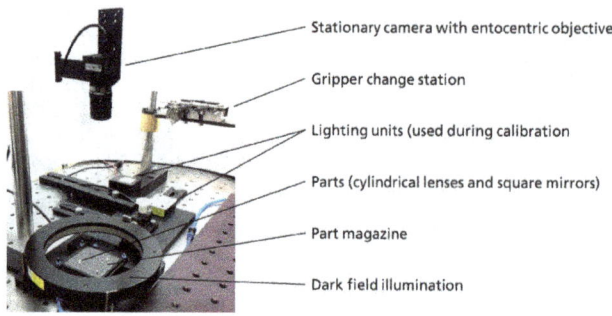

Figure 9. Setup for detecting and localizing randomly positioned optical components on a Gel-Pak® vacuum release tray using dark field illumination (ring light).

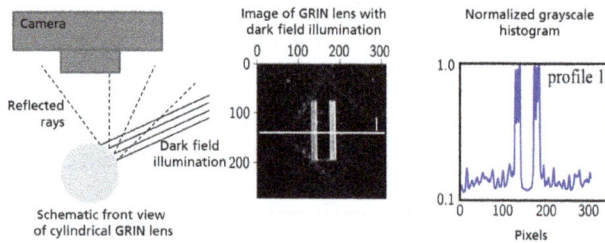

Figure 10. The center image shows the enhanced contrast achieved through dark field lighting. The right diagram presents a corresponding normalized intensity profile. The left drawing explains the separation of blobs using dark field illumination on a cylindrical lens (only one direction of illumination is illustrated): only the rays hitting a specific region of the cylindrical surface will be reflected into the camera. This phenomenon occurs on both sides of the GRIN lens so that there are two separated blobs in image processing.

also suitable to distinguish and group parts. Formed groups can finally be mapped to templates, which have to be configured only once for every new component type.

Part descriptions based on salient points are not suitable because of small and mostly homogenous surfaces, which do not offer many features.

The image segmentation is based on binary thresholding with a watershed algorithm to achieve accurate edges. In order to enhance the contrast between the Gel-Pak and mostly transparent optical components, dark field lighting has been introduced in the experimental setup. Occasionally, parts such as cylindrical GRIN optics lead to separated blobs, which have to be combined in a postprocessing step. This has been implemented as a heuristic rule that combines closely lying blobs (cf. the two deflections in the plot of Fig. 10). The separation of the blobs is caused by the dark field illumination and the reflection on the cylindrical surface of the GRIN lens.

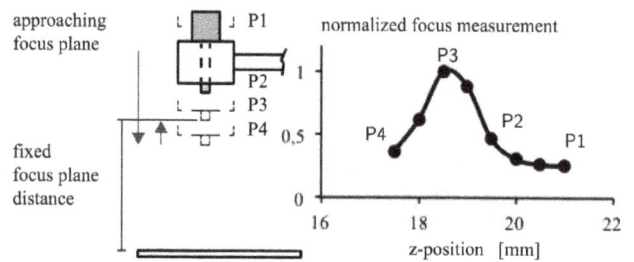

Figure 11. The left model illustrates the proceeding of a focus measurement. On the right side normalized focus measurements are plotted against the z coordinate of the robot. Measurements start above the focus plane, so the measurement points were taken from right to left.

4.2   Height measurement through variation of focus

Information on the stationary camera can only be used to obtain two-dimensional information about the position and orientation of a part. For a fully automated process, the height information has to be detected as well.

Due to the fact that the focus plane of the mobile camera is in a fixed and known distance to the lens, focus measurements can be utilized to determine the z coordinate of an investigated surface in comparison to an autofocus feature of a camera. Therefore, the positioner is moved in small steps towards the surface of the Gel-Pak. At each step the camera image is analyzed. In Nayar and Nakagawa (1990) and Firestone et al. (1991), different algorithms are presented to quantify the focus quality. The presented results are based on the Laplace operator (sum of second partial derivatives). The focus of an image correlates with the smoothness of edges in the image, which can be extracted with a Laplace filter. To weaken the effect of noise, the Laplacian of Gaussian filter is applied to the investigated image region. The focus is then quantified by the weighted average of the obtained pixel intensity. In Fig. 11 normalized focus measurements are plotted against the z coordinate of the robot. The focus plane is determined by the absolute maximum, which can be numerically calculated.

Autofocus algorithms must have a reliable and early abort criterion because the focus plane is tuned to be aligned with an attached gripping tool. The assigned micromanipulator allows pulling up the gripper 2 mm, which is generally enough vertical space for an automated positioning of the positioner. During the course of this work, no robust autofocus could be implemented. Only well-structured surfaces have led to acceptable results. Therefore, different approaches such as stereovision might be used in the future although this might increase the costs of the assembly solution.

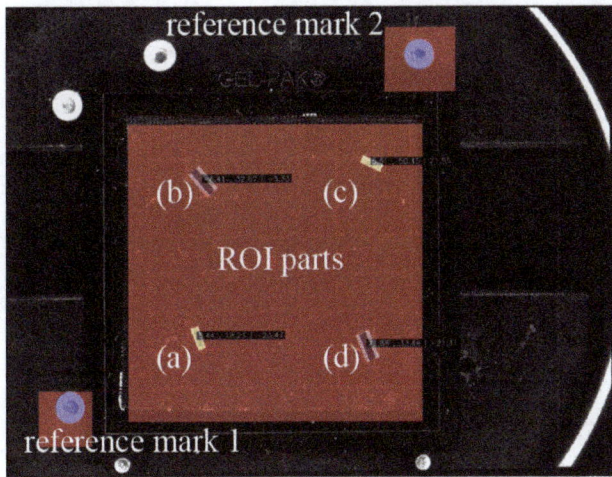

Figure 12. Image data of the stationary camera are shown. Three regions (ROI, red) are used to isolate details of interest. Identified parts **(a–d)** are grouped and colored by their type (GRIN, HR).

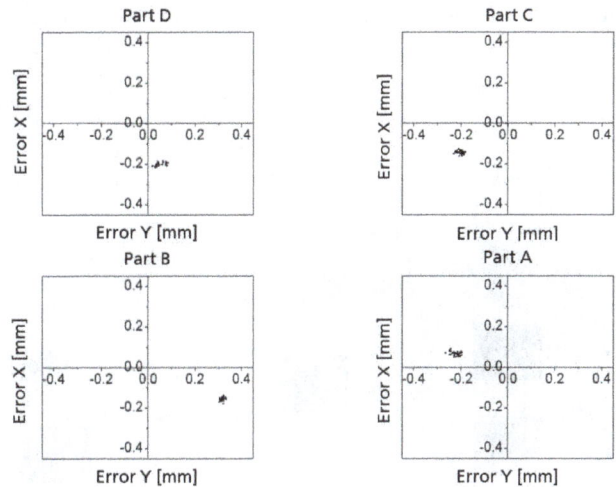

Figure 13. The identified parts of Fig. 12 have each been approached by the robot in a way that the image center of the mobile camera is overlaid with the center point of each part.

5 Evaluation of results

In the following, results regarding the camera calibration as well as the part localization processes will be presented. In the case of camera calibration, two measurements of 40 repetitions have been carried out. Between measurements the mobile camera has been unmounted from the mechanical interface and remounted again. Results are summarized in Table 1.

At a 6σ level, the scaling factor error accounts for an absolute error of less than 0.02 %. The 6σ level for the orientation was identified at 0.2112 degrees, which is a sufficient value for micro-optical part pickup. The 6σ level of the X–Y offset is below an error of 20 μm. According to these first results, calibration is sufficiently precise for the task of micro-optical part pickup. For more reliable results, more changeover scenarios need to be carried out. The quality of calibration is strongly determined by the repeatability of the positioning system. Calibration of the mobile camera should be carried out after each camera changeover.

Using a single calibration configuration of the stationary camera, part localization has been carried out. The localization of an individual GRIN lens for 40 times has led to 6σ levels of 1.5 μm, 2.2 μm for the X–Y offset.

Detected parts are grouped and colored for convenience as shown in Fig. 12. After calibrating the local coordinate system, the positioner is moved above each detected component. The image from the mobile camera is then processed in order to evaluate the achieved precision. The component center is therefore detected in analogy to the algorithm of the stationary camera and then compared to the image center. Exemplary results are given in Fig. 13. Positioning errors in X and Y direction are obvious. The results show that part of the error seems to be systematic depending on the corner of the

part magazine that was approached by the robot. All of the error offsets are within a range of 70 μm (most of them even within a range of 30 μm). One explanation for this behavior is that the robot used was in prototype stadium during the work and that kinematic transformations on the robot controller do not precisely correspond to the actual kinematic structure. The repeatability of the robot is sufficient for the pickup process.

The implemented image-based part localization and identification allows for a reliable pickup process for the investigated optical components. Currently, height information for each component type is provided by the operator or an underlying geometrical model. This ensures collision avoidance since the presented measurement via focus determination needs further investigation. The work has shown that a dependency exists between the surface structure of the optical component and the quality of the height measurement. For well-structured surfaces, the Laplacian approach leads to acceptable results. Also, the Tenengrad algorithm as mentioned in Nayar and Nakagawa (1991) performed well. Tenengrad is based on two Sobel operators calculating gradients in horizontal and vertical directions. A 6σ level of 28.2 μm (Laplace) and 18 μm (Tenengrad) has been achieved in individual cases. Figure 14 shows a single measurement run of the autofocus algorithm for a well-structured surface. For transparent parts or parts with large homogenous surfaces showing no structures, no reliable results have been achieved yet. Table 2 presents the results for a non-transparent and well-structured heating element. Smaller step sizes (e.g., of 0.1 mm) led to worse results for both algorithms.

Table 1. Results of camera calibration.

Calibration measure	Unit	max(6σ)
Scaling factor	[pixels mm^{-1}]	$6 \times 5.92 \times 10^{-3} = 35.52 \times 10^{-3}$
Orientation	[degrees]	$6 \times 0.0352 = 0.2112$
X–Y offset (x, y)	[pixel], [pixel]	$6 \times 0.553 = 3.318, 6 \times 0.601 = 3.606$

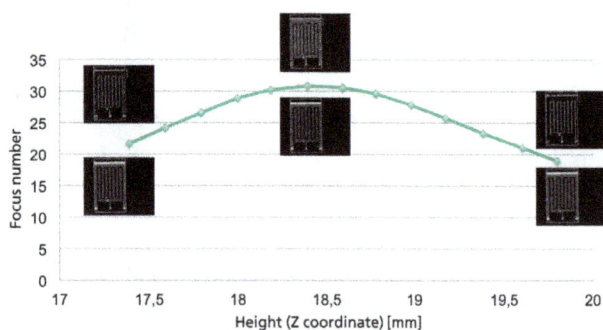

Figure 14. The plot shows a single measurement run determining the focus number calculated by the Tenengrad algorithm at a 0.2 mm step size. For statistical analyses, the z coordinate of the surface has been determined 40 times by an autofocus algorithm.

6 Summary and outlook

The paper presented a concept of chaining process steps where each step transforms the assembly state to the next more granular level and named it "chain of refined perception". This concept was motivated and conceptually embedded in the context of self-optimizing micro-optical assembly systems. Such systems strongly utilize model-based control architectures, which need continuous matching with measurement data. Model-based control is an enabler for automated planning and optimization algorithms such as path planning. Sensor integration is still required to meet the precision requirements.

In more detail, techniques necessary for a computer-vision-based feeding of optical components have been presented and evaluated. Implemented in manual laboratory processes, this allows for a convenient way to support operators. In an automated and self-optimizing scenario, it completes the chain of refined perception. A reliable calibration routine has been presented for identifying camera parameters such as perspective distortion and for determining the scaling factor between image pixels and real-world metrics. Another routine was depicted for calibrating a local coordinate system with respect to the positioner base coordinates. Such calibration allows for picking up randomly aligned optical components. This approach was enhanced by a strategy for identifying the z coordinate of a plane through a sequence of images collected by the mobile camera attached to the tool center point. Results of this work were presented by depict-

Table 2. Results of focus measurements.

Focus strategy	6σ level
Tenengrad (0.5 mm steps)	29.4 µm
Tenengrad (0.2 mm steps)	18.0 µm
Laplace (0.5 mm steps)	48.0 µm
Laplace (0.2 mm steps)	28.2 µm

ing the achieved positions in comparison with the ideal target positions.

The "chain and refined perception" will be established as an approach in further research activities focusing on the efficient planning and commissioning of flexible micro-optical assembly systems. Future work aims for a product-centric approach by establishing a formalized product description similar to the descriptions in Whitney (2004) for mechanical assemblies and by deriving the assembly execution logic automatically leading to drastically reduced planning and commissioning efforts.

Acknowledgements. Research in the Cluster of Excellence for "Integrative Production Technology for High-wage Countries" at RWTH Aachen University is funded by the German Research Foundation (DFG).

Edited by: R. Tutsch
Reviewed by: two anonymous referees

References

Bradski, G.: The OpenCV Library, Dr. Dobb's Journal of Software Tools, 2000.

Brecher, C. (Ed.): Integrative production technology for high-wage countries, Berlin, New York, Springer, 2012.

Brecher, C., Pyschny, N., Haag, S., and Lule, V.: Micromanipulators for a flexible automated assembly of micro optics, in SPIE Photonics Europe, International Society of Optics and Photonics, 2012.

Brecher, C., Pyschny, N., and Bastuck, T.: Design and Optimization of Flexure-Based Micro- manipulator for Optics Alignment, in: Proceedings of the 13th International Conference of the European Society for Precision Engineering and Nanotechnology, edited by: Leach, R., 27–31 May, 268–271, 2013.

Firestone, L., Cook, K., Culp, K., Talsania, N., and Preston Jr., K.: Comparison of autofocus methods for automated microscopy, Cytometry, 12, 195–206, 1991.

Haag, M. and Härer, S.: SCALAB. Scalable Automation for Emerging Lab Production, Final report of the MNT-ERA.net research project, 2012.

Laganière, R.: OpenCV 2 computer vision application programming cookbook, Packt. Publ. Limited, 2011.

Loosen, P., Schmitt, R., Brecher, C., Müller, R., Funck, M., Gatej, A., Morasch, V., Pavim, A., and Pyschny, N.: Self-optimizing assembly of laser systems, Prod. Engineer., 5, 443–451, 2011.

Miesner, J., Timmermann, A., Meinschien, J., Neumann, B., Wright, S., Tekin, T., Schröder, H., Westphalen, T., and Frischkorn, F.: Automated Assembly of fast-axis collimation (FAC) lenses for diode laser bar modules. High-power diode laser technology and applications VII, edited by: Zediker, M. S., San Jose, CA, 24 January 2009: SPIE (v7198), 71980G-71980G-11, 2009.

Nayar, S. and Nakagawa, Y.: Shape from focus: An effective approach for rough surfaces, in Robotics and Automation Proceedings, IEEE International Conference, 218–225, 1990.

Pierer, J., Lützelschwab, M., Grossmann, S., Spinola Durante, G., Bosshard, Ch., Valk, B., Brunner, R., Bättig, R., Lichtenstein, N., Zediker, and Mark S.: Automated assembly processes of high power single emitter diode lasers for 100W in 105 µm/NA 0.15 fiber module. High-power diode laser technology and applications IX, edited by: Zediker, M. S., 23–25 January 2011, San Francisco, California, United States, Bellingham, Wash: SPIE (v. 7918), 79180I-79180I-8, 2011.

Russell, S. J., Norvig, P., and Davis, E.: Artificial intelligence. A modern approach, 3rd Edn., Upper Saddle River, NJ, Prentice Hall, 2010.

Schmitt, R., Pavim, A., Brecher, C., Pyschny, N., Loosen, P., Funck, M., Dolkemeyer, J., and Morasch, V.: Flexibel automatisierte Montage von Festkörperlasern. Auf dem Weg zur flexiblen Montage mittels kooperierender Roboter und Sensorfusion, wt Werkstatttechnik online, Vol. 98, No. 11/12, 955–960, 2008.

Whitney, D. E.: Mechanical assemblies. Their design, manufacture, and role in product development, New York: Oxford University Press, 2004.

19

Looking at the future of manufacturing metrology: roadmap document of the German VDI/VDE Society for Measurement and Automatic Control

J. Berthold[1] **and D. Imkamp**[2]

[1]VDI/VDE GMA within VDI e.V., Dusseldorf, Germany
[2]Carl Zeiss Industrielle Messtechnik GmbH, Oberkochen, Germany

Correspondence to: J. Berthold (berthold@vdi.de)

Abstract. "Faster, safer, more accurately and more flexibly" is the title of the "manufacturing metrology roadmap" issued by the VDI/VDE Society for Measurement and Automatic Control (www.vdi.de/gma). The document presents a view of the development of metrology for industrial production over the next ten years and was drawn up by a German group of experts from research and industry. The following paper summarizes the content of the roadmap and explains the individual concepts of "Faster, safer, more accurately and more flexibly" with the aid of examples.

1 Metrology and production

Under the impact of global megatrends, manufacturing technology is faced with a number of different challenges. The topics of resource efficiency, of mastering new process technologies, of increasing flexibility and of transparency have a special significance in production today (Fig. 1).

At the same time the trend towards higher product quality continues uninterrupted. Globalization has made it possible for production to be linked on a worldwide basis, in which the exchange of information is becoming more and more important in securing quality. Much of this information, particularly that concerned with the state of the products and production processes, is obtained with the aid of metrology.

Against a background of discussions about the supply of energy and shortages in raw materials, the subject of resource efficiency plays an important role today in production as well. New methods in manufacturing can make a contribution to improving resource efficiency. This also requires the use of measurement technology since only this can supply the information needed for evaluating efficiency.

Manufacturing technology continues to be faced with challenges arising from customer desire for individually designed products and from fluctuating demand, which it counters with a greater degree of flexibility. This is accompanied by a considerable reduction in batch sizes, which can often only be managed by a more intensive use of metrology since lengthy production start-ups and pilot production runs can hardly be afforded any longer. At the same time more and more sectors of industry (such as aviation, medical products) are calling for a seamless documentation of the conformity assessment of all manufactured products, which is also impossible without a more intensive use of metrology.

The term "production metrology" is a natural one for metrology within the context of production, but this metrology is nevertheless also referred to as "manufacturing metrology" in German (Pfeifer and Schmitt, 2010; Dutschke and Keferstein, 2007). Figure 2 provides an overview of the more important fields of application for manufacturing metrology. These are also examined in the technical committees of the "manufacturing metrology" department of the VDI/VDE Society for Measurement and Automatic Control (GMA, 2012) where a working group has been formed which, in light of the aforementioned trends in manufacturing technology, has assigned itself the task of forecasting the future of manufacturing metrology (Fig. 3).

The results of this work have been published by the Verein Deutscher Ingenieure e.V. (VDI) under the title of "Manufacturing metrology 2020: a technology roadmap

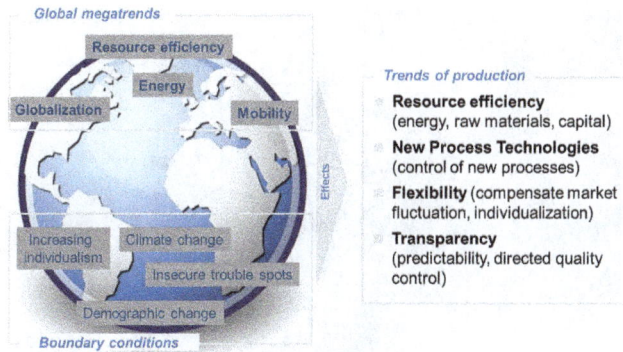

Figure 1. Global trends and trends in manufacturing technology (image source: WZL, RWTH Aachen).

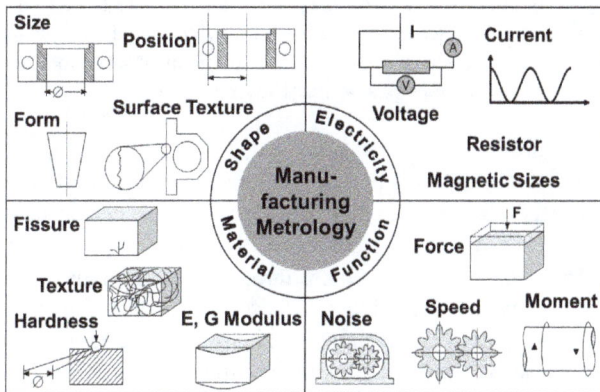

Figure 2. Fields of application for manufacturing metrology (derived from Pfeifer and Schmitt, 2010).

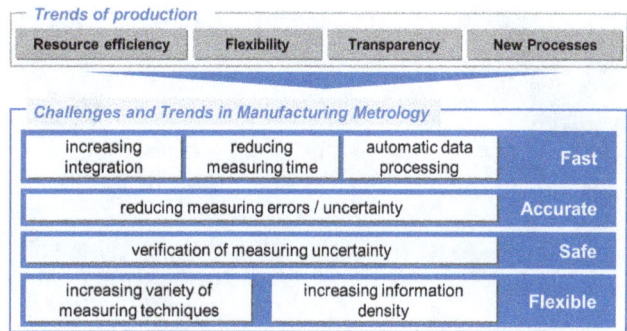

Figure 3. Production trends and their impact on challenges and trends in manufacturing metrology.

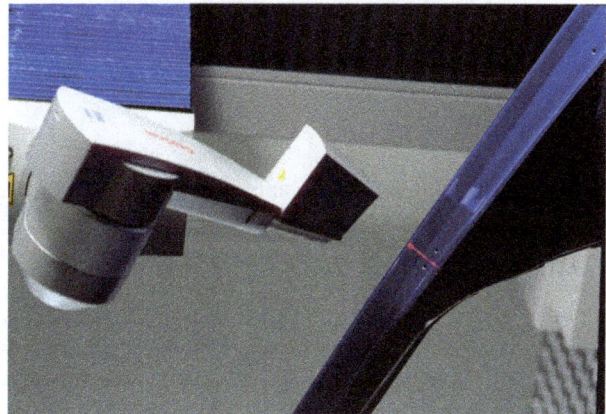

Figure 4. Faster metrology by optical capture of gap dimensions during inspection of automotive bodywork with triangulation sensor.

for metrology in industrial production" (VDI/VDE, 2011). Summaries have been presented nationally (Imkamp and Berthold, 2009, 2011; Schmitt and Imkamp, 2011) and internationally (Schmitt et al., 2011; Grzesiak and Imkamp, 2012). This present paper points out the main results of the work.

The challenges and trends in manufacturing metrology can be described with the terms "faster", "safer", "more accurately" and "more flexibly". The topics of accuracy and speed are in particular of central importance, as can also be gathered from other studies of metrology, such as, for example, the market study on 3-D metrology prepared by the Fraunhofer Society (Fraunhofer-Allianz, 2010) and the technology roadmap for process sensors in the chemical and pharmaceutical industry (VDI/VDE/NAMUR, 2009).

2 Faster

On the one hand, speed means the development and application of metrological procedures by which information about product quality can be obtained in a shorter time. Here it is less a matter of developing procedures basically from scratch

than of adapting a large number of known measurement principles for utilization in production. Optical methods play a significant part here (Leibinger and Tünnermann, 2012) (Fig. 4). On the other hand a tighter integration of metrology into production processes especially by means of automation will contribute to getting measurement results faster and using them more efficiently (Imkamp and Frankenfeld, 2009). In this way the times required for transportation to the measuring equipment can be reduced or even cut completely (Fig. 5). Furthermore, the information from measurements is directly available in production, thereby allowing the incorporation of control loops, for example. Regulation by means of an automated transmission of data can be implemented with a particularly high level of efficiency (Heizmann et al., 2009; Pfeifer and Imkamp, 2004).

3 More accurately

Demands relating to the accuracy of measurement technology are also increasing in conjunction with stricter quality requirements. This change affects procedures not only in

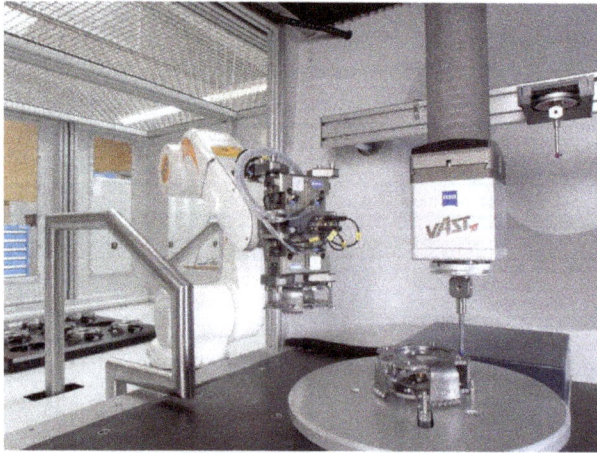

Figure 5. Faster metrology due to the automated integration of metrology into material flow with the aid of a robot.

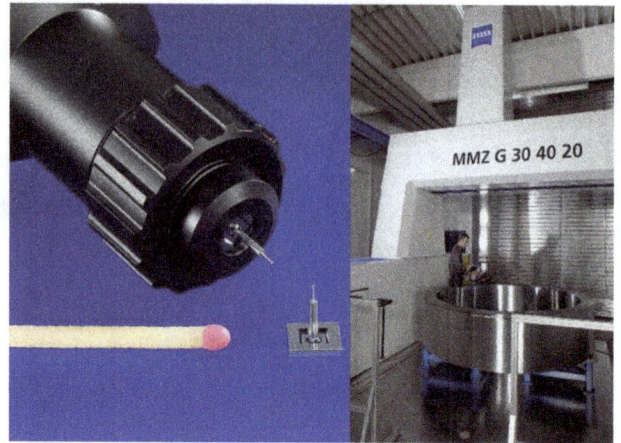

Figure 6. Accuracy in coordinate metrology in micro-scale, e.g. probe of a micro-scale part measuring device (Wiedmann et al., 2011), and macro-scale parts, e.g. measurement of large mechanical parts for wind energy systems (DeGlee, 2010).

macrometrology (Schmitt et al., 2009) but also the micro- and nanometrology used for capturing the product shape (Bosse et al., 2009) (Fig. 6).

In macrometrology, as tolerances become tighter, e.g. for drives in wind power systems (DeGlee, 2010), a greater accuracy of the measuring instruments is required. In this context it is worth noting that, in response to the requirements of industrial quality inspection regarding, for example, traceability, techniques from geodesy are being used more and more frequently in manufacturing metrology (Hennes, 2007). Furthermore progress in optical technology and fast, low-cost computation leads to wide-spread application of laser trackers and digital photogrammetry for coordinate metrology (Estler et al., 2002). In micrometrology higher levels of accuracy are required on account of increasing miniaturization (Porath and Seitz, 2005; Wiedmann et al., 2011). Figure 7 shows the order of magnitude of these trends.

Demands for greater accuracy are also to be found in the measurement of material properties (Frenz and Schenuit, 2009) and electrical characteristics (Naß and Berthold, 2010). In addition to optimization of the procedures themselves, the monitoring and correction of environmental influences is becoming more important in this context.

4   Safer

Determination of measurement uncertainty and taking it into consideration in the conformity assessment are becoming increasingly important. Standardized procedures for determining measurement uncertainty will become more established and will be applied at different levels of detail depending on the task in question. More effort in determining uncertainty will need to be justified for the calibration of standards than in the inspection of straightforward product characteristics. As regards production, simplified procedures will be-

come established. It is precisely with safety-related products such as, for example, in the aviation industry and in medical technology that an evidential document regarding the determination of measurement uncertainty and its inclusion in the inspection decision will become standard and product safety will improve (Imkamp and Sommer, 2009). In addition, the computer-aided simulation of measurement processes on the basis of the Monte-Carlo method (JGCM 101, 2008) for determining measurement uncertainty will become more important. In the meantime implementations have become available for different measurement methods, in most cases in the form of prototypes (Schwenke, 1999; Bai et al., 2002; Hiller, 2011; Schmitt et al., 2008). In the field of coordinate metrology, systems are also already on the market (Fig. 8) (Wäldele and Schwenke, 2002) which are used in particular in the calibration of individual standards, and normative publications are now also available (ISO/TS 15530-4, 2008; VDI/VDE 2617-7, 2006).

5   More flexibly

The wide variety of measurement methods used in production is increasing and with it the flexibility of metrology. On the one hand, techniques are used which holistically register the shape of a product. These include fringe projection and photogrammetry (Bauer, 2003). With computer tomography it is even possible to register structures which are not accessible from the outside (Benninger et al., 2009; Kruth et al., 2011). Used, for example, to locate defects in castings or for running dimensional plausibility checks, computer tomography systems today attain measurement times which permit their integration into the clock-pulse-controlled production process – in other words, in-line utilization (Schnell, 2011) (Fig. 9).

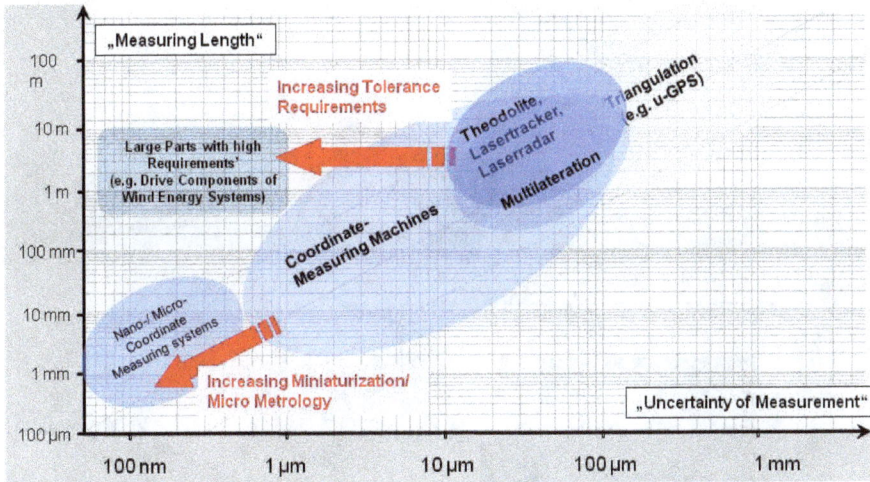

Figure 7. Tendencies in the development of accuracy (here quantified by "uncertainty of measurement") in the case of instruments used in length measurement (Schmitt et al., 2009).

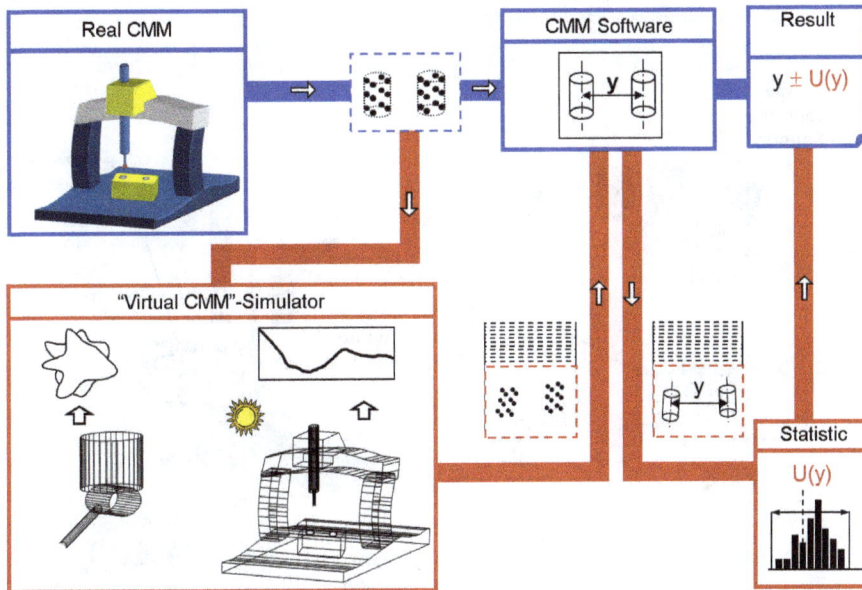

Figure 8. Determination of the measurement uncertainty in coordinate measuring machines by means of Monte-Carlo simulation: "Virtual CMM" and its connection to the instrument software (Wäldele and Schwenke, 2002).

On the other hand, different methods are being increasingly combined into measuring systems that are called multi-sensor measuring systems (Weckenmann et al., 2009; Imkamp and Vizcaino-Hoppe, 2007) (Fig. 10). The combination of results from several sensors is called sensor fusion (Heizmann et al., 2009). This boosts the flexibility of the systems. It does however also increase the complexity of the measuring systems and also the demands imposed on the user as regards training and the effort required in preparation for measurements.

6 Summary

In addition to the technical aspects we have described, the 2020 manufacturing metrology roadmap (VDI/VDE, 2011) will include future developments in the fields of the economic assessment of metrology and of training not only in institutes of higher education but also in the commercial sector (Wäldele, 2011). This topic has a special importance since the qualifications of measuring instrument operators have in many cases a great deal of influence on the accuracy of results and on their usefulness in evaluating and improving production.

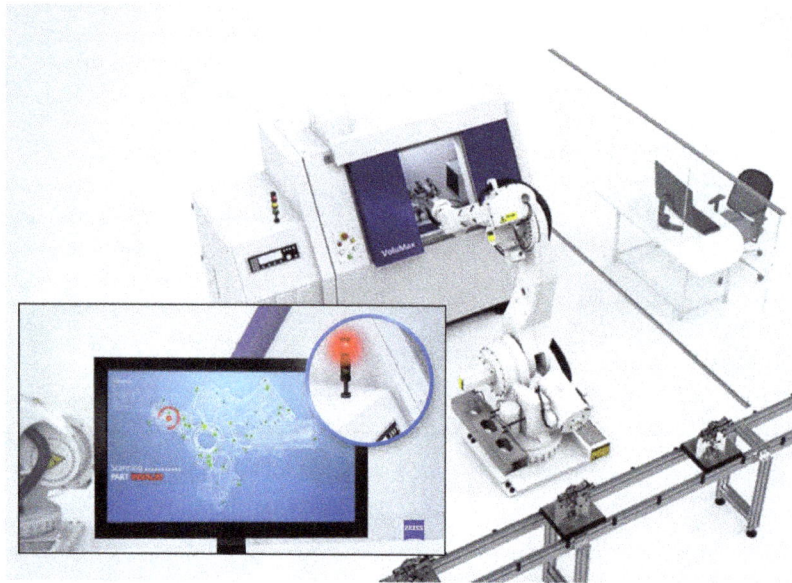

Figure 9. Computer tomography system for defect detection for in-line service.

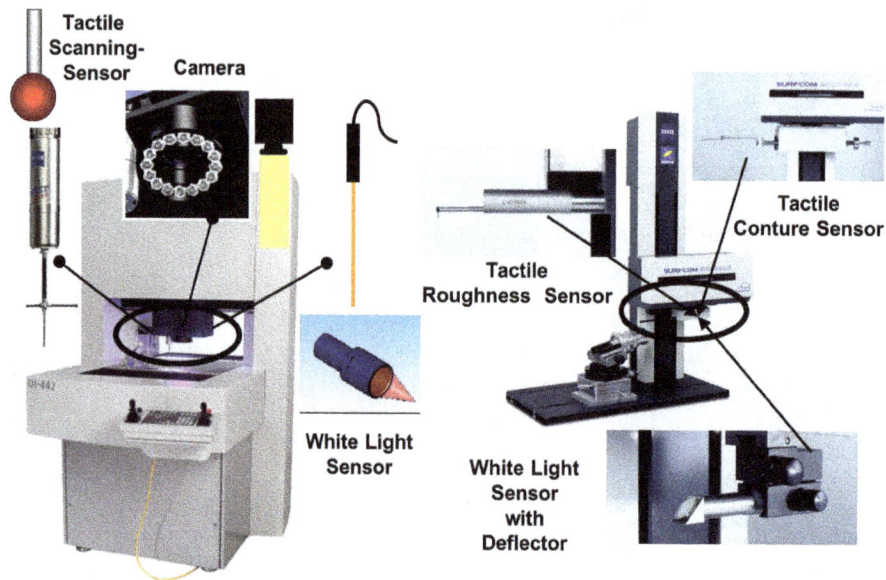

Figure 10. Multi-sensor coordinate measuring machine (left) with parallel sensors and multi-sensor surface profiler with interchangeable sensors (right).

Metrology will continue to grow in importance to industrial production. The increasing performance of metrology is reflected in its speed and levels of accuracy. At the same time it is becoming more flexible and can thus deliver more information about production. Mastering the uncertainty of metrology in production will contribute to making production more efficient and products safer.

Acknowledgements. The authors thank all members of the working group and all contributors of the final report (all from Germany): Rainer Bartelt; Mahr GmbH – Mahr Akademie Göttingen, Michael Heizmann; Fraunhofer Institute of Optronics, System Technologies and Image Exploitation (IOSB) Karlsruhe, Frank Lindenlauf, University of Applied Science Pforzheim, Harald Bosse, Klaus-Dieter Sommer and Frank Löffler; Physikalisch Technische Bundesanstalt (PTB) Braunschweig, Robert Schmitt, Philipp Jatzkowski and Susanne Nisch; Laboratory for Machine Tools and Production Engineering of RWTH Aachen University, Eduard Schenuit, Zwick GmbH & Co. KG Ulm, Rainer Tutsch, Technische

Universität Braunschweig, Franz Wäldele, Ausbildung Koordinatenmesstechnik e.V. Braunschweig, Stefan Kasperl, Fraunhofer-Entwicklungszentrum Röntgentechnik EZRT, Fürth.

This paper was published in German language as Imkamp, D., Schmitt, R., and Berthold, J.: Blick in die Zukunft der Fertigungsmesstechnik, tm – Technisches Messen, 79, 433–439, doi:10.1524/teme.2012.0251, 2012.

Edited by: R. Tutsch
Reviewed by: two anonymous referees

References

Bai, A., Bitte, F., and Pfeifer, T.: Der Einsatz von Simulationen zur Bestimmung der Messunsicherheit von Interferometern (English: Assessment of the Measurement Uncertainty of Interferometers by Means of Simulation), in: tm – Technisches Messen, Oldenbourg Industrieverlag, 69, 27–32, 2002.

Bauer, N. (Ed.): Leitfaden zu Grundlagen und Anwendungen der optischen 3D-Messtechnik, Vision 6, Fraunhofer Allianz Vision, Erlangen, 2003.

Benninger, R., Bleicher, M., and Berthold, J.: Mit Röntgenblick zum Allrounder – Roadmap Fertigungsmesstechnik 2020 (Teil 6), Qualität und Zuverlässigkeit QZ, 54. Jg., Nr. 10, 44–47, 2009.

Bosse, H., Koenders, L., and Schmitt, R.: Von Mikro zu Nano – Roadmap Fertigungsmesstechnik 2020 (Teil 3), Qualität und Zuverlässigkeit QZ, 54. Jg., Nr. 7, 28–31, 2009.

DeGlee, G.: Measuring for Wind Energy, in: Wind Systems (http://windsystemsmag.com), 42–47, October 2010.

Dutschke, W. and Keferstein, C. P.: Fertigungsmesstechnik, praxisorientierte Grundlagen, moderne Messverfahren, 7. Auflage, Teubner Verlag, Stuttgart, 2010.

Estler, W. T., Edmundson, K. L., Peggs, G. N., and Parker, D. H.: Large-scale metrology – An update, CIRP Ann.-Manuf. Techn., 51, 587–609, 2002.

Fraunhofer-Allianz Vision (Ed.): Marktstudie 3-D-Messtechnik in der deutschen Automobil- und Zulieferindustrie, Fraunhofer Verlag Stuttgart, Erlangen, 2010.

Frenz, H. and Schenuit, E.: Sinkende Toleranzschwelle – Roadmap Fertigungsmesstechnik 2020 (Teil 7), Qualität und Zuverlässigkeit QZ, 54. Jg., Nr. 11, 47–49, 2009.

GMA (Gesellschaft für Mess- und Automatisierungstechnik), http://www.vdi.de/fertigungsmesstechnik/, last access: 20 February 2012.

Grzesiak, A. and Imkamp, D.: Faster, Safer With More Accuracy and Flexibility – The VDI Roadmap Manufacturing Metrology, Xth International Scientific Conference, Coordinate Measuring Technique, University of Bielsko-Biala, Poland, 23–25 April 2012.

Heizmann, M., Beyerer, J., and Puente León, F.: Mehr Wissen durch Fusion von Sensordaten – Roadmap Fertigungsmesstechnik 2020 (Teil 2), Qualität und Zuverlässigkeit QZ, 54. Jg., Nr. 6, 35–39, 2009.

Hennes, M.: Potentiale der Ingenieurgeodäsie im Maschinenbau, 3. Dresdener Ingenieurgeodäsietag, Berufliche Weiterbildung (BWB) Industriemesstechnik, TU Dresden, Geodätisches Institut, 21–27, 1 June 2007.

Hiller, J.: Abschätzung von Unsicherheiten beim dimensionellen Messen mit industrieller Röntgen-Computertomographie durch Simulation, Dissertation, Universität Freiburg, 2011.

Imkamp, D. and Berthold, J.: Road to success – Der Weg der Fertigungsmesstechnik in die Zukunft, Qualität und Zuverlässigkeit QZ, 56. Jg., Nr. 9, 26–29, 2011.

Imkamp, D. and Berthold, J.: Schneller, sicherer, genauer – Roadmap Fertigungsmesstechnik 2020 (Teil 1), Qualität und Zuverlässigkeit QZ, 54. Jg., Nr. 5, 36–39, 2009.

Imkamp, D. and Frankenfeld, T.: Schnittstellen zur informationstechnischen Integration von Geräten der Fertigungsmesstechnik in die automatisierte Produktion, in: Tagungsband zur Automation 2009, Baden-Baden, VDI Verlag Düsseldorf, 16–17 June 2009.

Imkamp, D. and Sommer, K.-D.: Für eine sichere Fertigung, Roadmap Fertigungsmesstechnik 2020 (Teil 4), Qualität und Zuverlässigkeit QZ, 54. Jg., Nr. 8, 31–33, 2009.

Imkamp, D. and Vizcaino-Hoppe, M.: Mehr als die Summe der Sensoren – Optische Sensoren für Multisensor-Koordinatenmessgeräte, in: Tagungsband zur VDI Tagung Optische Messung technischer Oberflächen, Hannover (VDI Bericht 1996), VDI Verlag Düsseldorf, 9–10 October 2007.

ISO/TS 15530-4: Geometrical Product Specifications (GPS) – Coordinate measuring machines (CMM): Technique for determining the uncertainty of measurement – Part 4: Evaluating task-specific measurement uncertainty using simulation Ausgabe, June 2008.

JGCM 101:2008: Evaluation of measurement data – Supplement 1 to the "Guide to the expression of uncertainty in measurement" – Propagation of distributions using a Monte Carlo method, JCGM (Joint Committee for Guides in Metrology), available at: http://www.bipm.org/en/publications/guides/gum.html, 2008.

Kruth, J. P., Bartscher, M., Carmignato, S., Schmitt, R., De Chiffre, L., and Weckenmann, A.: Computed tomography for dimensional metrology, CIRP Ann.-Manuf. Techn., 60, 821–842, 2011.

Leibinger, P. and Tünnermann, A. (Ed.): Agenda Photonik 2020 des Programmausschusses für das BMBF-Förderprogramm Optische Technologien, Düsseldorf, available at: www.photonik2020.de, November 2010.

Naß, M. and Berthold, J.: Basis neuer Messtechnologien – Roadmap Fertigungsmesstechnik 2020 (Teil 8), Qualität und Zuverlässigkeit QZ, 55. Jg., Nr. 1, 53–55, 2010.

Pfeifer, T. and Imkamp, D.: Koordinatenmesstechnik und CAx-Anwendungen in der Produktion – Grundlagen, Schnittstellen und Integration, Carl Hanser Verlag, München (English: Pfeifer, T., Imkamp, D., and Schmitt, R.: Coordinate Metrology and CAx Applications in Industrial Production, Carl Hanser Verlag, München, 2006), 2004.

Pfeifer, T. and Schmitt, R.: Fertigungsmesstechnik, Oldenbourg Verlag, München (English: Pfeifer, T.: Production Metrology, Oldenbourg Verlag, München, 2002), 2010.

Porath, M. and Seitz, K.: Koordinatenmesstechnik für mikromechanische Bauteile: Herausforderungen und Lösungen, VDI-Tagungsband 1914, Koordinatenmesstechnik, Innovative Entwicklungen im Fokus des Anwenders, Tagung Braunschweig, VDI Verlag Düsseldorf, 15–16 November 2005.

Schmitt, R., Fritz, P., Jatzkowski, P., Lose, J., Koerfer, F., and Wendt, K.: Abschätzung der Messunsicherheit komplexer Messsysteme mittels statistischer Simulation durch

den Hersteller, in: VDI/VDE-Gesellschaft Meß- und Automatisierungstechnik – GMA (Ed.): Messunsicherheit praxisgerecht
bestimmen, Tagungsbericht: 4. Fachtagung Messunsicherheit, 12
und 13 November 2008 in Erfurt, Düsseldorf: VDI Wissensfo-
rum, 2008.

Schmitt, R. and Imkamp, D.: Wohin entwickelt sich die Fertigungsmesstechnik? – Roadmap Fertigungsmesstechnik 2020
der VDI/VDE-Gesellschaft Mess- und Automatisierungstechnik
(GMA), in: atp edition, Automatisierungstechnische Praxis, Nr.
6, 2011.

Schmitt, R., Jatzkowski, P., Nisch, S., and Imkamp, D.: Größer,
genauer und integrierter – Roadmap Fertigungsmesstechnik
2020 (Teil 5), in: Qualität und Zuverlässigkeit QZ, 54. Jg., Nr.
9, 31–33, 2009.

Schmitt, R., Nisch, S., Heizmann, M., Bosse, H., and Imkamp, D.:
Production Metrology – Future Trends and Challenges, in: Pro-
ceedings of the 10th International Symposium on Measurement
Technology and Intelligent Instruments (ISMTII-2011) Daejeon,
S. Korea, 29 June–2 July 2011.

Schnell, H.: Hochgeschwindigkeits-Computertomografie zur
schnellen, zerstörungsfreien und intelligenten Inspektion und
Prozessoptimierung von Aluminium-Gussteilen, Dissertation,
Universität Erlnagen-Nürnberg, 2011.

Schwenke, H.: Abschätzung von Messunsicherheiten durch Simulation an Beispielen der Fertigungsmesstechnik (Dissertation),
PTB-Bericht F36, 1999.

VDI/VDE-Gesellschaft Mess- und Automatisierungstechnik
(GMA), Editor: Fertigungsmesstechnik 2020, Technologie-
Roadmap für die Messtechnik in der industriellen Produktion,
VDI Verein Deutscher Ingenieure e.V., Düsseldorf, ISBN
978-3-00-034706-1, available at: www.vdi.de/44080.0.html,
April 2011.

VDI/VDE-Gesellschaft Mess- und Automatisierungstechnik
(GMA), NAMUR (Interessengemeinschaft Automatisierungstechnik der Prozessindustrie), Editor: Prozess-Sensoren 2015+,
Technologie-Roadmap für Prozess-Sensoren in der chemisch-
pharmazeutischen Industrie, VDI Verein Deutscher Ingenieure
e.V., Düsseldorf, NAMUR, Leverkusen, http://www.namur.de/
publikationen-und-news/fachinformationen/roadmap-sensorik/,
November 2009.

VDI/VDE-Richtlinie 2617 Blatt 7 Genauigkeit von Koordinatenmessgeräten – Kenngrößen und deren Prüfung – Ermittlung der
Unsicherheit von Messungen auf Koordinatenmessgeräten durch
Simulation (English: Accuracy of coordinate measuring machines – Parameters and their checking – Estimation of measurement uncertainty of coordinate measuring machines by means of
simulation), April 2006.

Wäldele, F. and Schwenke, H.: Automatische Bestimmung der
Messunsicherheiten auf KMGs auf dem Weg in die industrielle
Praxis, tm – Technisches Messen, Oldenbourg Industrieverlag,
69, 550–557, 2002.

Wäldele, F.: Die Wissenstankstelle für Fertigungsmesstechnik,
10 Jahre AUKOM, in: Quality Engineering, Konradin Verlag,
Leinfelden-Echterdingen, 6, 27 pp., available at: www.aukom-ev.
de, 2011.

Weckenmann, A., Jiang, X., Sommer, K.-D., Neuschaefer-Rube, U.,
Seewig, J., Shaw, L., and Estler, T.: Multisensor Data Fusion
in Dimensional Metrology, CIRP Ann.-Manuf. Techn., 58, 701–
722, 2009.

Wiedmann, W. K., Imkamp, D., and Bader, F.: Mikroteilemessgerät
F25 – Einsatzbereiche und Anwendungserfahrung, in: Tagungs-
band (VDI-Bericht 2133) zur 4. Fachtagung "Metrologie in der
Mikro- und Nanotechnik 2011 – Messprinzipien – Messgeräte –
Anwendungen", 25 und 26 Oktober 2011 in Erlangen, VDI Wis-
sensforum GmbH, Düsseldorf, 2011.

20

Towards assessing online uncertainty for three-phase flow metering in the oil and gas industry

M. P. Henry, M. S. Tombs, and F. B. Zhou

University of Oxford, Oxford, UK

Correspondence to: M. P. Henry (manus.henry@eng.ox.ac.uk)

Abstract. A new three-phase (oil/water/gas) flow metering system has been developed for use in the oil and gas industries, based on Coriolis mass flow metering. To obtain certification for use in the Russian oil and gas industries, trials have taken place at the UK and Russian national flow laboratories, NEL in Glasgow and VNIIR in Kazan, respectively. The metrology of three-phase flow is complex, and the uncertainty of each measurement varies dynamically with the operating point, as well as the metering technology, and other aspects. To a limited extent this is reflected in the error limits allowed in national standards, which may vary with operating point. For example, the GOST standard allows errors in the oil flow rate of ±6 % for water cuts of less than 70 %, which is increased to ±15 % for water cuts between 70 and 95 %. The provision of online uncertainty for each measurement, for example in accordance with the British Standard BS-7986, would be highly desirable, allowing the user to observe in real time variations in measurement quality. This paper will discuss how an online uncertainty assessment could be implemented in the Coriolis meter-based system.

1 Introduction

A Coriolis mass flow meter (Fig. 1) consists of a vibrating flowtube through which the process fluid passes, and an electronic transmitter. The transmitter maintains flowtube vibration by sending a drive signal to one or more drivers, and performs measurement calculations based on signals from two sensors. The physics of the device dictates that Coriolis forces act along the measurement section between the two sensors, resulting in a phase difference between the sinusoidal sensor signals. This phase difference is essentially proportional to the mass flow rate of the fluid passing through the measurement section.

The frequency of oscillation of the flowtube varies with the density of the process fluid. The frequency value can be extracted from the sensor signals (for example by calculating the time delay between consecutive zero crossings) so that the process density can be obtained. The flowtube temperature is also monitored to enable compensation for variations in flowtube stiffness.

Coriolis meters are widely used throughout industry. The direct measurement of mass flow is often preferred over volumetric-based metering, for whereas the density and/or volume of a material may vary with temperature and/or pressure, mass remains unaffected. This is particularly important in the oil and gas industry, where energy content and hence product value is a function of mass.

The exploitation of new technology, such as audio quality analog-to-digital convertors and digital-to-analog convertors (ADCs and DACs), and field-programmable gate arrays (FPGAs), has facilitated the development of new capabilities for Coriolis meters, such as the ability to deal with multiphase flows. Multiphase flow introduces highly variable damping on the flowtube, up to three orders of magnitude higher than in single phase conditions, requiring agile and precise drive control, which only the latest technology can provide. In addition, the mass flow and density measurements generated under multiphase flow conditions are subject to large systematic and random errors, for which correction algorithms must be defined and implemented.

There is great interest within the oil and gas industry for exploiting the new Coriolis metering technology in upstream applications, where the process fluids are inherently multiphase. A Coriolis meter measuring two parameters – mass

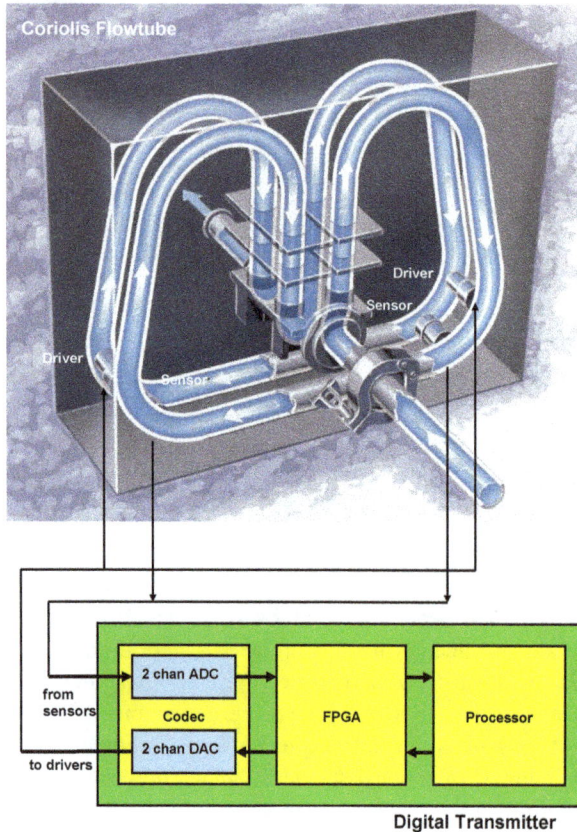

Figure 1. Coriolis mass flow meter: flowtube and digital transmitter.

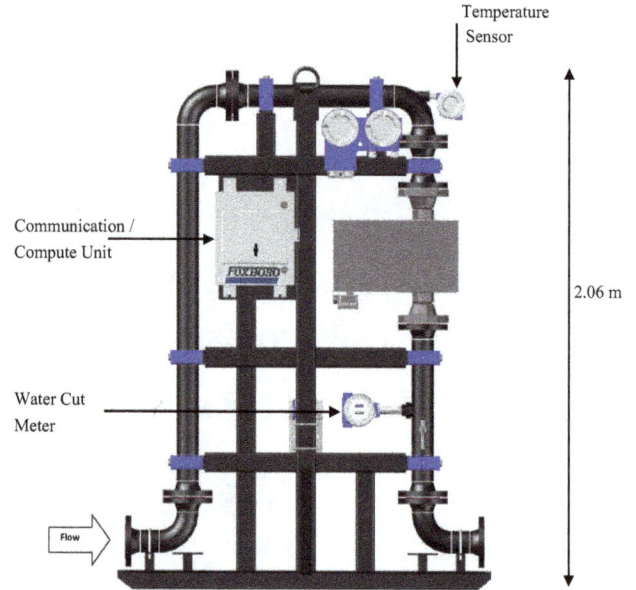

Figure 2. Net Oil and Gas Skid.

flow and density – is theoretically able to resolve a two-phase (liquid/gas) mixture. However, unless simplifying assumptions are made, a Coriolis meter cannot on its own resolve the general three-phase oil/water/gas mixture that characterises most oil well production. Including a third measurement, such as water cut (the proportion of water in the liquid mixture, typically scaled between 0 and 100 %), enables true three-phase metering to be achieved.

The term "Net Oil" is used in the upstream oil and gas industry to describe the oil flow rate within a three-phase or a liquid (oil/water) stream. A Net Oil & Gas Skid (from here on referred to as the Skid) measures the oil and gas flow rates, and hence also the produced water, in a three-phase produced fluid.

The Skid (Henry et al., 2013) has been designed by the authors and their industrial partners as a replacement for three-phase separator measurement systems conventionally used for well testing and production monitoring in the field. Figure 2 shows the design of the Skid. The pipework dimensions and internal diameter (50 mm) remain the same while a range of Coriolis meter inlet diameters from 15 to 50 mm can be fitted to match the flow rate of the wells to be monitored. The Skid has been successfully tested at the UK and Russian

national flow laboratories, and is currently undergoing field trials.

The metrology of three-phase flow is complex, and in reality the uncertainty of each measurement varies dynamically with the operating point, as well as the metering technology, and other aspects. The need to accommodate variations in measurement quality at different three-phase operating points is acknowledged to a limited extent in some national standards (e.g. GOST, 2008). However, it is argued in this paper that a truly dynamic uncertainty analysis of the three-phase measurements would facilitate extending the range of operating conditions under which guaranteed measurement performance could be provided. It is further argued that the best approach to constructing a three-phase flow uncertainty analysis is through the use of Monte Carlo modelling.

After describing the design and performance of the Skid, this paper will discuss the steps needed to provide an online assessment of the uncertainty of the three-phase measurements, conforming to the SEVA concept, as specified in the British Standard BS-7986 (BSI, 2005), as well as the international standard known as the GUM – the Guide to the Expression of Uncertainty in Measurement (JCGM, 2008a).

2   Net Oil & Gas Skid

The Skid (Fig. 2) is designed to condition the process fluid flow to minimise slip between gas and liquid via the rise and fall of the pipework, and by an integrated flow straightener in the horizontal top section. The Coriolis meter is positioned on the downward and outward leg of the Skid. Other instrumentation consists of a water cut meter and a pressure and temperature transmitter. The latter reads the pressure at the

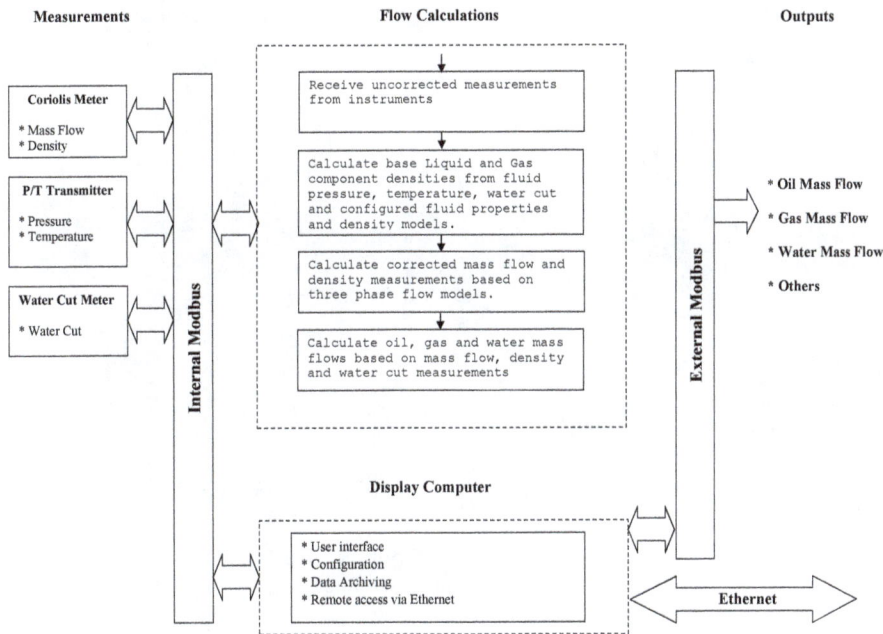

Figure 3. Hardware/software architecture of the Net Oil and Gas system.

inlet to the Coriolis meter and the temperature of an RTD (resistance temperature detector) sensor in a thermal well, positioned at the top of the Skid. The Communication/Compute Unit acts as a communication master for all the devices, using the Modbus industrial communications protocol, commonly used in the oil and gas industry. The Compute Unit performs three-phase flow measurement calculations based on the data received, provides a user interface (for providing, for example, gas and fluid density information) and also carries out data archiving. Real-time data is provided to the user's data acquisition system via a Modbus interface, with an update rate of 1 second.

The hardware/software architecture of the Skid is shown in Fig. 3. The Display Computer provides three communication interfaces: an internal Modbus for the Skid instrumentation, an external Modbus interface to provide measurement values to the user, and an Ethernet interface to enable remote configuration, monitoring and archival data retrieval. The Display Computer further provides a user interface to enable local configuration, data display, etc.

Figure 3 further shows an overview of the flow calculation algorithm. The uncorrected data from the instruments is gathered via the Modbus interface. Here, "uncorrected" refers to the effects of multi-phase flow: the mass flow, density and water cut readings are calculated based on their single-phase calibration characteristics. The liquid and gas densities are calculated based on the temperature, pressure and water cut readings and configuration parameters, based on data provided by the user. Corrections are applied to the Coriolis meter mass flow and density readings based on the three-phase flow measurement models. Finally, the oil, water and gas

measurements are calculated from the corrected mass flow, density and water cut.

The corrections to the mass flow and density readings are implemented using neural networks, based on internally observed parameters. One important parameter is the density drop, i.e. the difference between the pure liquid density (for a particular water cut value) and the observed density of the gas/liquid mixture. For example, Fig. 4 shows a 3-D visualisation of the observed density drop error against the observed mass flow and density drop, keeping other parameter values constant (e.g. the water cut is 52 %). Here a zero density drop indicates no gas present and, as would be expected, results in no density error. Models based on laboratory experimental data are used to provide online corrections for the mass flow and density readings.

Once models have been constructed and implemented in real-time software, formal trials can be carried out to test the resulting performance, and to demonstrate compliance with oil industry standards. For example, the Russian Standard GOST R 8.165 (GOST, 2008) has the following key specifications:

- Total liquid flow accuracy requirement ±2.5 %

- Total gas flow accuracy requirement ±5.0 %

- Total oil flow accuracy requirement dependent upon water cut:

- For water cuts < 70 %, oil accuracy requirement ±6.0 %

- For water cuts > 70 % and < 95 %, oil accuracy requirement ±15.0 %

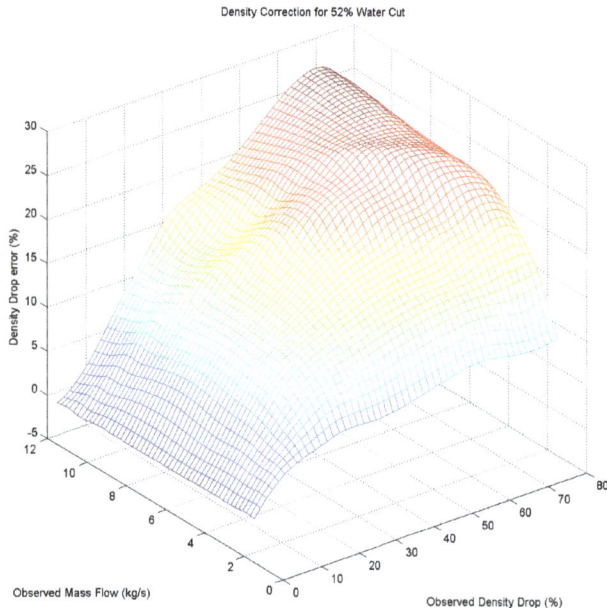

Figure 4. Density error induced by effects of three-phase flow.

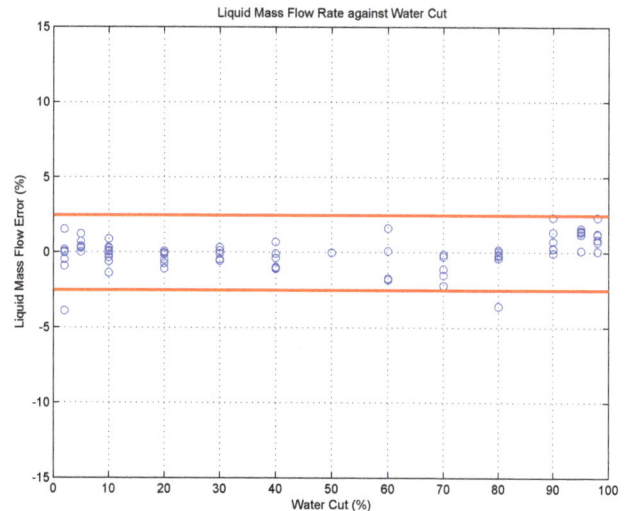

Figure 5. Total liquid mass flow rate errors against water cut. The target limit is ±2.5 % (red boundary).

– For water cuts > 95 %, no universal oil accuracy requirement is specified. Instead, the limit of the permissible relative error is to be specified in a measurement procedure to be approved and validated by the appropriate authorities.

These accuracy requirements are absolute, i.e. all test results must fall within these specifications. Trials have taken place on the Skid at the UK national flow laboratory, NEL, in Glasgow, and at the Russian national flow laboratory, VNIIR in Kazan. The resulting performance (Henry et al. 2013) matches the GOST requirements, and the Skid has been certified for use in Russia. For example, Fig. 5 shows the liquid mass flow errors from 75 formal trials at NEL, over the full range of water cuts, where the specified accuracy requirement is ±2.5 %.

3   Motivation for an online uncertainty analysis

Typically, formal trials at laboratories are carried out under steady state conditions. For example, in Fig. 5, each test result is based on a five-minute trial where all reference conditions are kept constant. The advantage of testing at steady state is that it reduces the uncertainty of the reference flow rates so that the performance of the Skid can be accurately assessed at specific operating points.

In practice, a desired accuracy performance can only be achieved over a limited range of conditions. For example, the maximum total liquid flow rate achievable through the Skid is likely to be determined by pressure drop considerations; conversely the minimum total liquid flow rate is likely to be constrained by the accuracy performance of the Skid at

low flow. With three-phase flow, there are many dimensions to consider in specifying the operating envelope for acceptable measurement uncertainty. For example, as the water cut increases towards 100 %, it becomes increasingly difficult to measure the absolute oil flow rate to within ±6.0 %; in this case the GOST standard varies the oil flow rate accuracy requirement with the water cut, as discussed above, but no such provision is made for the gas flow measurement, which is required to be accurate to within 5 % in all cases. As the gas volume fraction (GVF) tends to zero, it becomes increasingly difficult to meet this requirement.

For example, consider a mixture of pure water and gas, where the water density is taken as $1000\,kg\,m^{-3}$, the gas density at line temperature and pressure is $5\,kg\,m^{-3}$, and the GVF is 5 %. Then in every cubic metre of gas/liquid mixture, there are 950 kg of water, and only 250 g of gas; the GOST standard requires the latter to be measured to within ±12.5 g. To achieve this resolution for gas dispersed within 950 kg of water is extremely challenging, although this performance was successfully achieved in trials at NEL (Henry et al., 2013).

Testing performance with static flow conditions in laboratories can thus be used to set limits on the range of parameters over which the Skid can deliver the required accuracy performance. In practice, the accuracy of each of the oil, water and gas flow measurements will vary dynamically with the operating point (e.g. water cut, GVF and liquid mass flow rate), as well as other conditions (e.g. process noise).

Furthermore, real oil and gas wells often exhibit dynamic behaviour. For example, Fig. 6 shows data from a field trial in Russia over the course of a three-hour test. The upper graph shows the proportion by volume of free gas, oil and water in the produced fluid, while the lower graph shows the absolute volumetric flow rates. Here the well flow rate and composition show significant dynamic variation in water cut, GVF,

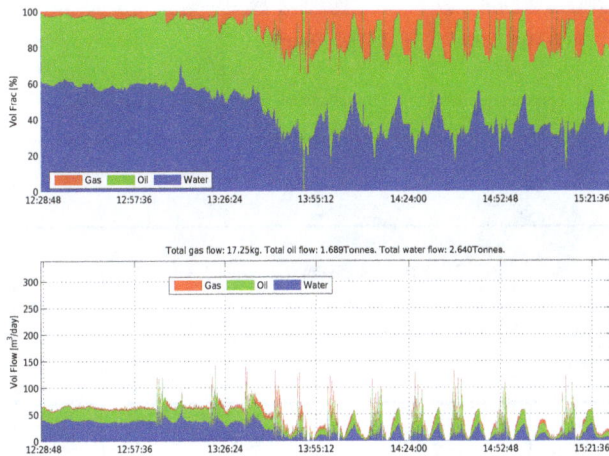

Total gas flow: 17.25kg. Total oil flow: 1.689Tonnes. Total water flow: 2.640Tonnes.

Figure 6. Oil, water and gas measurements from a Russian field trial.

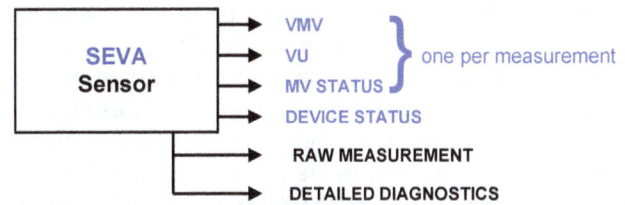

Figure 7. A SEVA sensor generating measurement and validity data.

4 Online uncertainty and SEVA

The Sensor Validation (SEVA) concept (Henry and Clarke, 1993) proposes a model of how a "self-validating" or SEVA sensor should behave, assuming the availability of internal computing power for self-diagnostics, and of digital communications to convey measurement and diagnostic data. This model has been incorporated into the British Standard BS-7986 (BSI, 2005). Note that a similar concept, "metrological diagnostic self-regulation", has been developed independently in Russia, and this has been incorporated into national standards (GOST, 2009, 2011). A generic set of metrics are proposed for describing measurement quality (see Fig. 7). For each measurement, three parameters are generated:

and liquid flow rate. One major advantage of the Skid over conventional separator technology is that it provides dynamic measurements, as opposed to simple totalised flows over a period of several hours. Data on the dynamics of flow are potentially useful to reservoir engineers for understanding the evolving state of the oilfield.

Conventionally, it is assumed that as long as the Skid operating conditions fall within the specification of the certification standard (e.g. GOST) throughout the entire well test period, then the measurement accuracy can be considered to be within the specified limits (e.g. 5 % for gas flow).

A more pragmatic and flexible approach would be to assert that, for a particular well test, as long as the operating conditions averaged over the duration of the test fall within the specification of the certification standard, then nominal accuracy can be assumed.

An alternative approach would be to provide a dynamic uncertainty analysis for each measurement value, as a function of the operating conditions, process noise and other influencing factors. With this approach, the overall uncertainty of each measurement is estimated, based upon its dynamic behaviour over the course of the well test period.

In particular, this approach might facilitate the demonstration of acceptable levels of uncertainty over wider ranges of operating conditions than for a purely static analysis. For example, if the liquid flow rate drops below the threshold for acceptable accuracy based on a static analysis, a dynamic uncertainty analysis may demonstrate that the contribution of this low flow to the overall uncertainty of the entire test period may be small, and that the overall well test total flow remains within specification.

Thus, developing a dynamic uncertainty analysis for the Skid may be able to demonstrate acceptable uncertainty performance over a wider range of operating conditions than is possible using static, laboratory-based verification.

- The Validated Measurement Value (VMV). This is the best estimate of the true quantity value of the measurand, calculated using an automated measurement procedure. Where diagnostic information indicates a known fault, the measurement procedure is adjusted to compensate for the fault. Typically such diagnostic information will be derived from one or more reference values (ISO, 1994; BIPM 2012) internal to the sensor, the Validated Uncertainty (VU). This is the measurement uncertainty, or probably error, of the VMV. For example, if the VMV is $4.31\,\mathrm{kg\,s^{-1}}$, and the VU is $0.05\,\mathrm{kg\,s^{-1}}$, then the sensor is claiming that the true measurement value lies between 4.26 and 4.36 kg s^{-1} with the stated level of coverage (typically $k = 2$, 95 % probability).

- The Measurement Value Status (MV Status). Given the requirement to provide a measurement, even when a fault has occurred, the MV Status indicates the generic fault state under which the current measurement value has been calculated.

For the purposes of this work, the most important aspect of the SEVA scheme is the generation of the Validated Uncertainty, a dynamic assessment of the uncertainty associated with each measurement value provided by the sensor. In the case of a complex instrument such as a Coriolis meter, the uncertainty of each measurement (e.g. the mass flow and density) is calculated separately within the instrument, and will vary dynamically with operating point, process noise and other parameters. Online uncertainty can be used for a variety of purposes, such as deciding on control system behaviour (e.g. whether to accept or reject the quality of the measurement value for the purposes of taking control decisions). Where measurements are combined (for example in forming mass balances or other higher-level calculations), the SEVA scheme proposes the provision of a higher-level uncertainty analysis, where the dynamic uncertainty of the input measurements are used in the calculation of the uncertainty of the resulting measurement. Consistency checking between redundant SEVA measurements has also been developed (Duta and Henry, 2005).

Here, it is assumed that dynamic assessments of the uncertainty of each measurement from the Coriolis meter, water cut meter and other sensors are available, and that these will be used to generate a corresponding online uncertainty assessment of the three-phase measurements of gas, water and oil flow.

5   Towards assessing online uncertainty for three-phase flow metering

In the Guide to the Expression of Uncertainty in Measurement or GUM (JCGM, 2008a), a number of techniques are described for calculating the uncertainty of an output variable from the values and uncertainties of input variables. In the case of a simple analytical relationship between inputs and output, formulaic expressions can be used. In more complex cases, where for example there may be a correlation between input variables and/or the functional relationship is not readily expressed algebraically, Monte Carlo modelling (MCM) is proposed as an alternative technique (JCGM, 2008b). Given the complexity of the three-phase flow calculations, which includes neural net models, MCM is proposed as the most appropriate means of assessing output uncertainty (Fig. 8) for the Skid.

In outline, with MCM the measurement calculation is carried out multiple times, where in each case the input variables are randomly selected based on their respective probability distributions. With a sufficient number of repeat measurements, it is possible to estimate the probability distribution of each output variable, and thereby to calculate a mean and coverage interval or uncertainty.

The GUM is primarily intended for static, offline analyses. In Sect. 7 of Supplement 1 (JCGM, 2008b), where the number of Monte Carlo trials M is discussed, it is suggested that one million simulations might be appropriate to ensure a good approximation of the distribution of the output variable Y. This is clearly unlikely to be feasible in an online Skid with a 1 s update rate. However, the following text appears applicable:

"If the model is complicated, ..., because of large computing times it may not be possible to use a sufficiently large value of M to obtain adequate distributional knowledge of the output quantity. In such a case an approximate approach

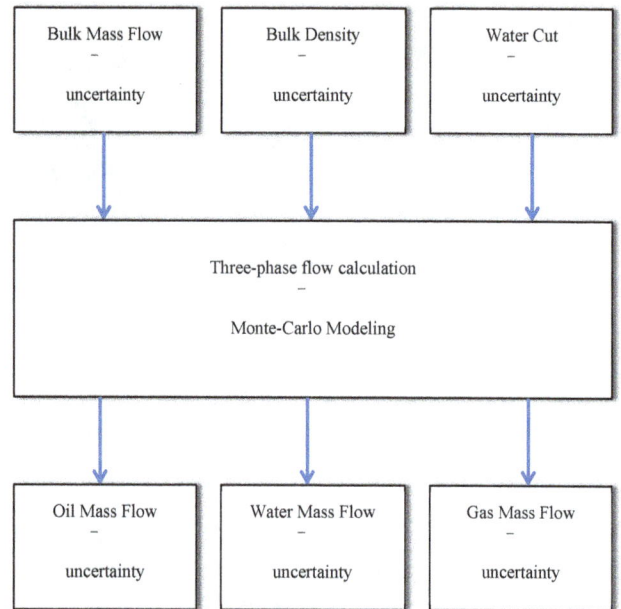

Figure 8. Monte Carlo modelling to estimate output uncertainties.

would be to regard [the distribution of Y] as Gaussian (as in the GUM) and proceed as follows. A relatively small value of M, 50 or 100, for example, would be used. The average and standard deviation of the resulting M model values of Y would be taken as its [mean and uncertainty] respectively."

The proposed approach will be implemented as follows:

- At the start of each new calculation period, mass flow, density, water cut, pressure, and temperature measurements are collected from the Skid instrumentation.

- Estimates of the uncertainties of each of these measurements are obtained either from the instruments themselves, or in the display computer.

- Simple Gaussian distributions can be assumed. The only likely correlations are between the mass flow and density measurements – all others can be assumed to be independent.

- Monte Carlo modelling simulation will entail performing 50–100 three-phase measurement calculations where the input parameters are randomly selected from their assumed Gaussian distributions.

- The resulting oil, water and gas mass flow rates are assumed Gaussian, so that the best estimate and uncertainty of each flow rate can be calculated from the MCM results.

- The totalised flow and its uncertainty are updated for each fluid type.

It will be challenging to validate the model especially based on only 50–100 points and to prove that this is representative. However, even with only 50–100 MCM calculations per measurement update, this approach will require a substantial increase in the computing power resources for the Skid if it is to be implemented in real time. A first step will be to implement an offline simulation of the MCM-based uncertainty analysis. Offline case studies based on field trial data will be used to evaluate the utility of online uncertainty analysis. The challenges of real-time implementation will be addressed once the benefits are demonstrated in simulation.

Acknowledgements. The authors thank Invensys Process Systems for funding this research, and our two reviewers for their helpful and insightful comments.

Edited by: R. Tutsch
Reviewed by: two anonymous referees

References

BIPM: BIPM, JCGM 200, Vocabulary of Metrology – Basic and General Concepts and Associated Terms, 3rd Edn., 2008 version with minor corrections, 2012.

BSI: BS7986:2005, Specification for data quality metrics for industrial measurement and control systems, British Standards Institute, 2005.

Duta, M. D. and Henry, M. P.: The fusion of redundant SEVA sensors, IEEE T. Contr. Syst. T., 13, 173–184, 2005.

GOST: R 8.615, Amended 2008: State system for ensuring uniformity of measurements. Measurement of quantity of oil and petroleum gas extracted from subsoil. General metrological and technical requirements, Federal Agency for technical regulation and metrology, 2008 (in Russian).

GOST: R 8.674-2009, State system for ensuring the uniformity of measurements. Intelligent Sensors and Intelligent Measuring Systems. Basic Terms and Definitions, 2009 (in Russian).

GOST: R 8.734-2011, State system for ensuring the uniformity of measurements. Intelligent Sensors and Intelligent Measuring Systems. Methods of metrological self-checking, 2011 (in Russian).

Henry, M. P. and Clarke, D. W.: The Self-Validating Sensor: Rationale, Definitions and Examples, Control Eng. Pract., 1, 585–610, 1993.

Henry, M. P., Tombs, M. S., Zamora, M. E., and Zhou, F. B.: Coriolis Mass Flow Metering for Three-Phase Flow, Flow Meas. Instrum., 30, 112–122, 2013.

ISO: 5725-1:1994, Accuracy (trueness and precision) of measurement methods and results – Part 1: General principles and definitions, 1994.

JCGM: JCGM 100:2008, Evaluation of measurement data – Guide to the expression of uncertainty in measurement, http://www.bipm.org/utils/common/documents/jcgm/JCGM_100_2008_E.pdf (last access: 17 April 2014), 2008a.

JCGM: JCGM 101:2008, Evaluation of measurement data — Supplement 1 to the "Guide to the expression of uncertainty in measurement" – Propagation of distributions using a Monte Carlo method, http://www.bipm.org/utils/common/documents/jcgm/JCGM_101_2008_E.pdf (last access: 17 April 2014), 2008b.

21

Prediction of bed-leaving behaviors using piezoelectric non-restraining sensors

H. Madokoro, N. Shimoi, and K. Sato

Faculty of Systems Science and Technology, Akita Prefectural University, 84-4 Tsuchiya Aza Ebinokuchi, Yurihonjo City, Akita, 015-0055, Japan

Correspondence to: H. Madokoro (madokoro@akita-pu.ac.jp)

Abstract. This paper presents a sensor system to predict behavior patterns that occur when patients leave their beds. We originally developed plate-shaped sensors using piezoelectric elements. Existing sensors such as clip sensors and mat sensors require restraint of patients. Moreover, these sensors present privacy problems. The features of our sensors are that they require no power supply or patient restraint. We evaluated our system using a basic experiment to predict seven behavior patterns. We obtained a result of predicted behavior patterns related to bed-leaving using only six sensors installed under a bed. Especially, our system can correctly detect behavior patterns of lateral sitting, which is a position that occurs when a patient tries to leave from the bed, and terminal sitting, which is the position immediately before bed-leaving. They were discerned from other behavior patterns.

1 Introduction

According to the National Population Census 2010 in Japan, the aging rate of the country is 23.1 % (NPC, 2010). This rate implies that Japan has entered a super-aging society, which is defined as the rate of the number of people aged over 65 yr old is greater than 20 percent of the total population. A report of National Institute of Population and Social Security Research estimated that a quarter of Japanese residents will be more than 65 years old in 2015 (IPSS, 2012). Along with the longevity of the society, labor shortages will become severe, especially at nursing-care facilities (Knokuchi, 2008). Few caretakers must care for numerous patients. For this situation, caretakers monitor patients inadequately, especially during sleep at night (Yamada et al., 2010). One approach to this problem is to use bed-leaving sensors that signal when patients leave from their beds. They can be used to prevent falling from their beds. The number of hospitals and nursing-care facilities using these sensors has increased (Imaizumi et al., 2010; Matsuda et al., 2003). Table 1 presents existing sensors to be used actually. We compare features of these sensors related to cost, detection speed, accuracy, privacy, and restrictiveness.

Actually, clip sensors are the lowest-cost sensors that can be introduced easily. This is a simple sensor attached to a patient's clothing (Tatsumi et al., 2007). According to protection of human rights, the usage of clip sensors has been prevented recently because it requires constraint of the wearer. For the performance of clip sensors, malfunctions and anomaly detections occur frequently because of the binary response, ON or OFF, used to detect bed-leaving behaviors. Regarding the reliability and perspective of management, clip sensors are insufficient to prevent falling from a bed completely. Moreover, accidents caused by binding of the neck in a cable have been reported (Tatsumi et al., 2007). We regard clip sensors as inadequate for use at clinical sites, although it is easy to introduce them at low cost.

Recently, mat sensors are widely used as a low-cost and convenient sensors that can be installed easily (Kondo et al., 2006). Haruyama et al. developed an alarm system to detect patients leaving from their beds using mat sensors (Haruyama et al., 2006). Medical and welfare suppliers released various mat sensors installed on a floor, a bed, or on rolling handrails. Mat sensors used on a floor are unnecessary for authentication for medical devices under the pharmaceutical law. Regarding the performance of detection, a problem

Table 1. Comparison of characteristics of existing bed-leaving sensors.

Sensor type	Cost	Timing	Accuracy	Privacy	Restriction	Reference
Clip	Low (up to $100)	Fast	Low	High	Yes	Tatsumi et al. (2007)
Bed mat	Low (up to $200)	Fast	Low (70.1%)	Low	No	Kondo et al. (2006)
Floor mat	Low (up to $200)	Slow (after standing)	Low	Low	No	Haruyama et al. (2006)
Handrail	Low (up to $300)	Fast	Low	Low	No	Haruyama et al. (2006)
Camera	High	Fast	High	High	No	Seki et al. (2002)
Infrared	High	Fast	High	High	No	Hirasawa et al. (2008)
Supersonic	High	Fast	High (95.1%)	High	No	Shimizu et al. (2009)
Strain gauge	High	Fast	High (99.0%)	Low	No	Uezono et al. (2010)

of a delay remains because of the response after sitting at the end of the bed, although such sensors are easy to produce and to sell. Moreover, sensor responses are apparent when medical staff members such as a nurse or a medical doctor walk on the mat. To distinguish the responses of patients and medical personnel is a challenging problem for signal pattern recognition. Sensors rolled over handrails not only obstruct a view of a bed, but also present a risk of removal of a sensor when a patient finds it and feels negatively about being restrained. Furthermore, false detection occurs when patients leave their bed without gripping a handrail. Mat sensors installed on a bed can detect bed-leaving with higher reliability than other mat sensors. However, existing mat sensors are actuated by a binary response similar to that for clip sensors. Early detection is not realized, especially in the initial stage of bed-leaving behavior.

Large-scale systems using numerous sensors of various types have been proposed for prediction at the initial stage to measure behavior patterns in detail. Shimizu et al. proposed a bed-leaving detection system using ultrasonic array sensors (Shimizu et al., 2009). They evaluated their system at a hospital as a demonstration experiment. Hirasawa et al. proposed a method to expose infrared rays to the upper part of the bed as a system to prevent falling accidents (Hirasawa et al., 2008). Uezono et al. proposed a large-scale monitoring system for detecting bed-leaving behavior patterns using 96 strain gauges assigned for a reticular pattern (Uezono et al., 2010). These large-scale sensor systems can realize higher accuracy and more stable detection than low-cost sensors, such as clip sensors or mat sensors, can. However, these sensors are not put into practical use because of their cost. Moreover, high expenditures are necessary to replace a bed or for construction for installation whenever these systems are improved for practical use in a market.

Using a camera as a bed monitoring sensor can provide a low-cost system. Moreover, it can obtain much information for a subject. However, it is a challenging task to predict behavior patterns obtained from images, even when state-of-the-art computer vision technologies are used. For this method, medical staff members must observe images di-

rectly. It is impossible to monitor numerous subjects simultaneously with a few operators. Moreover, we must consider aspects of human rights and quality of life (QOL). Especially, it is impossible to recognize behavior patterns related to bed-leaving using only sensor responses, even when detailed analyses are conducted, because behavior patterns differ among people (Seki et al., 2002). Moreover, monitoring using a camera imposes a mental load on patients because they feel as though they are under surveillance all day and all night.

For solving these problems, this paper presents an non-restraining sensor system using piezoelectric films. We aim at reducing the amount of data used for predicting and minimizing incorrect recognition given the minimum number of sensors employed. Our sensor requires no electric power for detection because it uses piezoelectric elements. Moreover, we developed an integrated system to send data obtained from sensors between a capturing device and a monitoring terminal computer using a close-range wireless module. We developed a microcomputer board in the device to remove noise from sensor data. We evaluated our proposed system at an environment that represents a clinical site. We obtained a result to determine and to predict seven behavior patterns related to bed-leaving. Moreover, we obtained a decision rule of bed-leaving from histograms for actual application to a clinical site.

2 Proposed sensor system

2.1 Sensor

High-performance and functional sensors of various types were used for existing bedside monitoring systems for targeting expensive care or medical treatments (Haruyama et al., 2006; Shimizu et al., 2009; Hirasawa et al., 2008; Uezono et al., 2010). In contrast, we designed a system providing low cost and user-friendliness for practical use. Our originally developed plate-shaped sensors can be installed easily under a bedsheet.

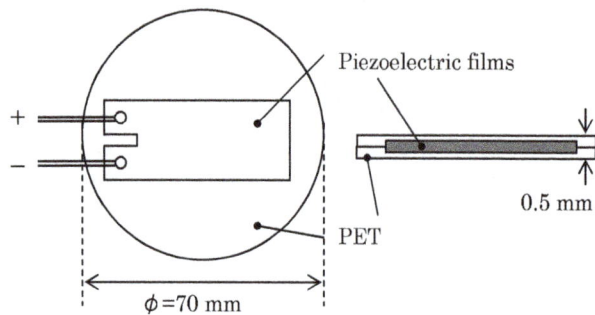

Figure 1. Design structure of prototype sensor.

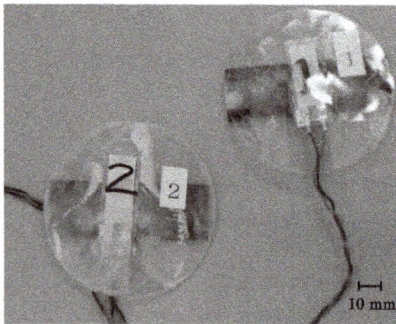

Figure 2. Overview photograph of prototype sensor.

Table 2. Specifications of Piezoelectric film DT2-028K/L (Measurement Specialties Inc.).

Parameter	Value
Minimum impedance	$1.0 \times 10^3 \; \Omega$
Output voltage	10 mV–100 V
Stress constant (g_{31})	$216 \; \text{mV m}^{-1}$
Capacity	$380 \; \text{pF cm}^{-2}$
Young's modulus	$2-4 \times 10^6 \; \text{Nm}^2$
Voltage resistance	$80 \; \text{V} \, \mu\text{m}^{-1}$
Tension strength (T_B)	$140-210 \times 10^3 \; \text{N m}^{-2}$
Tension strength (T_Y)	$30-55 \times 10^3 \; \text{N m}^{-2}$
stretch strength (S_B)	2.5–4.0 %
stretch strength (S_Y)	2.0–5.0 %
Operating temperature	0–70 °C

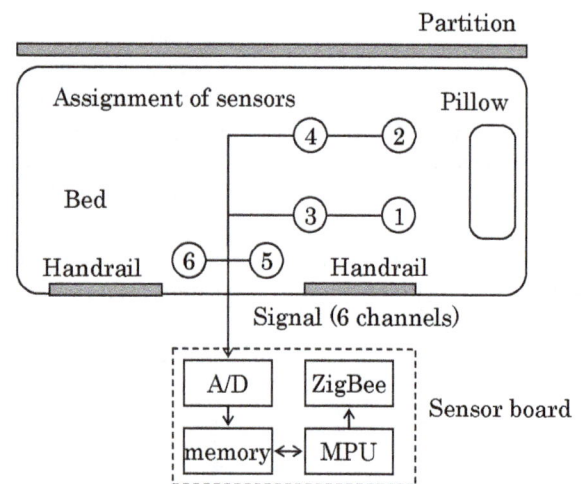

Figure 3. Block diagram of our system and assignment of sensors on a bed.

As a prototype, we developed a sensor using piezoelectric film DT2-028K/L by Measurement Specialties Inc. Table 2 presents specifications of the DT2-028K/L film (Measurement Specialties Inc., 2009). We fixed a piezoelectric film between two polyethylene terephthalate (PET) plates of laminated polyester. The polyester and PET plate sizes were, respectively, 125 µm and $\phi 70 \times 0.5$ mm. Figures 1 and 2 respectively depict the design structure and an overview photo of our prototype sensors.

Output voltage is generated from the bent piezoelectric films when a subject transfers body weight on the bed. This sensor can measure recursively because the reference potential is offset when the bending stops. Moreover, the strength of weight according to changes of the body is obtainable linearly because the bend of the piezoelectric film and output voltage has a relation of proportionality. Furthermore, piezoelectric films are less troublesome and provide no false operations because they have simple wiring without electric power supply for measurements. Additionally, we can provide a low-cost system requiring no maintenance related to replacement of a battery.

2.2 System structure

Figure 3 portrays the entire structure of our sensor system that we originally developed for this study. Our system comprises six plate-shaped sensors installed on the bed and a microprocessor board that can obtain output voltage from each

sensor for sending to a monitoring terminal computer. The output voltage is generated from these sensors when piezoelectric films are bent to receive body weight changes with movements of a subject caused by rolling or rising on a bed. The assignment of six sensors is S1 and S2 for the shoulder part, S3 and S4 for hip part, and S5 and S6 for the terminal part. We assigned these six sensors referring to the literature related to the development of a monitoring system for users of welfare care beds (Imaizumi et al., 2010; Matsuda et al., 2003).

For this assignment, a subject is available on the bed if responses from S1, S2, S3, and S4 are given alternately. In this case, the system can recognize that a subject is sleeping or rolling on the bed. In contrast, the system can detect a subject attempting to try to leave from the bed when responses are given from S5 or S6. Moreover, the system judges that a subject left from the bed completely if no sensor gives any response. This is the boundary to determine complete leaving or estimated leaving. Moreover, it can be estimated that a

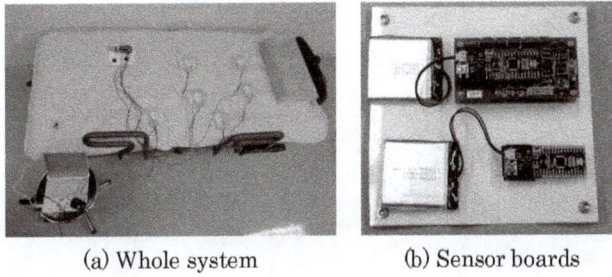

(a) Whole system   (b) Sensor boards

Figure 4. Photographs of our experimental environment, microprocessor board, and monitoring tool.

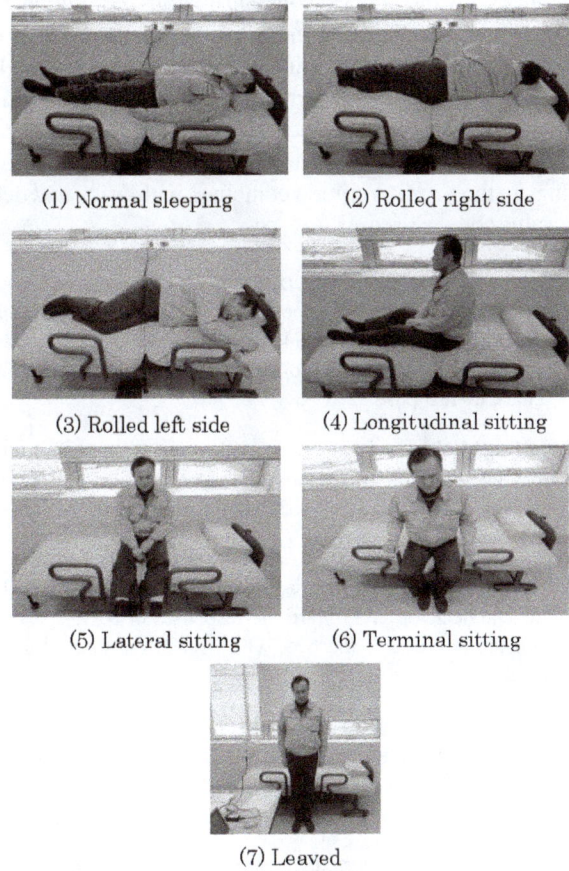

(1) Normal sleeping   (2) Rolled right side

(3) Rolled left side   (4) Longitudinal sitting

(5) Lateral sitting   (6) Terminal sitting

(7) Leaved

Figure 5. Behavior patterns for detection and recognition of bed-leaving.

subject becomes unconscious if no sensors has any response. Both are severe situations. We consider that our system can provide attention to a nurse or a medical doctor if our system is combined with a hospital information network system such as a nurse call network.

The primary feature of our system is that it realizes monitoring using non-restraining sensors. We consider QOL for a patient to live life normally. Our system requires no supervision using infrared cameras or constraining sensors such as clip sensors. Moreover, we can create a low-cost system using piezoelectric films as sensors that can function with little trouble or missed operations, and with remarkable characteristics for pressure resistance.

Figure 4b depicts the exterior of our capturing board equipped with a microprocessor and a wireless module. This board obtains output voltage from each sensor for wireless communication. For this study, we developed this board using Open Source Hardware ArduinoFIO. With consideration of power consumption, we used short-range wireless communication standard ZigBee communicated with a monitoring terminal computer. The input of this board is four channels. We used two boards for six sensors.

We set thresholds for measurement values from each sensor to reduce the communication traffic and the total amount of useless data for bed-leaving prediction. Moreover, we developed a function to analyze monitoring data as behavior logs. We originally developed software to recognize positions or movements of a subject from output patterns in each sensor used for monitoring by a terminal computer.

2.3 Detection and recognition algorithms

The target behaviors for recognition or prediction of bed-leaving consist of three patterns: sleeping, attempted leaving, and complete leaving. For this study, we also attempt to classify detailed behavior patterns from the responses of six sensors. The behavior of sleeping consists of three patterns: upside sleeping, left-side sleeping, and right-side sleeping. The behaviors of attempted leaving consist of three patterns (Mogi et al., 2011). The first behavior pattern is the status of sitting on the bed to the longitudinal side. For this position,

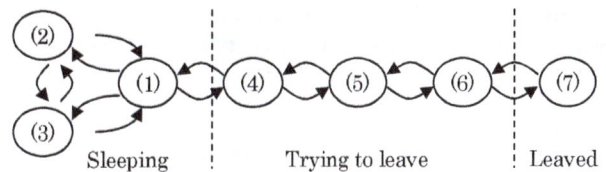

Figure 6. Status transition in respective behavior patterns.

subjects are trying to rise from the bed. We designate this as longitudinal sitting. The second behavior pattern is the status of sitting on the bed to the lateral side. For this position, subjects are trying to move to the terminal of their bed after turning the body from longitudinal sitting. We designate it as lateral sitting. The third behavior pattern is the status of sitting on the terminal of the bed. For this position, subjects are trying to leave the bed. We designate it as terminal sitting. The decision target for this study is to produce seven patterns with the status of complete leaving.

For the recognition of these behavior patterns, we set thresholds of fuzzy values and a definite value to the output voltages from sensors. Our system automatically calculates the number of fuzzy values from a histogram. For output related to changes of local parts of the body, a piezoelectric film generates voltage proportionally according to the weight of loading. Moreover, the number of definite values is calculated automatically from the histogram in the case where body weight is used because high voltage is output. The following are the decision criteria combined with fuzzy and definite values of each sensor.

1. Normal sleeping. In this status, a subject is sleeping on the bed normally to the upper side of the body. The decision criteria are that the responses of S1, S2, S3, or S4 have intermittent outputs more than fuzzy values, whereas S5 and S6 installed near the terminal of the bed have no response.

2. Rolled right side. In this status, a subject is rolling over to the right side. The decision criteria are the responses of S1 and S3 installed on the right side of the bed, which reach the thresholds of definite values.

3. Rolled left side. In this status, a subject is rolling over to the left side. The decision criteria are the responses of S2 and S4 installed on the left side of the bed, which reach the thresholds of definite values.

4. Longitudinal sitting. In this status, a subject is sitting longitudinally on the bed after rising. The decision criteria are the responses of S3 and S4, installed on the hip of a subject, which reach the thresholds of fuzzy values.

5. Lateral sitting. In this status, a subject is sitting of the lateral of the bed after turning the body from the longitudinal position. The decision criteria are the responses of S3 and S4 and S5 and S6, installed on the terminal of the bed, which reach the thresholds of fuzzy values.

6. Terminal sitting. In this status, a subject is sitting of the terminal of the bed trying to leave the bed. The decision criteria are that the responses of S1, S2, S3, and S4 have no outputs, and that S5 and S6 reach the thresholds of definite values.

7. Leaved. This status is determined as a subject leaving from the bed. Alternatively, a subject is unconscious. The decision criterion is that no sensor has any output.

Figure 6 portrays the status transitions in each pose. In the status of longitudinal sitting, subjects have no direct path for bed-leaving. They will return to normal sleeping. In the status of lateral sitting, subjects will move to leave from their bed because they move to turn the body to the terminal. Therefore, our system must determine the status of the lateral sitting immediately for the prediction of a subject leaving the bed. Moreover, our system rings an alarm immediately if a subject moves their behavior to the status of longitudinal sitting. We consider that our system can protect patients from injury or accidents caused by falling from bed, judging by their status before bed-leaving.

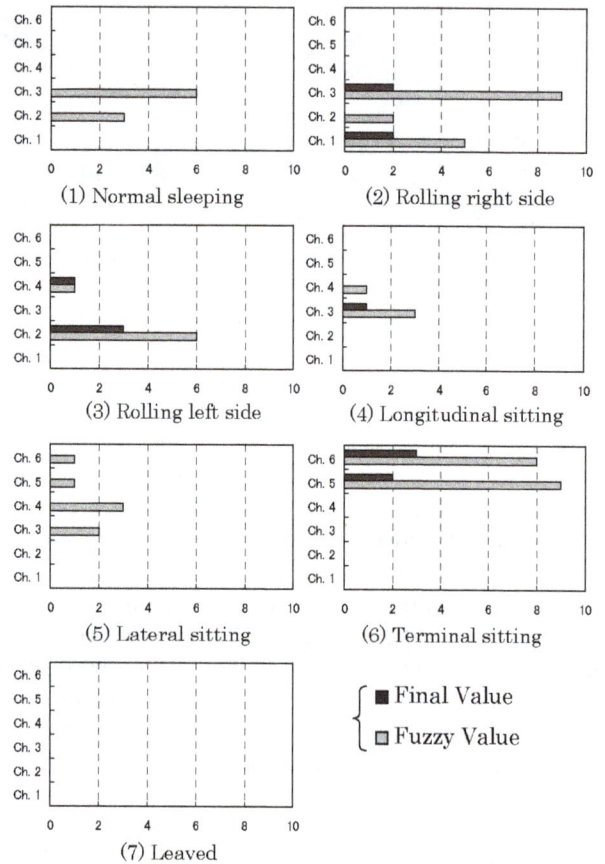

Figure 8. Frequency distributions of fuzzy and definite values in respective behavior patterns.

Table 3. Thresholds of fuzzy values and final values in respective sensors [V].

Sensor	S1	S2	S3	S4	S5	S6
Fuzzy Value	1.0	1.0	1.0	0.5	0.2	0.2
Definite Value	2.0	2.0	2.0	1.0	0.5	0.5

3 Experimental results

We created an experimental environment similar to a clinical site for evaluation of our developed sensor system. The subject is a man in his 50s. We obtained datasets of 10 iterations of stimuli to confirm the reproduction of responses. Figure 5 depicts images related to each status. The person shown in this figure is a subject for this experiment. The subject repeated each status according to the order shown in Fig. 6. Herein, the action period in each status is 20 s. The sampling rate is 50 Hz.

Figure 7 depicts time-series transitions of output voltages in each status. Subsequently, we calculated histograms of fuzzy values and definite values to set thresholds in each channel. Table 3 shows thresholds that we used for this

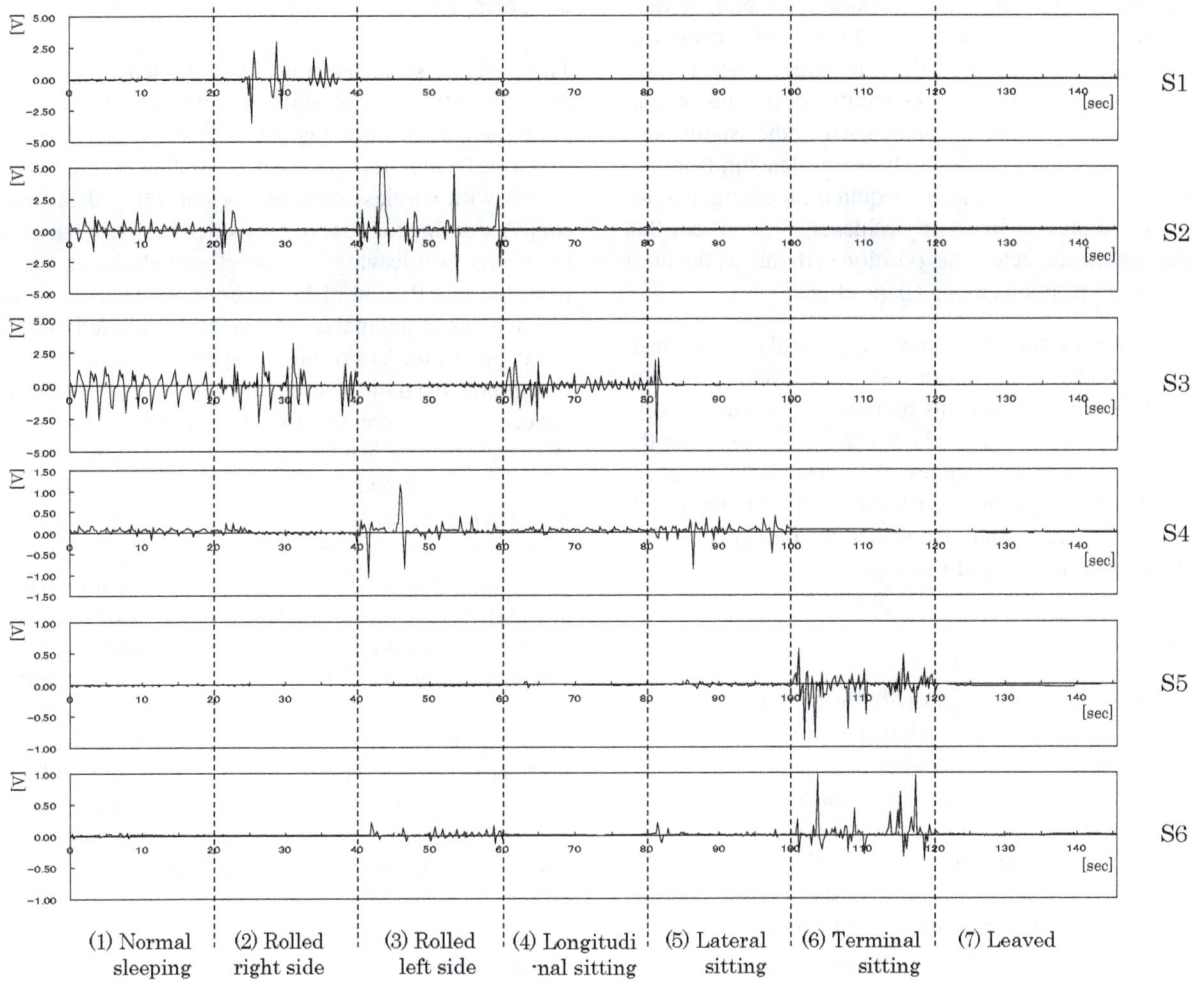

Figure 7. Time-series changes of the sensor output in respective behavior patterns.

experiment. We optimized these thresholds based on results of our preliminary experiments of 10 times. Figure 8 depicts histograms of fuzzy definite values and in each status.

1. For the status of normal sleeping, definite values were not apparent from all sensors; fuzzy values were apparent on S2 and S3. We considered that fuzzy values are generated by slight differences of body pressure according to the sleeping status. Moreover, we considered that the reason for a lack of output from S1 and S4 are influences of the setting of thresholds or the sampling intervals we used for this experiment.

2. For the status of rolled right side, fuzzy and definite values were apparent on S1 and S3, which were installed the left side of the bed. We consider that output voltage reached a definite value because the body weight was concentrated to the lateral side of the body. The fuzzy value can be recognized twice on S2. We consider that this effect results from body movements.

3. For the status of rolling to the left side, fuzzy and definite values were apparent on S2 and S4, which were installed on the right side of the bed. This result shows steady responses because other sensors had no output.

4. For the status of longitudinal sitting, output values were apparent on S3 or S4 installed under the hip of a subject. Other sensors had no output. This result shows the possibility of prediction of a subject to rise and to sit on the longitudinal of the bed.

5. For the status of lateral sitting, fuzzy values were apparent on S1, S2, S3, and S4. The status of this position can be recognized from the output of these four sensors, although definite values were not apparent because of the distribution of the body weight, detected as pressure. This status is the most important to achieve early detection of bed-leaving.

6. For the status of the terminal sitting, the outputs of fuzzy values and definite values were apparent on S5 and S6.

The fuzzy value and definite value from both sensors were, respectively, eight and two times. This result can be inferred as indicating that the subject tried to execute bed-leaving because the number of outputs of both values are numerous compared with other status. The terminal position immediately before leaving from the bed is a state for a subject required an emergency response to prevent injury or accidents. We consider that our system can detect this position certainly as the most important position among six positions.

7. After leaving the bed, our system can determine bed-leaving because no sensor had output values. Herein, the difference between this status and the status of sleeping on the upper side in (1) can be recognized from the distribution of fuzzy values. We consider that this status signals a high emergency state similar to the terminal status. This result means that our method achieved steady detection of bed-leaving.

4 Discussion

To detect the status of a subject with long-term and continuous monitoring, we must detect which behaviors in the basic six poses corresponded to a dataset of time-series measurements. From the experimentally obtained result, no fuzzy or definite value on S5 and S6 was observed as a feature in the status of sleeping or rolling of a subject. Moreover, both values showed a distribution according to the status of longitudinal sitting after rising or aiming to bed-leaving from lateral sitting to terminal sitting after turning the body to the lateral side. Therefore, we specifically examined the boundary between attempted leaving and complete leaving for analysis of monitoring data. Consequently, we can realize reduction of the total amount of measured data and protection of false recognition. We regard this is the trigger point that is useful to predict bed-leaving through monitoring.

For prediction of bed-leaving, we obtained a result from Fig. 8 to show a markedly different distribution between the distribution of the histogram of sensors and behavior patterns in the case of the sitting of a subject near the terminal of the bed. For complete leaving, we consider that our results present the possibility of comparative and steady judgments compared with expensive sensors because of a lack of fuzzy response and definite values from all sensors. Compared with the prediction system proposed by Uezono et al. (2010), which used 96 sensors of bending gages, our system can realize detection and prediction of six behavior patterns of bed-leaving only used six sensors: one-sixteenth the number used in their system.

5 Conclusions

This paper presented an non-restraining sensor system to predict behaviors of bed-leaving. We developed a prototype of plate-shaped sensors using piezoelectric films. Moreover, we developed a monitoring system consisting of microprocessor boards with wireless modules for collecting data from sensors. According to the seven classified behavior patterns used to predict bed-leaving, we developed decision algorithms from the distribution of histograms. We evaluated our system at an experimental environment constructed in reference to a clinical site. As a result, our system can recognize behavior patterns from the histograms of fuzzy and definite values. Especially, we were able to detect salient behavior features of lateral sitting when trying to leave and terminal status for leaving from the bed compared with other behavior patterns. Moreover, we obtained results enabling us to discern sleeping on the bed with rolling to the left and right side.

For our future work, we must detect behaviors that trigger our system. We must also combine our sensors with other sensors such as acceleration sensors. Moreover, we will obtain steady detection to expand the application range of our method to increase the number of subjects. We would like to apply our system to clinical information network systems such as nurse call networks and electric medical records. Moreover, we would like to apply our system to care facilities or single senior homes for security and safety observation that maintains QOL and privacy together.

References

Haruyama, K., Tanaka, K., Kobayashi, S., Yasuoka, K., Uchibori, A., and Oka, M.: Development of Getting Up Detection and Report Device using Power Line Communication and Mat-Sensor, Trans. Institute of Electrical Engineers of Japan Part. D, 126, 1507–1513, 2006.

Hirasawa, K., Matsumura, N., Kanemaru, N., and Abe, K.: Falling Accident Prevention System Used for a Clinical Site and Care Facility, Technical Journal of Nippon Telegraph and Telephone, 20, 32–35, 2008.

Imaizumi, K., Iwakami, Y., and Yamashita, K.: Availability of Monitoring System for Supporting Healthcare of Elderly People, Japanese Journal of Applied IT Healthcare, 5, 63–64, 2010.

National Institute of Population and Social Security Research (IPSS), Estimated Population in the Future of Japan, 2012.

Inokuchi, K.: The Labor Shortage in Care Workplace and Employment Intention of the Young, Proc. Kanazawa Univ. Graduate School Human and Socio-Environmental Studies, 15, 69–84, 2008.

Kondo, S., Kamiya, C., Miyamoto, H., Toriyama, Y., Mimura, E., and Tsuchida, F.: Availability of Sensor Mats to Detect Leaving for Protection of Falling Accidents from Bed, Trans. Japanese Association of Rural Medicine, 55, 245, 2006.

Matsuda, H., Yamaguchi, A., and Arakawa, T.: Monitoring System of Living Activities for Elderly People, National Technical Report, 82, 4–8, 2003.

Measurement Specialties Inc.: Datasheet ofDT Series Elements with Lead Attached, Rev. 1, 2009.

Motegi, M., Matsumura, N., Yamada, T., Muto, N., Kanamaru, N., Shimokura, K., Abe, K., Morita, Y., and Katsunishi, K.: Analyzing Rising Patterns of Patients to Prevent Bed-related Falls (Second Report), Trans. Japan Society for Health Care Management, 12, 25–29, 2011.

National Population Census 2010, Ministry of Internal Affairs and Communications, 2010.

Seki, H. and Hori, Y.: Detection of Abnormal Action Using Image Sequence for Monitoring System of Aged People, Trans. Institute of Electrical Engineers of Japan Part. D, 122, 1–7, 2002.

Shimizu, M., Dugawara, K., Ozaki, F., Hama, Y., Nishimura, M., and Yoshino, H.: Development of Detection System of Getting out of Bed with Ultrasonic Array Sensor (Part 3), Trans. Society of Life Support Technology, 21, 9–16, 2009.

Tanaka, M.: A case example of application of remote monitoring for single living senior persons of biological signal monitoring technology using high sensitive presser sensor made in Finland, Trans. Japanese Telemedical and Telecare Association, 3, 231–233, 2007.

Tatsumi, T., Kanemoto, K., and Yagi, N.: How to Use Efficient Tentomushi – Considering the Length of Stride and Height for Protecting False Operation, Proc. Japanese Nursing Association, 38, 144–146, 2007.

Uezono, T., Kubo, A., Nakajo, M., Uekaraseta, A., and Uchida, T.: Application Study of Bed leaving Prediction System, Proc. Conference of Kagoshima Prefectural Industrial Technology Center 2010, 34–35, 2010.

Yamada, R., Takashima, M., Sato, Y., Ito, W., Ito, T., and Asanuma, Y.: Evaluation and Prevention of Inpatient Falls – A study using a classification system based on situational criteria, Proc. Akita Univ. School of Health Sciences, 18, 144–150, 2010.

Permissions

List of Contributors

P. Bartscherer
Robert Bosch GmbH, Robert-Bosch-Platz 1, 70839 Gerlingen-Schillerh¨ohe, Germany

R. Moos
Bayreuth Engine Research Center (BERC), University of Bayreuth, 95447 Bayreuth, Germany

C. Coillot
LPP/CNRS/UPMC/Ecole Polytechnique, Route de Saclay, 91128 Palaiseau, France

J. Moutoussamy
LPP/CNRS/UPMC/Ecole Polytechnique, Route de Saclay, 91128 Palaiseau, France

G. Chanteur
LPP/CNRS/UPMC/Ecole Polytechnique, Route de Saclay, 91128 Palaiseau, France

P. Robert
LPP/CNRS/UPMC/Ecole Polytechnique, Route de Saclay, 91128 Palaiseau, France

F. Alves
LGEP/CNRS/Paris XI, 11 rue Joliot Curie, 91192 GIF sur Yvette, France

M. Busek
Fraunhofer IWS Dresden, Dresden, Germany

M. Nötzel
Fraunhofer IWS Dresden, Dresden, Germany

C. Polk
Fraunhofer IWS Dresden, Dresden, Germany

F. Sonntag
Fraunhofer IWS Dresden, Dresden, Germany

A. Nocke
Solid-State Electronics Laboratory, Technische Universität Dresden, Dresden, Germany

C. Coillot
Laboratoire Charles Coulomb, BioNanoNMRI group, University Montpellier II, Place Eugene Bataillon, 34090 Montpellier, France

J. Moutoussamy
LPP/CNRS/UPMC/Ecole Polytechnique, Route de Saclay, 91128 Palaiseau, France

M. Boda
SubSeaStem, 25 rue des Ondes, 12000 Rodez, France

P. Leroy
LPP/CNRS/UPMC/Ecole Polytechnique, Route de Saclay, 91128 Palaiseau, France

N. Orlowski
Research Centre for BioSystems, Land Use and Nutrition (IFZ), Institute for Landscape Ecology and Resources Management (ILR), Justus-Liebig-University Giessen (JLU), Giessen, Germany

H.-G. Frede
Research Centre for BioSystems, Land Use and Nutrition (IFZ), Institute for Landscape Ecology and Resources Management (ILR), Justus-Liebig-University Giessen (JLU), Giessen, Germany

N. Brüggemann
Forschungszentrum Jülich GmbH, Institute of Bio- and Geosciences – Agrosphere (IBG-3), Jülich, Germany

L. Breuer
Research Centre for BioSystems, Land Use and Nutrition (IFZ), Institute for Landscape Ecology and Resources Management (ILR), Justus-Liebig-University Giessen (JLU), Giessen, Germany

K. Nörthemann
Humboldt-Universität zu Berlin, Brook-Taylor-Str. 2, 12489 Berlin, Germany

J.-E. Bienge
Johann Heinrich von Thünen-Institut, Alfred-Möller-Straße 1, 16225 Eberswalde, Germany

J. Müller
Johann Heinrich von Thünen-Institut, Alfred-Möller-Straße 1, 16225 Eberswalde, Germany

W. Moritz
Humboldt-Universität zu Berlin, Brook-Taylor-Str. 2, 12489 Berlin, Germany

M. Seifert
Fraunhofer IWS, Dresden, Germany

K. Anhalt
Physikalisch-Technische Bundesanstalt, Berlin, Germany

C. Baltruschat
Physikalisch-Technische Bundesanstalt, Berlin, Germany

S. Bonss
Fraunhofer IWS, Dresden, Germany

B. Brenner
Fraunhofer IWS, Dresden, Germany

K. Retan
ZF Friedrichshafen AG Graf-von-Soden-Platz 1 88046 Friedrichshafen, Germany

A. Graf
ZF Friedrichshafen AG Graf-von-Soden-Platz 1 88046 Friedrichshafen, Germany

L. Reindl
IMTEK Lehrstuhl für Elektrische Mess- und Prüfverfahren Georges-Köhler-Allee 106, 79110 Freiburg, Germany

A. Zuzuarregui
CEIT and Tecnun (University of Navarra), Paseo Manuel Lardizábal No. 15, 20018 San Sebastián, Spain

S. Arana
CEIT and Tecnun (University of Navarra), Paseo Manuel Lardizábal No. 15, 20018 San Sebastián, Spain
CIC microGUNE, Goiru Kalea 9, 20500 Arrasate-Mondragon, Spain

E. Pérez-Lorenzo
CEIT and Tecnun (University of Navarra), Paseo Manuel Lardizábal No. 15, 20018 San Sebastián, Spain
CIC microGUNE, Goiru Kalea 9, 20500 Arrasate-Mondragon, Spain

S. Sánchez-Gómez
Department of Microbiology and Parasitology, University of Navarra, Irunlarrea 1, 31008, Spain

G. Martínez de Tejada
Department of Microbiology and Parasitology, University of Navarra, Irunlarrea 1, 31008, Spain

M. Mujika
CEIT and Tecnun (University of Navarra), Paseo Manuel Lardizábal No. 15, 20018 San Sebastián, Spain
CIC microGUNE, Goiru Kalea 9, 20500 Arrasate-Mondragon, Spain

R. Zeiser
Department for Microsystems Engineering, University of Freiburg, Freiburg, Germany

T. Fellner
Department for Microsystems Engineering, University of Freiburg, Freiburg, Germany

J. Wilde
Department for Microsystems Engineering, University of Freiburg, Freiburg, Germany

B. Fabbri
Department of Physics and Earth Science, University of Ferrara, Via Saragat 1/c, 44122 Ferrara, Italy

S. Gherardi
Department of Physics and Earth Science, University of Ferrara, Via Saragat 1/c, 44122 Ferrara, Italy

A. Giberti
Department of Physics and Earth Science, University of Ferrara, Via Saragat 1/c, 44122 Ferrara, Italy
MIST E-R S.C.R.L., Via P. Gobetti 101, 40129 Bologna, Italy

V. Guidi
Department of Physics and Earth Science, University of Ferrara, Via Saragat 1/c, 44122 Ferrara, Italy
MIST E-R S.C.R.L., Via P. Gobetti 101, 40129 Bologna, Italy
CNR-INO – Istituto Nazionale di Ottica, Largo Enrico Fermi 6, 50124 Firenze, Italy

C. Malagù
Department of Physics and Earth Science, University of Ferrara, Via Saragat 1/c, 44122 Ferrara, Italy
CNR-INO – Istituto Nazionale di Ottica, Largo Enrico Fermi 6, 50124 Firenze, Italy

P. Sharma
Instrumentation & Control Group, Indira Gandhi Centre for Atomic Research, Kalpakkam-603102, India

N. Murali
Instrumentation & Control Group, Indira Gandhi Centre for Atomic Research, Kalpakkam-603102, India

T. Jayakumar
Metallurgy & Materials Group, Indira Gandhi Centre for Atomic Research, Kalpakkam-603102, India

S. H. Jang
Department of Nanomechanics, Tohoku University, Sendai, Japan

Y. Shimizu
Department of Nanomechanics, Tohoku University, Sendai, Japan

S. Ito
Department of Nanomechanics, Tohoku University, Sendai, Japan

W. Gao
Department of Nanomechanics, Tohoku University, Sendai, Japan

S. Hoche
Chair of Brewing and Beverage, Bio-PAT (Bio-Process Analysis Technology), Freising, Germany

M. A. Hussein
Chair of Brewing and Beverage, Bio-PAT (Bio-Process Analysis Technology), Freising, Germany

T. Becker
Chair of Brewing and Beverage, Bio-PAT (Bio-Process Analysis Technology), Freising, Germany

D. Griefahn
Institute of Production Management and Technology (IPMT), TU Hamburg-Harburg, Denickestrasse 17, 21073 Hamburg, Germany

J. Wollnack
Institute of Production Management and Technology (IPMT), TU Hamburg-Harburg, Denickestrasse 17, 21073 Hamburg, Germany

W. Hintze
Institute of Production Management and Technology (IPMT), TU Hamburg-Harburg, Denickestrasse 17, 21073 Hamburg, Germany

S. Haag
Fraunhofer Institute for Production Technology IPT, Aachen, Germany

D. Zontar
Fraunhofer Institute for Production Technology IPT, Aachen, Germany

J. Schleupen
Fraunhofer Institute for Production Technology IPT, Aachen, Germany

T. Müller
Fraunhofer Institute for Production Technology IPT, Aachen, Germany

C. Brecher
Chair at the Laboratory for Machine Tools and Production Engineering (WZL) at RWTH Aachen University, Aachen, Germany

T. Bley
Centre for Mechatronics and Automation Technology (ZeMA), Saarbruecken, Germany

E. Pignanelli
Centre for Mechatronics and Automation Technology (ZeMA), Saarbruecken, Germany

A. Schütze
Lab for Measurement Technology, Department of Mechatronics, Saarland University, Saarbruecken, Germany

M. P. Henry
University of Oxford, Oxford, UK

M. S. Tombs
University of Oxford, Oxford, UK

F. B. Zhou
University of Oxford, Oxford, UK

Y. Wang
Institute Cluster IMA/ZLW & IfU Institute, RWTH Aachen University, Germany

D. Ewert
Institute Cluster IMA/ZLW & IfU Institute, RWTH Aachen University, Germany

T. Meise
Institute Cluster IMA/ZLW & IfU Institute, RWTH Aachen University, Germany

D. Schilberg
Institute Cluster IMA/ZLW & IfU Institute, RWTH Aachen University, Germany

S. Jeschke
Institute Cluster IMA/ZLW & IfU Institute, RWTH Aachen University, Germany